Viruses
and the Evolution
of Life

Viruses
and the Evolution
of Life

Luis P. Villarreal

Center for Virus Research and
Department of Molecular Biology
and Biochemistry
University of California, Irvine
Irvine, California

ASM
PRESS

WASHINGTON, D.C.

Address editorial correspondence to ASM Press, 1752 N St. NW, Washington, DC 20036-2904, USA

Send orders to ASM Press, P.O. Box 605, Herndon, VA 20172, USA
Phone: (800) 546-2416 or (703) 661-1593
Fax: (703) 661-1501
E-mail: books@asmusa.org
Online: www.asmpress.org

Library of Congress Cataloging-in-Publication Data

Villarreal, Luis P.
 Viruses and the evolution of life / Luis P. Villarreal.
    p. ; cm.
 Includes bibliographical references and index.
 ISBN 1-55581-309-7 (hardcover)
 1. Viruses—Evolution. 2. Host-virus relationships. 3. Viruses—Ecology.
4. Evolution (Biology)
 [DNLM: 1. Viruses. 2. Evolution. QW 160 V727v 2005] I. Title.

 QR392.V554 2005
 579.2'13—dc22

                              2004013977

10 9 8 7 6 5 4 3 2 1

To my sons, Joseph and Alexander Villarreal

Contents

Preface

Are viruses alive? Often students have asked this question with an almost childlike innocence. Yet this simple question raises fundamental issues. Although a precise scientific definition of life is an elusive thing, it is generally agreed that life is a state of being bounded by birth and death. From a biological perspective, living organisms are thought to also require a degree of biochemical autonomy, allowing metabolism to produce the molecules and energy needed to sustain the state of life. This level of autonomy appears to be essential to most definitions. Viruses are inherently nonautonomous, symbiotic molecular parasites and were defined as such in the early 1950s. And parasites inherently challenge our views of autonomy. Viruses are parasitic to essentially all biomolecular aspects of life. Thus, by this reasoning, we might conclude viruses are simply nonliving parasites of living metabolic systems. But life can also be thought of as a potential for continued life, not simply ongoing metabolism. A metabolically active sac, devoid of genetic potential for propagation, is not considered alive. Thus, a seed might not be considered alive but might be considered to have maintained in a "nonliving" state the potential for life. This potential is itself born from a living cell. However, a seed can also be destroyed or killed. In this regard, viruses resemble seeds more than live cells. They have a certain living potential, and they can be killed, but they do not attain the more autonomous state of life. Some have referred to this situation as a "kind of borrowed life" (4).

Why does it matter? Is this simply a philosophical question, the basis of a lively and heated rhetorical debate with little real consequence? Viruses are essentially chemical entities. If we consider them alive, are we also defining life as a chemical state? The issue of whether viruses are considered alive is not simply of interest philosophically. It is important in the context of evolutionary biology and understanding how all life works. If viruses are indeed alive, evolutionists must factor them into the forces that create life and drive evolution. Viruses would need to be considered a thread in the fabric of life, and the enormously high rates of viral reproduction and variation would define the leading edge of all evolving entities. This would be a hugely important viral role in evolution.

However, viruses are not alive. At least, that is the broadly held view of most evolutionary biologists. Hence, we need not consider their contributions to the origin of species and the maintenance of life, and they have been dismissed from the tree of life. They are like environmental or chemical toxins that simply kill off less-fit hosts: a nasty part of the natural habitat that must be overcome. This view is expressed by Margulis and Sagan in *Acquiring Genomes: a Theory of the Origins of Species* (2), which is based on the premises that the minimal unit of life is the cell and that the simplest of cells has 500 genes. Because viruses lack the means of producing their own proteins, they behave as chemicals, not as living cells. That this is indeed a broadly held view is attested to in the *Encyclopedia of Evolution* (3), which in its 1,205 pages dedicates just over 4 pages to viruses and considers them mainly as degenerate elements derived from the host.

The concept of a virus as "simply a chemical" is well supported. Ever since 1935, when Wendell Stanley and colleagues first crystallized tobacco mosaic virus, this chemical concept has been ingrained. Before that, it was thought that viruses did indeed represent the simplest of all living and genetic principles. Ironically, this earlier view—that viruses embody the basic genetic and replication mechanisms of life—was in fact stupendously successful in the development of the field of molecular biology. Viruses were first used to unravel the basic genetic definitions and principles of how all life works. The foundations of our experimental systems for understanding the molecular mechanisms of living organisms are based on what we learned from virus experiments. In the 1950s, these viral studies laid the foundations of all the molecular biology that followed. Molecular biology has since defined all the basic life processes as essentially chemical in nature. We can now crystallize many, perhaps most, of the essential components of life, including ribosome subunits, the RNA-protein complexes responsible for protein synthesis. It is probably for this reason that most molecular biologists do not ponder too much the question of whether viruses are alive. It would seem equivalent to pondering if a gene or protein is alive: a non-question of sorts.

Dead hosts can make virus. However, individual subcellular constituents (e.g., mitochondria, DNA, RNA, genes, enzymes, and membranes) are not considered alive. These cellular systems represent chemical complexity levels equivalent to those of many viruses. Cells are considered alive only when these constituents remain functionally interactive, and viruses become living constituents only when part of infected living cells. Thus, the viral living "state of being" is dependent on the live host (and is hence a borrowed life). Or at least so it would seem. Yet here too, viruses blur our very definitions of both life and death. This is because some dead viruses can bring themselves back to life and also because some viruses can grow in otherwise dead host cells, such as cells that have no nucleus (metabolically active sacs). This capacity to grow in dead hosts is most apparent in unicellular hosts, many of which live in the oceans and are subjected to constant UV killing. In the cases of bacteria and photosynthetic cyanobacteria and algae, the hosts are often "killed" by UV inactivation, but some virus infections restore a living state and allow more virus to be made. These viruses have or encode enzymes that repair various molecules and aspects of host machinery, restoring the needed genetic and biochemical capacity. In the case of photosynthetic cyanophages, one very sensitive component of the cyanobacteria is the enzyme that functions as the photosynthetic center. This enzyme can be "overcharged" with light, destroying photosynthesis and cellular metabolism. However, some cyanophages encode more rugged viral versions of this photosynthetic enzyme that are much more resistant to UV killing. The host genome is larger, and more easily targeted and killed by UV light, than those of most viruses. In the oceans, much cellular turnover is due to such UV killing. It stands to reason that some viruses would take advantage of these "living dead" cells.

Dead virus can make virus. UV inactivation is a common laboratory method used to kill or disinfect viruses. Amazingly, such dead viruses can sometimes come back to life. This is due to a well-established process known as multiplicity reactivation. If an individual cell is infected by more than one "dead" virus, the multiple viral genomes can complement the damage and reassemble, by genetic recombination, to form a whole virus. This is group selection from the "dead." It is exactly such a reassembly capacity that allows us to create artificial recombinant viruses in the laboratory. Viruses are the only known biological entities that have this capacity to animate their dead, a type of "phoenix" phenotype. A dead (or defective) virus need not remain dead. Thus, neither the partial destruction of cellular metabolism nor its potential for life need preclude the "life" potential of a virus. What, then, is life: a metabolic state or a genetic potential? Viruses clearly cross the boundaries that we use to define these concepts, and they force us to question our definitions. They appear to capture the chemical principles of life but exist often in inanimate states.

So are viruses alive or not? We can resolve the different views by considering the principle of emergent complex properties. Both life and consciousness are examples of emergent complex systems. Both require a critical level of complexity or interaction to achieve their respective states. A neuron, by itself or even in a ganglion, is not conscious. The complexity of a whole brain is needed for consciousness. However, even an intact human can be biologically alive but brain dead. Similarly, neither cellular nor viral individual genes or proteins are by themselves alive. An enucleated cell would be "brain dead" in that it would lack the full critical and cooperative complexity of cellular life. Life itself has an emergent, complex, and cooperative character, but it is made from the same fundamental building blocks that make a virus. Hence, the molecular and evolutionary views can be reconciled if we accept this view of viruses as nearly alive but lacking the essential complexity of life. This recognition that viruses are nearly alive can be found in the first reported isolation of a virus by Martinus W. Beijerinck, who referred to them as *"contagium vivum fluidum"* (infectious living fluid). In his famous 1929 essay "The Origin of Life" (1), J. B. S. Haldane also noted that viruses were a step beyond enzymes on the road to life, nearly but not fully alive, lacking the cellular systems that cooperate to make life.

What is really important, however, is to understand how viruses affect the Darwinian process of evolution. How do viruses alter the principles of evolutionary biology and the origin of living systems? We can accept the recent proposition that viruses "belong to biology because they possess genes, replicate, evolve, and are adapted to particular hosts, biotic habitats, and ecological niches" (4). However, it will surprise most evolutionary biologists to find out that the majority of known viruses are in fact persistent and inapparent, not pathogenic (toxic). Many such viruses have their own, ancient evolutionary history, dating to the very origin of cellular life. For example, the UV repair enzymes noted above (those involved in excision and resynthesis of damaged DNA, ligation of broken DNA, repair of oxygen radical damage, etc.) are unique, highly conserved virus-specific genes. Viruses create unique genes in very large numbers. The largest virus, a DNA virus called mimivirus, has more genes than the smallest cell, and these genes are mostly unique. Many believe that such viral genes are "stolen" during the process of horizontal transfer from one host to another. But this view is both oversimplified and generally incorrect, as we have previously noted (V. DeFilippis, L. P. Villarreal, S. L. Salzberg, and J. A. Eisen, Letter, *Science* **293**:1048, 2001). Virus genomes can permanently colonize their host, adding viral genes to host lineages. Some destructive viruses force life to change, and this change is also reflected in genomes. Now that we are in a genomic era, viral remnants and footprints in all domains of life are obvious and the evidence of past viral colonization is clear. All living populations appear susceptible to virus-induced disease, and, given current events, there seems to be no end in sight to this phenomenon.

From prebiotic chemical replicators to unicellular life to human populations, viruses affect life's outcome and give an ever-changing shape to the fitness landscape, often determining which organisms will survive. But viruses themselves also evolve and create new viral genes and species at sometimes astonishing rates. New viruses, such as human immunodeficiency virus type 1, may in fact be the only authentically new species whose emergence has actually been witnessed by modern science. Viruses matter to life. Consider the oceans, the vast cauldron from which all modern life has evolved. Current estimates are that the oceans harbor a population of 10^{31} viral particles, most of which is killed by UV radiation and turns over every day. We know that some of this viral gene pool finds its way into cellular genomes as prophage. How has this enormous, dynamic, and ancient gene pool impacted the evolution of life? At this time we can only guess, but my guess is that it represents the equivalent of the background cosmic microwave radiation; it is the still hot remnant of the biological big bang of creation that continues to contribute to the evolution of prokaryotes, which make up the immense black hole of our planet's biodiversity. Viruses are an inherent and, I would suggest, essential ingredient in the web of life (5; D. P. Mindell and L. P. Villarreal, Letter, *Science* **302**:1677, 2003). They represent the fast-evolving boundary between the living world and the chemical world. It is time to acknowledge and study the role of viruses in the web of life, be they living or not.

References

1. Haldane, J. B. S. 1929. The origin of life. *Rationalist Annu.* **3.**

2. Margulis, L., and D. Sagan. 2002. *Acquiring Genomes: a Theory of the Origins of Species*, 1st ed. Basic Books, New York, N.Y.

3. Pagel, M. D. (ed. in chief). 2002. *Encyclopedia of Evolution.* Oxford University Press, Oxford, United Kingdom.

4. van Regenmortel, M. H. V., and B. W. J. Mahy. 2004. Emerging issues in virus taxonomy. *Emerg. Infect. Dis.* **10:**8–13.

5. Villarreal, L. P. 1999. DNA virus contribution to host evolution, p. 391–420. *In* E. Domingo, R. G. Webster, and J. J. Holland (ed.), *Origin and Evolution of Viruses.* Academic Press, San Diego, Calif.

Acknowledgments

I gratefully acknowledge the following people for providing critical feedback on the drafts of this book as well as for providing informative conversation: Ed Wagner, Frank Ryan, Maria Eugenia Gonzalez, Esteban Domingo, Marilyn Roossinck, Roland Davis, Gail Kurath, Huw Davies, Victor DeFilippis, Keith Gottlieb, Francisco Ayala, Paul Gershon, and Nancy Beckage.

I give special thanks to my mentors: to John Holland for planting the seed in my mind, 30 years ago, concerning the importance of viral persistence in virology and to Paul Berg for conveying his skeptical, rigorous, but imaginative style in evaluating science.

I thank the following individuals for helpful conversations, reprints, and comments on the topics covered in this book: Walter Fitch, Jose Almendral, Jim Van Etten, Karla Kirkegaard, Curtis Suttle, David Prangishvili, Clyde Dawe, Jay Levy, Robin Harris, Rosemary Rochford, Anna Moroni, Gerhard Thiel, and Nancy Reich. I also thank J. Lederberg for an early conversation that confronted some of the assumptions in this book and promoted a deeper analysis.

I thank Joanna Boerner for excellent assistance in graphic illustration and copyediting of the manuscript drafts. I also thank Tony Velazquez for excellent assistance with illustrations as well as Allison Kanas and the Center for Virus Research for excellent administrative support.

I thank the National Science Foundation for the support of a Presidential Award for Mentoring Minorities in Science (HRD 003099), which allowed me some time to work on this book.

Overall Issues of Virus and Host Evolution

Introduction

This book seeks to present the evolution of viruses from the perspective of the evolution of their hosts. Since viruses infect essentially all life forms, the book broadly covers all life. Such an organization of the virus literature differs considerably from the usual pattern of presenting viruses according to either the virus types or the types of host disease with which they are associated. This book presents broad patterns for the evolution of life and evaluates the role of viruses in host evolution as well as the role of the host in virus evolution. This book also seeks to broadly consider and present the role of persistent viruses in evolution. Although we have come to realize that viral persistence is indeed a common relationship between virus and host, it is usually considered as a variation of a host infection pattern and not the basis from which to organize our thinking on virus-host evolution. Most students of microbiology or molecular biology will be familiar with the virus families organized according to their replication strategies or to the diseases they cause. Such classical textbook organization generally includes a section, often at the end, in which some issues or observations concerning the evolution of a particular virus are presented. However, this presentation pattern is inevitably narrow and fails to address broader issues or integrate our thinking about virus-host evolution. For students of evolutionary biology, the importance of viruses to the evolution of life will be a new topic. As discussed below, evolutionary biology has generally failed to consider the contribution that viruses have made to the evolution of life. Some of the reasons are historical, but mainly this is due to the view that viruses do not represent living entities and thus cannot be significant components of or contributors to the tree of life. Yet viruses do have the characteristics of life: they can be killed, they can become extinct, and they adhere to the rules of evolutionary biology and Darwinian selection. In addition, viruses have enormous impact on the evolution of their hosts. Viruses are ancient life forms, their numbers are vast, and their role in the fabric of life is fundamental and unending. They represent the leading edge of evolution of all living entities, and they must no longer be left out of the tree of life.

Definitions

The concept of a virus has old origins, yet our modern understanding or definition of a virus is relatively recent and directly associated with our unraveling of the nature of genes and nucleic acids in biological systems. As it is important to avoid the perpetuation of certain vague and sometimes inaccurate views of viruses, below I discuss some definitions as they apply to modern virology.

Historical writings about viruses can be traced back several thousand years. However, all historical descriptions are in reference to specific and/or recognizable diseases caused by viruses. The very name *virus* stems from the concept of a poison or illness that appears to move through the air. In an early account of the 430 B.C. plague in Athens, which was likely due to a viral epidemic, Thucydides carefully described the epidemic. Although a clear progression of symptoms was presented, with respiratory disease followed by rashes, gastrointestinal symptoms, and central nervous system symptoms, scientists today cannot now be certain of which virus might have been responsible, and to this day it remains a point of contention. My own assessment of the timing and pattern of symptoms indicates that the epidemic clearly resembles those seen with distemper (paramyxovirus) in domestic dogs. However, distemper is not a currently recognized human disease. Other historical descriptions are sufficiently distinct for modern virologists to be more confident of the virus involved, such as infections with smallpox virus. However, the main theme has remained that viruses are invisible agents of disease and enemies that cause harm to the host. While viral transmissibility and immunity were described in these early writings, the first proposal that these agents might be invisibly small entities was written by Girolamo Fracastoro in 1546. Although the transmission potential was used to develop a vaccine against smallpox by Edward Jenner in 1798, it was not until the 1800s that the germ theory of disease finally prevailed following experimental evaluations by Friedrich Henle, Louis Pasteur, Joseph Lister, Robert Koch, and others. That viruses were so small that they could be filtered through ceramic filters which would not pass bacteria was determined by Friedrich Loeffler and Paul Frosch in 1898 and by Martinus Beijerinck in 1899. Thus, in early definitions viruses were considered to be very small agents of host disease. This was the only view of viruses that was available to Theodosius Dobzhansky and others, who developed the new synthesis of evolutionary biology with genetic theory and the origin of species in the 1930s. Viruses were also first crystallized around this time (by W. Stanley in 1935). The ability to crystallize suggested a very chemical-like nature for viruses and reinforced the view that they were acellular replicators of disease, belonging outside of the tree of life. However, also around this time additional lines of research, by Max Delbrück and others, on viruses that destroy bacteria began to unravel the genetic nature of viruses. Yet the modern definition of a virus as a molecular genetic parasite, which was first articulated by S. E. Luria in an essay published in *Science,* awaited the advent of molecular biology in the 1950s. It was also around this time that it became clear in molecular terms that some viruses, such as temperate phage, were silent and could be genetically maintained for long periods by the host. Both the viruses causing disease and the silent viruses colonizing the host genome were defined as molecular genetic parasites. It soon became clear that if a virus could be a silent, host-associated genetic element, then the evolutionary histories of virus and host could be tightly entwined. Evidence that some viruses could silently persist in their host but still emerge from the host genome had actually been reported earlier (1909, Rous sarcoma virus; 1915, temperate phage). But without an understanding of the molecular genetic nature of viruses, these observations had little influence on the understanding of virus or host evolution. It was not until many viral genomes were finally sequenced, beginning in the 1970s, and phylogenetic methods for the analysis of sequence similarity were developed (neighbor joining and parsimony) that inferences concerning the evolutionary history of viruses could be drawn. Finally, with the sequencing of many host genomes in the 1990s, it became clear that all host genomes, from bacterial to human, have been strongly affected by viral colonizing activity. Thus, the needed information has finally been assembled to allow us to evaluate virus and host evolution together and connect these two elements of the tree of life.

Before other general issues of virus evolution are considered, it is important to define some terms used in this book.

Virus.   A molecular genetic parasite that uses cellular systems for its own replication. Note that this definition makes no reference to the molecular identity of the viral entity. Nor does it specify viral genes or their role in replication, or the specific viral life cycle. This is in order to allow the inclusion of both traditional viruses that transmit predominantly via extracellular means, hence making virions of specific molecular structure, and viruses that transmit through the host genome or other inapparent means, including defective viruses. In this context, a defective virus is a virus whose replication is conditional upon replication of another virus.

Defective virus.   A viral genome or particle that lacks sufficient instructional elements to code for its own replication and depends on another virus for such functions.

Virus species.   The traditional definition of a species for an organism is an interbreeding population that shares gene flow. As viruses have no sexual exchange process, a virus species must be defined by its lineage. A virus species is thus "a polythetic class of viruses that constitute a replicating lineage and occupy a particular ecological niche," as stated by M. van Regenmortel in 2000. Thus, a virus species is fundamentally a related lineage. The characteristic of occupying a particular niche, however, is problematic for those viral species that are known to jump species, adopt an alternative lifestyle, or occupy different niches.

Symbiosis.   The state of two previously separate living entities living together in one organism. This definition includes the persistent virus that has colonized a host, and it does not distinguish between mutually beneficial and parasitic states. All cohabitation relationships that are defined below, such as viral persistence, are considered symbiotic. For an in-depth consideration of the definition of symbiosis and its role in evolution, see the writings of F. Ryan in Recommended Reading.

Viral persistence.   The capacity of a virus to be maintained in an individual host organism while the ability of the virus to be transmitted to other host organisms or offspring of the host is also maintained. Persistence can be maintained regardless of the host immune response. This definition includes both latent and chronic infections. A latent infection involves periods, sometimes extensive, in which no virus is made in the host. In contrast, a chronic infection produces a steady level of virus progeny. This definition also includes genomic or defective viruses, which can be efficiently transmitted to host offspring or another host in the presence of the appropriate helper virus. The term *persistence* is sometimes used in other contexts in virology. For example, the ability of a virus to persist outside a cell in the environment is called persistence. Such uses are not included in this definition of viral persistence.

Acute viral infection.   A type of virus infection associated with the replication and production of an amplified number of viral progeny in which the capacity of the virus to continue to replicate is transient and is not maintained in an individual host. Ongoing virus replication is limited either by the death of the host cell or by the immune response of the host organism. In terms of cell culture-based studies of acute viruses, the cell destruction wrought by these acute viruses has been the basis of one quantitative method of virology—the plaque assay. In the plaque assay, a continuous region of cell death caused by the spread of virus progeny from a single infected cell is observed. This assay has been the basis for measuring biologically active virus. In contrast, persistent viruses often fail to lyse cells and can be much more difficult to measure. Table 1.1 summarizes the distinctions between acute and persistent viruses.

Fitness.   The characteristics that endow an organism or virus with the capacity or high probability for continued life or the capacity for its offspring to persist and continue life. Fitness is a conditional, or relativistic, concept.

Table 1.1 Distinctions between acute and persistent life strategies of viruses[a]

Acute life strategy	Persistent life strategy
Virus must find new host during reproductive period in order to continue infectious cycle	Virus maintains the capacity for continued or episodic reproduction in individual hosts (reservoirs)
Requires sustained generation of new host: host immune response to transient virus replication prevents same virus from continued replication in the same host	Includes lysogeny and latency, and some chronic infections—one virus can have both states
Applies to most epidemic viral diseases	Not selfish: requires persistence gene functions (accessory genes)
Examples of DNA viruses: T4 phage, PBCV-1 of algae, human smallpox virus; examples of RNA viruses: human poliovirus, measles virus, human influenza virus	Applies to many DNA viruses: herpesvirus, adenovirus, papillomavirus, polyomavirus, parvovirus, TT virus
	Includes some RNA viruses: coronaviruses, arenaviruses, rodent-specific hantaviruses, HTLV, foamy viruses, and endogenous retroviruses of primates

[a]PBCV-1, *Paramecium bursaria Chlorella virus 1*; HTLV, human T-cell lymphotropic virus.

It depends greatly on the competition that is present at the time of selection. Thus, the concept of "fitness space" also has a conditional quality. As described below, fitness or fitness space can be very difficult to measure experimentally because measurements are usually based on relative rates of reproduction.

Lateral gene transfer. The movement of genetic information from one lineage of organism to another isolated lineage of organism, for example, the movement of genes from a bacterial genome to a eukaryotic genome.

Viral emergence. The sudden appearance of a novel viral epidemic in a particular host organism.

Types and Classification of Viruses

Viruses are generally classified according to the type of nucleic acid in their genome (double-stranded DNA [dsDNA], single-stranded DNA [ssDNA], dsRNA, or ssRNA), the replication strategy of the genome (minus-strand RNA, positive-strand RNA, circular genome, linear genome, segmented genome, or retrotranscription) and the morphology of the virus particle (capsid size and type, virion assembly, type and number of membranes, and nuclear or cytoplasmic assembly). These features are generally maintained during virus evolution. Historically, viruses were first classified by the diseases they caused, but this led to much confusion because a single virus can be responsible for an array of disease (or no disease) states in different hosts. In addition, several nonrelated viruses can produce similar disease states. For example, viruses that induce hepatitis (liver swelling and resulting jaundice) were called hepatitis viruses. Yet we know that there is no relationship between hepatitis A virus (a positive-strand ssRNA virus) and hepatitis B virus (a pararetrovirus with a DNA genome). Morphological classification according to appearance under an electron microscope was more successful. However, this too proved inadequate because distinct viral species can be morphologically identical. In 1971, David Baltimore proposed a viral classification scheme based on genome type, polarity, and organization. Current classification also includes sequence similarity and gene organization to assign and differentiate viral species.

Viruses vary substantially in genome size and content. Defective viruses can be as small as several hundred nucleotides and not contain any open reading frames (ORFs). Satellite viruses and dependoviruses are a bit larger but may contain as few as one gene. Most viruses contain between 10 and 20 genes and have genomes that range between 5 and 25 kbp. The largest viruses are dsDNA viruses found in bacteria (*Bacillus megaterium* phage, 670 kbp), microalgae (*Pyramimonas*-specific virus, 560 kbp), and amoebae (mimivirus, 670 kbp). Mimivirus is the largest and most recently discovered (in 2003). Its genes are clearly related to those of the phycodnaviruses and poxviruses. Yet it is so large that it will not pass through 0.2-μm-pore-size filters. Mimivirus contains over 900 ORFs—a number that exceeds that in some free-living cells. Eighty percent of the mimivirus ORFs are unique to the virus.

The overall diversity of viruses is hard to estimate since so many have not been characterized. The current virus database contains about 3,600 viral species. This corresponds to about 30,000 virus strains and subtypes. Analysis of the current collection suggests that ssRNA viruses are the most diverse types, followed by dsDNA viruses, dsRNA viruses, and finally ssDNA viruses. However, these numbers are likely to be highly biased due to sampling limitations, as scientists have historically focused their studies on the viruses of *Escherichia coli,* humans, and domesticated animals and plants. Clearly, relatively unstudied habitats are known to exist. These are anticipated to have enormous populations of certain virus types not included in the database. For example, about 20,000 species of polydnaviruses (genomic DNA viruses of parasitoid wasp species) are estimated to exist, and about 10^{31} mostly unclassified dsDNA virus particles are found in the oceans. Thus, the current tally of virus species is likely to be an enormous underestimation. Viruses of humans are the best studied, and their number can be estimated to be on the order of less than 1,000 human-specific, exogenous viruses (about 100 rhinoviruses, 100 papillomaviruses, 40 adenoviruses, and smaller numbers of herpesviruses, polyomaviruses, parvoviruses, and various RNA viruses). The human genome also harbors a large number (thousands) of endogenous retroviruses, most of which appear to be inactive. It is difficult to know if these numbers of viruses are representative of other host species or if humans host an unusually larger number of viruses. One thing does seem clear: viral species greatly outnumber host species.

Virus Habitat

Not only do viruses have ecological habitats in the usual sense, such as oceans, soils, etc., but also they have host- and tissue-specific habitats that are very distinct, such as bacterial versus fungal versus animal hosts. Each of these habitats tends to have its own specific viral ecology. For example, bacterial cells differ in many basic ways from eukaryotic cells: bacteria have cell walls, they lack nuclei and mitochondria, and they carry out mixed transcription and translation. The most common and diverse of the bacterial viruses, including those of cyanobacteria, are large-tailed phages containing dsDNA that resemble the phage lambda. Some of these viruses integrate into the host's DNA. Large-tailed DNA viruses are essentially absent from metazoans, and large DNA viruses do not normally integrate into metazoan chromosomes. Unicellular eukaryotic green algae also show a particular viral ecology in that their viruses tend to be large-tailed dsDNA viruses. As these organisms are the most abundant cells in the oceans, it is likely that they are the hosts for the large numbers of tailed phage-like viral particles found in the oceans (10^8 to 10^{11} particles per liter). There are additional host order-associated differences in virus occurrence. For example, higher plants are observed to support a very large number of positive-strand ssRNA viruses, which are uncommon in many other host orders, including those of *Bacteria*. Conversely, mammals support infection with herpes- and retroviruses, both of which are absent from higher plants. Most filamentous fungi are persistently infected with some form of dsRNA virus, whereas mycoplasmas tend to support infection with ssDNA viruses. Fish and bats support the infection of many rhabdoviruses (minus-strand ssRNA viruses), which are rare in avian species. Overall, we see broad but well-maintained patterns of virus-host relationships. These patterns also apply to isolated host populations. For example, diverse virus types, such as those specific for algae, fish, or mammals, are often distinguished by being adapted to either New World or Old World populations of their host species. It is assumed that the various hosts provide specific habitats that favor or allow only certain types of viruses to succeed.

Host Evolution

It seems most likely that cellular life initially evolved in the oceans. It also seems that there must have existed a prebiotic, acellular system that preceded the evolution of cells. Viruses are defined as obligate intracellular parasites, leading some to conclude that viruses must have evolved after the evolution of the first cellular life forms. However, as is discussed in chapter 2, viruses are simply molecular genetic parasites. And as such, they are capable of parasitizing any replication system, including other viruses as well as prebiotic systems. Thus, there is a reason to think that even prior to the evolution of cellular life forms, molecular genetic parasites may have already existed. Evolutionary biologists typically monitor host evolution by examining the homology of organismic physical characteristics. While this methodology can be applied to bacteria, which retain much homology (such as with cell wall morphology), this process presents a problem for the study of virus evolution. Figure 1.1 shows a summary of the current understanding of the relationship between host species and virus diversity. As suggested above, it is likely that the sampling for viral species counts is highly biased. In addition, it can be assumed that the sampling of host species also strongly underrepresents the prokaryotes, since they are likely to represent the greatest biodiversity of the planet.

Fossil records indicate that cellular life as the prokaryote started about 4 billion years before present (ybp). It is often suggested that this first cellular life form was the common progenitor to all life, and it has been called the last universal common ancestor (LUCA). Recent sequence analysis of the major domains of extant life forms suggests that the number of genes in common to all life is surprisingly small—only about 360. These genes are

Figure 1.1 The broad pattern of host evolution. The distribution of classified virus families among host groups is depicted. The size of each pie slice corresponds to the number of virus families. As shown here, each host group has a characteristic virus type: dsDNA for algae, *Archaea*, and *Bacteria*; dsRNA for fungi and protozoa; ds-DNA for invertebrates; negative-strand [(−)] ssRNA for vertebrates; and positive-strand [(+)] ssRNA for plants. Furthermore, the life strategies of infecting viruses also tend to correspond with host groups. For example, the viruses of hyperthermophiles and filamentous algae both have a very strong tendency to be chronic or persistent, and not lytic. The data presented here are derived from the universal database of the International Committee on Taxonomy of Viruses (http://www.ncbi.nlm.nih.gov/ICTVdb). RT, reverse transcription.

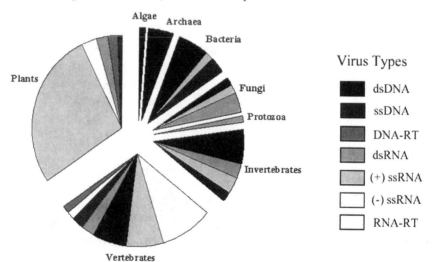

thought to be descended from the LUCA, but curiously, replication proteins are not included in this conserved set. The *Archaea* and *Bacteria* are believed to have diverged early from the LUCA, nearly 4 billion ybp. The cyanobacteria appear to have been the next major cell type to have evolved, about 2.6 billion ybp. These three groups of prokaryotes all currently have distinct and characteristic viruses that are discussed in chapter 3. However, common to all these prokaryotes are the tailed phage, which appear to have evolved prior to the divergence of the host cells. The earliest eukaryotes, the first unicellular algae, appear to have evolved between 2.2 billion and 1.8 billion ybp. After the emergence of these algae, there was a period of relative stasis, and for more than 1 billion years, cellular life appears to have evolved slowly and changed little. At the end of this period, for reasons that remain unknown, living systems appear to have acquired a method for evolutionary creativity that resulted in the Cambrian explosion—a burst of new species. Many scientists feel that some functional process of genetic innovation must have been acquired at this time to allow for the transition to rapid evolution. However, this unknown system of genetic novelty was neither the acquisition of sex nor that of gametes, as both are thought to have arisen following the Cambrian explosion.

The Cambrian explosion was immediately preceded by the origination of filamentous algae. The earliest kingdom to diverge after unicellular algae was the Diplomonadida/Parabasalia, which includes *Trichomonas* and *Giardia* species. All are species with dimorphic nuclei—two nuclei that separate genes according to germ line or soma function. Other kingdoms of microscopic organisms to have diverged relatively early were the Ciliophora/Sporozoa (e.g., *Tetrahymena* and *Plasmodium*) and Euglenophyta/Mastigophora (e.g., *Euglena, Leishmania,* and *Trypanosoma*). The Viridiplantae lineage appears to have descended from green algae and given rise to green plants (e.g., *Arabidopsis*, Solanaceae, and *Chlamydomonas*). Another divergence gave rise to fungi (e.g., *Saccharomyces* and *Schizophyllum*), which split off to form metazoans. In metazoans, a basal divergence produced *Caenorhabditis,* followed by the divergence of protostomes from deuterostomes and then insects from vertebrates.

As discussed above, the Cambrian explosion of species occurred about 545 million ybp, leading to the immense increase in evolution that has led to all modern life forms. This explosion in the number of species has mainly been observed in the fossil record as the abrupt appearance of numerous skeletal forms (trilobites, molluscs, and echinoderms), which were totally absent from the preceding fossil record. As discussed in chapters 5 and 6, the evolution of these skeletal animal forms is also correlated with the likely emergence of numerous types of viruses. The very first ocean animals had evolved prior to this explosion; these were flat, boneless, eyeless, mouthless, and brainless filter feeders that became extinct at the Cambrian period. Also prior to the Cambrian period, there were no predators of these early animals. In addition, various species of algae that had existed for long periods also became extinct. The mechanisms that could account for this mass planktonic extinction remain unknown. Interestingly, C. Emiliani proposed in 1982 that the mass extinction of planktonic ocean species during evolution may have been due to selective sweeps by lytic virus infections.

In terms of modern life forms, the evolution of fungi marks a most important event, as it was directly involved in the origination of animals and indirectly involved in, but central to, the origination of terrestrial plants. About 450 million ybp, plant and animal life emerged from the oceans onto the land. Fungi had acquired characteristics that were able to withstand the desiccation and sunlight of this new, harsh environment. Fungal symbiosis with plants appears to have allowed the plants to create root systems with the ability to pull in water and nutrients from soil, in addition to photosynthesis, which produced carbon-based energy sources for the fungi. There is also evidence that the emergence of life onto land had a major effect on life in the oceans. The oceans are generally food-poor habitats that are perpetually in a state of famine and tend to resemble deserts. Land occupation by life appears to have significantly increased the flow, or runoff, of nutrients into the oceans, thereby increasing the ability of this habitat to support diverse life forms. In fact, current estimates are that land-based species represent about 50 times the biomass of the combined oceanic species. On land, modern plants represent a majority of this biomass. Although, unlike for the oceans, scientists are hard-pressed to estimate the combined viral load on land species, it is known to be high for land plants (see chapter 7). In addition, the fungi of land plants also support many types of viruses. Consider, for example, the Douglas fir, which can host 2,000 species of fungi, most of which themselves host dsRNA viruses (see chapter 5).

How Viruses Evolve

There are several major difficulties that apply to the study of virus evolution. Viruses leave no fossils in the geological record; thus, we have no outside reference to calibrate the time of possible viral evolution events. Another problem is that viruses clearly have numerous origins and are thus polythetic. Hence, we cannot easily fit them into one congruent tree of life. The polythetic character corresponds mainly to the specific genome replication strategy of the individual virus families, each of which seems to have evolved from a distinct common ancestor. For example, all small dsDNA viruses (papillomaviruses or polyomaviruses) appear to be related to each other and probably have a common origin. However, some viral groupings, such as the positive-strand ssRNA viruses, are so large and diverse that available sequence data do not support the view that they evolved from one common ancestor. In this case, it appears there may have been several origins for the positive-strand riboviruses. There is also the problem that viruses are too simple to show homology in the classic sense. We cannot monitor homologous traits suitable for phylogenetic analysis in virus evolution. Yet it is still clear that viruses do have lineages and evolutionary relationships. Viruses generally conserve information regarding replication proteins and *cis*-genomic signals for replication. Although high mutation rates can obscure this information, consensus sequences at important protein domains are generally conserved and provide useful phylogenetic data. Viruses also conserve replication strategy and gene order. Viral morphology and morphogenesis are other generally conserved features of a virus family. All these traits can be used to deduce virus lineage. However, that lineage may not be linear, and different parts of viruses can have different evolutionary histories. In fact, it is now generally accepted that, at least for bacterial DNA viruses, evolution has occurred mainly by the high-level recombination of subgene domains. An example of this problem is seen with the lambdoid phage. Although these viruses have clearly similar life strategies and morphologies, and many are capable of recombining with each other, no one gene, including core replication genes, is conserved among all these phage. The high rates of recombination within their genomes appear to have erased any record of sequence information that could have indicated their lineage. This makes it very difficult to discern evolutionary relationships among this phage family.

Another way to think of virus evolution is to consider it from the perspective of host evolution. As discussed below, some viruses do show a strong tendency to coevolve with their hosts. In addition, as mentioned above, viruses often show broad associations with their specific host orders. Figure 1.2 shows a "tree of life" dendrogram that includes all major lines of currently living organisms. A few points follow here but are developed in greater detail in the subsequent chapters. First, as the labels in the figure indicate, host lineages have their own peculiar sets of virus that they tend to support. Also, there is a tendency for species to evolve to higher complexity, but also to evolve to support more viruses. Finally, the LUCA is colored in blue, since as discussed in subsequent chapters, it may have a strong viral origin.

Deciphering the ultimate origin of viruses seems unattainable. If they are as old as all life, then their high rates of evolution have erased any useful record of their lineage or age, so their antiquity can only be inferred and not deduced. Historically, various scenarios have been proposed to explain viral origins. Viruses with RNA genomes represent the only extant entities that use RNA for the purpose of storing genetic information. Thus, it has been proposed that riboviruses may trace their origins to the RNA world, the era prior to DNA genomes. Although it seems logical, we currently have no way to evaluate this idea. Negative-strand viruses in particular seem not to have any host analogues for their genome structure or replication in any known cell type. For example, there is no cellular equivalent of the negative-strand viral replicase, yet all negative-strand viral replicases seem to be distantly related. Thus, they may trace their origins to a time before cells. In contrast, although the positive-strand RNA viruses all have replicases with similar structures, leading some to suggest a common origin, subsequent

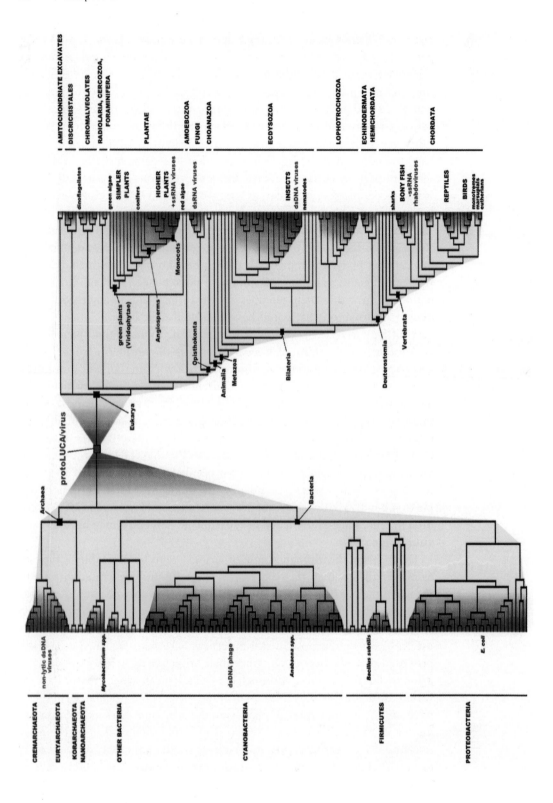

Figure 1.2 Tree-of-life dendrogram of currently accepted evolutionary relationships of all living organisms, with an overlay of color to show how the lineages support viruses. Prokaryotes are on the left and eukaryotes are on the right. These domains are connected by the LUCA, which is also identified as being viral. Highly common virus types for particular organisms are in blue. The color gradient (blue to yellow, with blue representing the most virus) identifies viral diversity for that group of organisms. Yellow indicates low viral diversity, and blue indicates high viral diversity. The LUCA is identified as being mainly viral.

sequence analysis has not supported the existence of a common ancestor. These viruses appear to exist in "supergroups" that have more related replicases. Although each of these supergroups may have a common ancestor, there is little support for linking these higher-level organisms. Another idea is that viruses represent escaped bits of host genomes, such as an origin of replication and a corresponding polymerase or primase along with genome-binding or -coating proteins. This idea has been proposed for various DNA viruses, especially those that use host-like replication processes. However, phylogenetic analysis does not support a cellular origin for any of the DNA viruses. For example, all evaluated eukaryotic viral DNA families appear to have originated from progenitor viruses, some of which can be traced to bacterial phage, not host cell ancestors. Another related proposal has been that retroviruses may also have evolved from host DNA sequences. Accordingly, various retroviruses (such as transforming retroviruses) appear to have originated from sequences found in the host genomes. Thus, it has appeared that there was some support for this idea of escaped host elements. Also, individual viral genes sometimes show strong similarity to homologous host genes, supporting the idea that these are host-derived genes. However, the emergence of a full virus, not just one specific gene, from the host genome has not been observed. Most all of the examples used to argue this idea can be traced to the emergence of endogenous and sometimes defective retroviruses that colonized the host. Thus, they cannot really be said to have originated from host elements. Another historical view was that viruses were degenerate unicellular life forms that had lost some genes and had become obligate intracellular parasites. This idea has been repeatedly applied to the large DNA viruses, such as the poxviruses, which physically resemble bacteria and have complex genomes. However, with the sequencing of these viral genomes, it has become clear that they did not originate from bacterial or other unicellular genomes. They all have originated from well-established viral lineages. Thus, the most supported view is that most viral lineages are old, originating independently from host replication systems, and that several independent viral origins have occurred.

Mathematical Biology, Host Populations, and Virus Evolution

Mathematical biology is the application of mathematical descriptions to biological issues. The relationship between a virus and its host can and has been mathematically modeled. Current models can be traced to the initial developments of predator-prey models by Vito Volterra. The premise is that a virus behaves much like a predator towards its host. Virus growth is dependent on consumption of its host, such that the virus and host population densities will be linked in a predictable way. This results in a differential equation called the Lotta-Volterra equation. In addition, if coefficients are added for the transmission efficiency of the virus, natural and virus-induced host death rates, survival from infection and subsequent immunity, and the birth of new host organisms, mathematical models can be developed that predict the outcome of viral growth or epidemics. With these models, we can understand the epidemic behavior of a virus with respect to host population density and host immunity, as first done for smallpox epidemics by the famous mathematician Daniel Bernoulli. When a virus is initially introduced into a dense and nonimmune host population, an epidemic called a "virgin soil epidemic" results. This is, for example, what happened to the human population of the New World with the introduction of the smallpox virus by the Spanish conquistadors. The characteristics of a virgin soil epidemic are summarized in Table 1.2. A summary of New World epidemics is shown in Table 1.3. However, once the initial sweep of the virus has occurred in the population, a different virus-host dynamic is established, depending on the various coefficients. Essentially all surviving adults will have been infected by the virus and will be immune to subsequent infection. This creates a situation in which only newborn (or immigrating) hosts will not be immune, and they will constitute the main hosts for virus propagation. The virus will thus become a childhood disease and endemic to the host population. However, in order to produce a sufficient number of host offspring to maintain the chain of virus transmission, the host population needs to be sufficiently large and interacting. For example, it is estimated that with various acute human viruses, single populations of 50,000 can maintain some

Table 1.2 Characteristics of virgin soil epidemics

Initial epidemic sweep will "R select" (fast replicators) for rapidly transmitting (pathogenic) variants.

Virus introduction and spread can readily precede human migration or European (infected) "colonizations" of New World. Only need contact between infected individual and susceptible in a threshold population to result in fire-like virus spread.

Epidemic will not disappear as susceptible population condenses or recovers and becomes immune. Epidemic will become "K selected" (reduced rate of spread) for endemic infection of newborns. Overall population recovery (birthrate) will be affected by the increased child mortality.

Table 1.3 Summary of New World epidemics

Date	Epidemic
Early 1500s	Epidemic of unknown viral agent in New Mexico's Charma Valley. Eighty percent of Native American village sites abandoned in early 16th century.
1513	Smallpox or measles epidemic in Florida introduced by Juan Ponce de León's attempt to colonize, which was thwarted by hostile natives. The colonists were ill. Native American populations crashed in early 16th century.
1517	Probably swine influenza epidemic, introduced by Christopher Columbus's second voyage. By 1517, the Arawak Indian population crashed.
1520	Smallpox epidemic in Mexico, introduced by Pánfilo de Narváez. Aztec elites, leaders, and military died in large numbers. A broader epidemic ensued.
1524–1525	Epidemic of unknown viral agent in Peru. Incan emperor, relatives, and subjects died prior to Francisco Pizarro's arrival.
1540	Epidemic of unknown viral agent in South Carolina and Mississippi Valley, introduced by entourage of Hernando de Soto. Large towns in South Carolina and dense Native American populations and complex societies suffered significant losses. By the 1700s, French settlers reported low Native American population densities. Village site counts indicate population crash in early and mid-16th century.

viral diseases. In the context of modern human (postagricultural) populations, these numbers are easily attained. Thus, the advent of human civilization, with its rise in population density, has allowed for the maintenance of acute viral diseases. For example, European populations during the Middle Ages sustained childhood infections with acute viruses such as smallpox and measles viruses. An outline of the history of smallpox and measles epidemics is shown in Table 1.4. However, throughout most of human evolution, human populations were composed of hunter-gatherers, groups that were

Table 1.4 Histories of smallpox and measles epidemics

Smallpox
  Not compatible with early human (preagricultural) population structure
  Earliest accounts from India, in Sanskrit medical text, and China, 1122 B.C.
  Not in ancient Greece, Rome, and probably Egypt
  Entered Europe via Islamic North African expansion to Spain; epidemics in Syria (A.D. 302) and Mecca (A.D. 569)
  Reintroduced to Europe via crusaders
  Disease descriptions of more mild illness limited to children (Spain in the 1400s)

Measles
  Plague of Athens (436 B.C.) described a distemper-like epidemic with high mortality
  Probable epidemics in Rome and then China in A.D. 165 and A.D. 251
  Considered a normal process of development; all adults were survivors of childhood infection in Europe

much smaller and could not have sustained the transmission chain of essentially any acute virus. Given that most terrestrial host organisms do not live in large, dense populations, virus-host evolution is significantly restricted.

The predator-prey mathematical models have been highly successful and have accurately predicted the outcome of various human viral epidemics, including the human immunodeficiency virus (HIV) pandemic in Africa. They have also allowed us to understand that distinct selection conditions can apply to the same virus in the same host when that host inhabits different populations or immunity structures. For example, in virgin soil epidemics, virus is selected for rapid spread. This presents a selective situation that closely resembles the growth-dominated selection of newly introduced virus species in island ecologies. However, in an already infected population, the virus is selected for infection of the immunocompromised host, which resembles the equilibrium-based selection of established viruses in island ecologies.

Viral Populations, Quasispecies, and the Fitness Landscape

A distinguishing characteristic of virus populations is the capacity of their genomes to be highly variable. Some virus populations, such as HIV-1, are so diverse that it is estimated that there may be no two genomes in a population that are identical. Furthermore, this diversity can be generated very rapidly. One measurement suggested that about 10,000 variants could arise from one round of a single cytotoxic-T-cell infection, even when a cloned, genetically homogeneous HIV template is used to initiate that round of infection. This diverse genetic character of viral populations has been called *quasispecies*. The term *quasispecies* was originally developed by Manfred Eigen and Peter Schuster as part of their study of the chemical basis for the origin of life to explain the diversity of chemical replicators that would result from an inaccurate replication process. They coined the term *quasispecies* to name an error-prone process of chemical replication that results in a population of related but nonidentical chemicals. However, the term *species* has introduced some confusion to the concept of quasispecies because this term has a different meaning in a biological context, which is "an interbreeding population of living organisms." As they apply to viruses, these two uses of the term *species* sometimes overlap, since a genetically diverse virus population may also represent a population that exchanges genetic information. In fact, it has been demonstrated that quasispecies of some viral populations can have a distinct relative-fitness profile, suggesting that evolutionary pressure may operate on quasispecies level. The absolute amount of sequence variation in a viral genome, however, can be enormous and has been called "hyperastronomical." In the case of HIV, with its genome length of 10,000 nucleotides, the total number of possible sequence variants is $10^{6,020}$. This number is large beyond comprehension. For example, compare it to the estimated number of protons in the universe, about 10^{300}. This vast variation in sequence can be thought of in terms of sequence space, where each variant occupies a coordinate in highly dimensional space. The distance between variants corresponds to the smallest number of point changes (Hamming distance) connecting the two variants. Although actual viral quasispecies can be large populations, they are relatively small compared to the total possible sequence space. For example, all of the combined HIV sequence variants from all people on Earth that have had HIV might correspond to 10^{60} variants. Because a real quasispecies virus would only correspond to a very small fraction of this potential sequence space, it can be thought of as occupying a relatively small cloud of the entire sequence space.

Fitness Landscape and Error Catastrophe

Evolution of a sequence often involves the adaptation of the sequence to higher fitness. Other parameters, such as the sampling restrictions (genetic bottlenecks) that are inherent to small numbers of a virus and are needed to establish successful virus transmission and neutral drift, also lead to population-based sequence changes. The pathway by which one sequence can evolve to another, more fit sequence can be thought of as a fitness landscape that exists in sequence space. Because viruses can have very high error rates (10^{-4} per viral genome base, compared to 10^{-9} per host genome base) in addition to their small genomes and rapid generation times, they have the capacity to move rapidly through fitness landscapes. In fact, it is estimated that HIV-1 can evolve 1,000,000 times faster than its host. The rate of sequence change, although very high, often appears to be essentially constant with time. Because of this, it can be used as a "molecular clock" to estimate times of evolutionary sequence divergence. However, the molecular clocks for RNA viruses are so fast that some have argued that essentially all RNA viruses are recently evolved (about 10,000 ybp). The very high adaptability of an RNA or retrovirus is dependent on a high error rate. Inherent in the quasispecies theory is the existence of an error threshold above which the error rate is so high that no viable virus will be reproduced. This threshold is called "error catastrophe." It is thought that without error correction systems, RNA genomes would have restricted lengths of about 30,000 nucleotides due to this threshold. Some inhibitors of virus replication have been demonstrated to increase replication error rates and push viral systems into error catastrophe.

Fitness

Fitness is considered to be the genetic feature that increases the survival and reproductive capacity of an organism. Fitness thus dictates the viable landscape in sequence space that quasispecies would follow to attain higher fitness. However, although scientists can readily agree on the concept of fitness, it is surprisingly difficult to define in measurable terms. Two terms that are commonly used to measure fitness experimentally are relative fitness and

Table 1.5 Biological characteristics of acute and persistent virus life strategies

Acute virus life strategy	Persistent virus life strategy
No persistence in individual host	Persistence in individual host
Often disease associated	Seldom causes acute disease; often inapparent
High mutation rates (RNA viruses)	Genetically stable
Virus replicates in more than one species	Virus is highly species specific
Virus does not show coevolution with host	Virus often shows coevolution with host
Transmission is horizontal	Transmission is often from parent to offspring (vertical) or through sexual contact
Highly dependent on host population structure	Less dependent on host population structure
Seldom evolves to persistence	Often the source of emerging acute disease in new host species

reproductive ratio. Relative fitness is the ratio of the number of progeny for an individual virus relative to the average number of progeny expected for the population. For reproductive ratio, this term can be defined for viruses as the number of newly infected host organisms that arise from one initially infected host organism. Thus, fitness is normally measured in terms of virus progeny or reproduction. This concept of fitness applies to an acute virus. Table 1.5 summarizes the characteristics of a virus with an acute life strategy and a virus with a persistent life strategy. As discussed below, persistent viral infections require an altered fitness definition which may need to be broadened from definitions based simply on progeny.

Persistence and Fitness

The foundation of all the previously described mathematical models has been that viruses behave as predators towards their hosts. This is certainly well supported in the disease-based epidemic models that have been considered, such as smallpox, measles, poliovirus, influenza virus, etc. Given the very origins of the conception of viruses as disease-causing agents, this predator-prey relationship is fully justified. I have previously defined the acute viral life strategy as one in which virus replication does not persist in the individual host. All these predator-prey models and examples fit this acute-lifestyle definition. However, I have also previously defined the persistent life strategy of a virus as one in which virus is maintained in the individual host. Examples of viruses that persist in individual hosts are numerous and span the entire spectra of virus and host types. In contrast to the acute lifestyle, even an initial consideration does not appear to support the idea that viral persistence has a predator-like relationship with its host. As a rule, persistent infections are inapparent and generally asymptomatic.

They do not destroy their hosts. Instead, they involve mechanisms or strategies that ensure the maintenance and stability of the viral genome within the host. In addition, the persistent virus may compete with and exclude other genetic parasites. Persistence is usually not associated with high-level or maximal production of progeny. The more typical outcome is that a small amount of progeny is produced, which is all that is needed to transmit infection to either new hosts or progeny hosts. In some cases, such as for temperate phage, the virus behaves as a genetic element of the host genome, so its reproduction is dependent on host reproduction. This is clearly not the relationship of a predator to its prey. Table 1.6 presents a summary of the characteristics of fitness for a persistent viral life strategy.

What, then, defines the fitness of a persistent virus? I have noted that reproductive ratios are often used to measure acute-virus fitness. These ratios are dimensionless metrics and thus lack a temporal component. It is not how long the virus or its host persists that is important, but rather how many successful offspring the virus makes. Yet, upon reexamination of the general definition of fitness, it is apparent that the survival of the juvenile individual, not just the number of offspring, is significant. A long-lived individual host that is stably infected may be the only remaining host organism left after a selective sweep has exterminated its kin. In this case, the fittest virus is the one infecting the surviving host, whether or not many or any virus progeny were produced. Consider, for example, the fitness of a long-term persistent virus, such as human herpes zoster virus. After initial infection and establishment of persistence, the virus can spend up to 50 years as a very low-level, persistent (latent) infection in a single ganglion, producing no progeny. However, to be fit, it must be able to reactivate with high probability after a particular and long duration and make enough virus to infect a new host, such as a grandchild of the original host, thereby reestablishing another generation of long-lived, persistent infection.

Table 1.6 Characteristics of fitness and persistence

Fitness is the genetic contribution of a trait to an individual's continued life and to the survival of its descendants.

Persistence must increase survival time of offspring or host to allow attainment of high transmission probability, not simply maximize virus reproduction. In other words, persistence is time and condition specific. Examples are lambda and varicella-zoster virus.

Requires a mechanism or strategy (genes) to ensure persistence and reactivation and transmission—a highly selected phenotype. In other words, persistent viruses are not selfish DNA, and defectives can mediate persistence.

Not dependent on host population structure; stable during long-term evolution; congruent with host; appears to be commensal

A temperate phage in its host bacteria can show similar temporal stability. The temperate lambda phage that infects *E. coli* can be passed for hundreds of bacterial generations, seldom, if ever, reactivating. But with the proper environmental conditions, such as UV-induced DNA damage, the virus will reactivate with high probability in almost all cells. It seems in this circumstance that temporal stability is part of viral fitness, as is retention of the capacity to sense the appropriate environmental signals. Despite these examples, our understanding of persistence is poor, due to both experimental and theoretical difficulties. Persistence presents a real problem for mathematical models and confounds viral epidemiology as it is currently applied to the study of viral epidemics. As a current example, epidemic models were recently developed and presented for the severe acute respiratory syndrome (SARS) epidemic. These models appeared to have accurately predicted the containment of the epidemic. However, if SARS had established a persistent infection in some patients, the models would have failed to address the situation and also failed to predict the outcome.

Persistence, Populations, and Evolution

In general, it seems clear that persistent viruses have a different population structure than acute viruses. It also seems clear that they have different evolutionary patterns and relationships to their hosts than acute viruses. Persistent viruses tend to show much greater genetic stability, which can be observed on an evolutionary timescale. Persistent viruses also show a pattern of coevolution with their hosts. This coevolution is consistent with the fact that persistent viruses tend to be highly host species specific. Both of these issues are examined in greater detail in subsequent chapters.

The genetic stability of persistent viruses during infection was initially noticed with many persistent DNA viruses (herpesviruses, adenoviruses, polyomaviruses, and papillomaviruses). Infections with these viruses tend to be much more homogeneous and do not to show the quasispecies population structure discussed above. The viral genetic stability is such that it can be used to monitor the migration and even the evolution of its host. It was initially assumed that because many persistent infections were due to DNA viruses, error correction mechanisms prevented the generation of the genetic diversity that is characteristic of acute virus populations. However, there are now numerous examples of species-specific, persistent infections with various types of RNA viruses that also show genetic stability and coevolution with their hosts. This has been observed for hantaviruses and coronaviruses in their native rodent hosts, as well as rhabdoviruses in their bat hosts and influenza viruses in their waterfowl hosts. Because all of these RNA viruses have been experimentally demonstrated to rapidly generate diverse genome populations in laboratory settings, it is clear that these viruses have high rates of error in RNA replication. Yet they usually maintain homogeneous populations in nature and show molecular clocks that are much slower than those observed in acute viruses. These slower molecular clocks are equivalent to the slow molecular clocks of their hosts' genomes. The basis for this stability has not been determined. It is possible that to persist for long periods, the virus simply colonizes specific cells in small numbers and that the resulting homogeneous progeny virus represents relatively few replication rounds from these small numbers of colonizing genomes. This issue needs further evaluation.

In terms of host population structure, persistent viruses differ from acute viruses and their hosts. Persistent viruses are not very dependent on host population densities and can be found to be highly prevalent in nongregarious host populations. In terms of humans and their association with human-specific viruses, preagricultural human populations were likely infected with most of these persistent human viruses, and even primate relatives of humans harbor most of these types of viruses. Thus, the persistent virus-host relationship is stable on an evolutionary timescale and is not dependent on host population densities. This viral life strategy is highly prevalent in natural host populations of essentially all orders. However, this virus-host relationship necessitates that the persistent virus have a process of transmission that is closely linked to host biology. For example, persistent viral infections tend to be transmitted from old to young hosts, during sex and birth or by some other process that is inherent to the host life strategy (such as with milk-borne viruses for mammals). Thus, persistent viruses are less associated with population structures than are acute viruses. This means that persistent viruses must have the capacity to sense biological or temporal cues to ensure a high probability of transmission at opportune times. Persistence also requires that the viruses have mechanisms to ensure maintenance of the virus within the host and prevent elimination. As is discussed later, maintenance can sometimes be ensured by the use of a system called "addiction modules," as seen in various unicellular hosts. Prevention of virus elimination requires mechanisms that counteract host immunity systems, as well as sometimes involving mechanisms that suppress competition by other genetic parasites. The main point to emphasize is that persistence requires some phenotype or strategy in order for the virus to be fit, to attain temporal stability, and to ensure transmission. This clearly differentiates the concept of persistence from that of selfish DNA or genes, since by definition, selfish DNA has no phenotype for the host. A summary of gene functions associated with persistence is shown in Table 1.7.

Table 1.7 Gene functions associated with persistence

Immunity—addiction modules (restriction or modification, toxin-antitoxin)

Silencing functions (small RNAs—latency-associated transcript, defectives, methylation, repressors)

Induction and sensing functions (reproduction-linked expression, receptors, expression factors, lytic virus activation)

Immune regulators (signal transduction, innate, adaptive; e.g., cowpox virus)

Finally, it needs to be noted that a persistent virus in one host can be an acute virus in another host species or host population. Some persistent viruses are able to jump species or shift hosts to become acute viruses in new hosts. Because persistence provides an inherently more stable evolutionary relationship between virus and host, this means that most acute viral diseases will have originated and adapted from a persistent state. This situation can generally explain viral emergence from a reservoir (persistent) host.

Organizational Overview

The chapters which follow examine the evolution of viruses from the context of host evolution. Chapter 2 begins with issues related to prebiotic evolution by considering simulations of early evolution that use computers and chemical replicators to model the prebiotic origins of life. In that chapter, the possible viral role in these simulations and the effect on the outcome are considered. In chapter 3, prokaryotes and their viruses are discussed. *Bacteria, Archaea,* and cyanobacteria all are presented from the perspective of viruses specific to these host groups. In this discussion, emphasis is put on persistent (temperate) viruses and how such viruses affect host evolution. Along the way, various dilemmas of evolutionary biology (especially the acquisition of complexity) are noted and evidence of viral involvement in these situations is considered. The bacterium-phage literature is very rich and detailed, and it is hoped that the reader will not get lost in the necessary evaluation of experimental details. As described in later chapters, bacteria are the most adaptable of all cellular life forms, and it has become clear that they evolve mainly by infectious processes. Therefore, the chapter includes a discussion of why an infectious process was not apparently maintained in eukaryotes. Chapter 4 gives special attention to a major dilemma in evolutionary biology: the origin of the eukaryotic nucleus. The relationship between the origin of the nucleus in unicellular algae and viral parasites is examined in some detail. Since fungi were so crucial to the emergence of animal and plant life from the oceans, I address the relationship of fungi to their viruses in chapter 5. The early life of the oceans is thought to have been crucial for the evolution of all higher life forms. Thus, in chapter 6, I evaluate what is known concerning viruses of aquatic animals and the early evolution of metazoans, maintaining the theme of also considering what is known about persistent viruses and genetic parasites that have colonized the host genome. Terrestrial plants, insects, and their viruses appear to have evolved together to a large degree. Consequently, the chapter that presents plants and insects (chapter 7) has the unusual organization of considering the evolution of plants, insects, and their viruses all together. This trinity will hopefully allow the reader to see the viral threads that link these hosts. The final chapter (chapter 8) is the longest and addresses the vertebrate hosts that have most often been the subject of virus evolution studies, the terrestrial vertebrates. However, this chapter is presented from an evolutionary perspective, initially addressing those animals that were first to evolve and diverge and also considering the viruses that infect them. Like the chapter on viruses of bacteria, this chapter also attempts to summarize a detailed and rich literature, so the reader is presented with much specific information. Although some of these viruses and hosts are indeed well studied, this organization also makes clear the major gaps in our knowledge, such as of the monotremes and their viruses that are so poorly studied. I end the book with a consideration of human and primate evolution and the study of their viruses. Attention is paid to the evaluation of what makes us distinct from our primate brethren and the viral associations of this difference.

Throughout this book, I also consider those viral agents and their defectives that have colonized host genomes and attempt to evaluate the relationship of these agents to host and viral evolution. From bacteria to humans, clear patterns exist that are frequently ignored in the traditional presentation of evolution. In this regard, I pay attention to those sequences within chromosomes that have been much less studied, such as the human Y chromosome, and why they are so colonized by endogenous retroviruses. Too often, such sequences are dismissed as junk, ignoring any virological inferences. Table 1.8 summarizes the relationship between host orders of increasing functional and genetic complexity to the colonization of their genomes by parasitic agents. Table 1.9 compares the common genetic characteristics of human host chromosomes to various viral lineages and the putative LUCA. The immense capacity of viruses for creating genetic novelty has often been ignored but is factored into the subsequent chapters. How virus colonization has driven host evolution is explored.

Other topics addressed in many chapters of this book include examination of natural biological populations with respect to their viruses, as well as discussion of the virus-related experiences reported by practitioners of biology: the brewer, the farmer, the fisherman, and all those that have had practical experience with large populations of living organisms and have witnessed the consequences of virus infections. Natural or field studies are

Table 1.8 Evolution of complexity and acquisition of parasitic genomic agents

Species	Haploid genome (bp)	Gene length (kbp)	No. of genes	Gene density (kbp/gene)
Escherichia coli	4.2×10^6	1.2	2,350	1.8 (insertion sequence and prophage [90% coding])
Saccharomyces cerevisiae	1.3×10^7	1.4	6,100	2.1 (DNA Ty and retroposons)
Drosophila melanogaster	1.4×10^8	11.3	8,750	16 (DNA transposons and retroposons [3 and 10%])
Homo sapiens	2.91×10^9	16.3	40,000	76 (*Alu*, SINEs, LINEs[a] [<2% coding], endogenous retroviruses)

[a]SINEs, short interspersed repeat elements; LINEs, long interspersed repeat elements.

Table 1.9 Common characteristics of host and viral chromosomes[a]

Chromosome	Characteristics
Human chromosome 21	225 genes; 50,000 repeat elements; 7,000 DNA elements; 2,000 ERV elements
Viral genomes	
T4	274 genes (69 essential), 140 of questionable function (rIIa); 42 similar to entries in GenBank
CMV	220 genes, 33 conserved in other herpesviruses or hosts, most genes are novel (e.g., UL 144, persistence)
WSSV	184 genes, 11 similar to entries in GenBank (e.g., DNA replication core proteins, T-Ag)
LUCA	184 genes, not including DNA replication proteins; this number is well beyond the error threshold for RNA genomes

[a]ERV, endogenous retrovirus; CMV, cytomegalovirus; WSSV, white spot syndrome virus; T-Ag, T antigen.

significantly underrepresented in the virology literature but are essential in order to evaluate the relevance of our many laboratory-developed viral models of actual virus-host relationships. It is important to understand the realities of virus-host relationships in an ecological context and not simply to consider virus-host relationships from the perspective of diseases caused by viruses or from laboratory models, for which most viruses are highly selected for specific biological characteristics. Ever since human populations first recognized that viruses exist and can cause diseases in humans and their domesticated plants and animals, disease eradication has been the main perspective and goal of virological studies. However, in order to better understand how viruses affect the evolution of all living organisms, scientists now need to include those numerous observations in which disease is not involved. Thus, the absence or rarity of viral disease in large sea urchin farms from Japan, for example, should receive equal attention as a natural demonstration of viral-host fitness and evolutionary success. It is the absence of such balanced observations and our failure to consider what such observations imply for the evolution of life that have led to the currently one-sided view of virus-host relationships. It is the intent of this book to provide the less popular, but more biologically relevant, perspective of persistent viruses.

The discussions in this book often stem from the consideration of simple, childlike questions, such as, "Where do viruses come from?" "Why do some viruses persist?" "Why do some viruses make us sick, but not others?" The reality is that most viruses do not cause disease in their native hosts. So more questions will follow, such as, "Why not, and why only sometimes?" Viruses will be seen to be fundamental, present at the dawn of life and present today as all species continue to differentiate. Pathogenesis can be looked at as resulting from a failed state of persistence. New host orders may result from successful viral persistence. Evolution itself may depend on this colonization process in order to create genetic complexity. It is hoped that this style of organization will stimulate many other such questions and possible answers and serve to remind us that we really have much to learn about what viruses are and what they do to life on our world. Viruses are part of this world and have an evolutionary power that is immense and unmatched by any other living entity. But how virus evolutionary power applies to host evolution is a topic in need of much study. We have a strong cultural bias regarding the concept of "virus." Our history and suffering have led us to view viruses as the hidden enemy, evil entities that simply seek to destroy life, and many books have been written with titles along these lines. But perhaps we have simply ignored those situations where destruction is not the outcome. And sometimes, it appears that the destruction wrought by a specific virus is the outcome of infecting a nonideal host. For example, humans have been lethally infected by viruses found in African monkeys (HIV), waterfowl (influenza virus), Gambian rats (monkeypox virus), or civet cats or their prey (SARS virus), all of which harbor these very viruses as benign persistent infections. Do the viruses of these species really exist only so that they can adapt to destroy us? What about the silent viruses all humans harbor but which seldom, if ever, cause disease in us or any other organism, such as TT virus, papillomaviruses, and polyomaviruses? Of what species are they the enemy? Clearly the perception of virus as merely the enemy of life is simpleminded and faulty.

In closing this introduction, it is worth considering a perception put forward by S. E. Luria, who helped articulate the modern definition of a virus as infective genetic material in the paper "Bacteriophage: an essay on virus reproduction" (*Science* **111**:507, 1950). Later that decade, when considering the role viruses might play in host evolution of the host, he wrote:

> . . . may we not feel that in the virus, in their merging with the cellular genome and re-emerging from them, we observe the units and process which, in the course of evolution, have created the successful genetic patterns that underlie all living cells?
>
> S. E. LURIA,
Virus Growth and Variation, 1959

I would amend this perception to include that it is the viruses that can persist in their host cells that have left their indelible mark on and assisted in the evolution of the cells of all life.

Recommended Reading

Historic Accounts and Definitions

Brock, T. D. 1961. *Milestones in Microbiology*. Prentice-Hall, Englewood Cliffs, N.J.

Brock, T. D. 1999. *Milestones in Microbiology: 1546 to 1940*. ASM Press, Washington, D.C.

Lipsitch, M., M. A. Nowak, D. Ebert, and R. M. May. 1995. The population dynamics of vertically and horizontally transmitted parasites. *Proc. R. Soc. Lond. B* **260**: 321–327.

Luria, S. E., E. Kellenberger, B. D. Harrison, W. Schafer, G. K. Hirst, A. Isaacs, J. M. Hoskins, M. G. P. Stoker, H. Rubin, H. B. A. P. Maitland, P. D. Cooper, E. S. A. A. Anderson, J. A. Niven, S. F. Janet, C. Morgan, and H. M. Rose. 1959. *Virus Growth and Variation: Ninth Symposium of the Society for General Microbiology.* Cambridge University Press, London, England.

van Regenmortel, M. H. V., and B. W. Mahy. 2004. Emerging issues in virus taxonomy. *Emerg. Infect. Dis.* 10:8–3.

Mathematical Biology

Doolittle, W. F., and C. Sapienza. 1980. Selfish genes, the phenotype paradigm and genome evolution. *Nature* 284:601–603.

Kerszberg, M. 2000. The survival of slow reproducers. *J. Theor. Biol.* 206:81–89.

Lipsitch, M., M. A. Nowak, D. Ebert, and R. M. May. 1995. The population dynamics of vertically and horizontally transmitted parasites. *Proc. R. Soc. Lond. B* 260:321–327.

Nowak, M. 1991. The evolution of viruses. Competition between horizontal and vertical transmission of mobile genes. *J. Theor. Biol.* 150:339–347.

Nowak, M. A., and R. M. May. 2000. *Virus Dynamics: Mathematical Principles of Immunology and Virology.* Oxford University Press, Oxford, England.

Szathmary, E. 1988. A hypercyclic illusion. *J. Theor. Biol.* 134:561–563.

Szathmary, E. 1992. Viral sex, levels of selection, and the origin of life. *J. Theor. Biol.* 159:99–109.

Issues of Evolutionary Biology

Cracraft, J., and M. J. Donoghue (ed.). 2004. *Assembling the Tree of Life.* Oxford University Press, Oxford, England.

Doolittle, W. F., and C. Sapienza. 1980. Selfish genes, the phenotype paradigm and genome evolution. *Nature* 284:601–603.

Giske, J., D. L. Aksnes, and B. Forland. 1993. Variable generation times and Darwinian fitness measures. *Evol. Ecol.* 7:233–239.

Kyrpides, N., R. Overbeek, and C. Ouzounis. 1999. Universal protein families and the functional content of the last universal common ancestor. *J. Mol. Evol.* 49:413–423.

Maynard Smith, J., and E. Szathmáry. 1995. *The Major Transitions in Evolution.* W. H. Freeman Spektrum, Oxford, England.

Orgel, L. E., and F. H. Crick. 1980. Selfish DNA: the ultimate parasite. *Nature* 284:604–607.

Ryan, F. 2002. *Darwin's Blind Spot.* Houghton Mifflin Co., Boston, Mass.

Evolutionary Virology

Domingo, E., R. Webster, and J. J. Holland (ed.). 1999. *The Origin and Evolution of Viruses.* Academic Press, San Diego, Calif.

Holland, J. J. 1998. The origin and evolution of viruses, p. 11–21. In W. W. Topley, C. S. Wilson, L. H. Collier, A. Balows, and M. Sussman (ed.), *Topley & Wilson's Microbiology and Microbial Infections*, 9th ed. Oxford University Press, London, England.

Mindell, D. P., J. S. Rest, and L. P. Villarreal. 2004. Viruses and the tree of life. *In* J. Cracraft and M. J. Donoghue (ed.), *Assembling the Tree of Life*. Oxford University Press, Oxford, England.

Nowak, M., and P. Schuster. 1989. Error thresholds of replication in finite populations: mutation frequencies and the onset of Muller's ratchet. *J. Theor. Biol.* **137:** 375–395.

Szathmary, E., and L. Demeter. 1987. Group selection of early replicators and the origin of life. *J. Theor. Biol.* **128:**463–486.

Villarreal, L. P., V. R. Defilippis, and K. A. Gottlieb. 2000. Acute and persistent viral life strategies and their relationship to emerging diseases. *Virology* **272:**1–6.

Viral Biology and Ecology

Cooper, J. 1995. *Viruses and the Environment*, 2nd ed. Chapman & Hall, New York, N.Y.

Hurst, C. J. 2000. *Viral Ecology*. Academic Press, San Diego, Calif.

Oldstone, M. B. 1998. Viral persistence: mechanisms and consequences. *Curr. Opin. Microbiol.* **1:**436–441.

Insights from Simulated Evolution

Introduction

All biological systems, including viruses, are essentially systems that store, copy, and express information. Because these basic attributes can also apply to man-made systems of information, including computer programs, it seems logical to consider that theoretical models derived from artificial computer-based or simulated systems could provide insight into some of the basic principles of biological information systems. This hope, that simulations are biologically informative, has been the motivation to develop and evaluate a large array of chemical- and computer-based models that attempt to emulate the evolution of biological systems. These models seek to create a bottom-up solution to problems of the early evolution of life. And it is hoped that if they are correct, then basic features of living information systems will also be elucidated, including the development of the complex (nonlinear or non-additive) behaviors that lead to the emergence of the complex characteristics in living organisms. The capacity for more complexity to emerge from less complex systems is a basic feature of evolving, living systems that is currently poorly understood. A common, and sometimes compelling, argument against these artificial computer and chemical models is that the fact that something can be modeled, or a model exists that has good internal consistency or behaves in complex ways, does not necessarily make the model applicable to the real biological world. It thus becomes incumbent on the model builders to show that the systems that have been developed have clear relationships to authentic biological processes. However, when we try to model very early events in evolution, such as prebiotic evolution, we have few if any solid facts that can be applied to evaluate the validity of the simulations. In this case, the simulations themselves may be one of our only sources of insight as we attempt to reconstruct the process from which life and viruses emerged. The concept here is that if we are indeed able to understand some of the more theoretical aspects of early evolution, we may be able to understand and predict the emergent properties leading to living systems. This chapter presents some of these insights of simulated evolution and attempts to evaluate the relevance of these simulations to extant biological systems and processes that we can now observe.

Viruses: Parasites of the Prebiotic World

What is the ultimate origin of the virus?

It is generally accepted that prior to the evolution of cellular life forms, there must have existed a period in which precellular, chemical, lifelike forms (or autocatalytic replicators) existed as the predecessors of cellular life. Such replicators would essentially have been chemical entities that were able to catalyze and template their own synthesis from existing substrate molecules which were present in the primordial soup or had spontaneously generated. The study of chemical replicators, described below, thus attempts to create models of catalysis from which prebiotic characteristics can be determined. Because all existing life now uses nucleic acid-based genetic information and protein-based catalysis, it seems likely that the prebiotic replicators that led to extant life forms would also have been based on nucleic acid-related chemistries. The main molecule for the storage of genetic information in extant life forms, DNA, is a rather chemically inert and stable molecule, which could not perform the needed catalysis. In contrast, RNA is known to function as the genetic material of viruses and also as a catalyst, in which case it is referred to as a ribozyme. Accordingly, one of the more accepted views of the prebiotic world asserts that autocatalytic RNA was the principal molecule for both information storage and catalysis. This situation constitutes the prebiotic "RNA world." However, there is one problem with this reasoning. Although RNA is known to be catalytic, especially with respect to the cleavage of RNA bonds, it is not efficient at polymerizing RNA during replication of an RNA template. Thus, RNA appears to lack one of the basic features required for a prebiotic replicator. Even so, inefficient RNA-based RNA polymerization may still have sufficed in the prebiotic world to allow preliving replicator systems to get a foothold, since there would have existed no competing or more efficient replicators. If so, the subsequent development and evolution of the much more efficient protein-based catalysts may be viewed as the emergence of a much faster, parasitic protein replicator that imposes on RNA templates and replicators. It has been assumed that with the emergence of protein catalysis and the ensuing faster RNA replication, replicators solely based on RNA became essentially extinct. RNA viruses and viroids may represent the sole descendants of this prebiotic world, in that only they retain an RNA-based genome.

It is often assumed that, in the precellular world, viruses were absent. This is because viruses are obligate parasites of cellular hosts and are dependent on host-specific systems for replication (e.g., protein translation and energy generation systems). Accordingly, viruses presumably could have come into existence only after the genesis of the cellular life forms needed to support their replication. However, according to the definition of a virus as simply a molecular genetic parasite, any genetic replicator, even noncellular prebiotic ones, would be susceptible to parasitic replicators or viruses. The tendency for replicators to become parasitized, and even for the parasitic replicators themselves to become parasitized, is a well-established phenomenon in virology. The parasites of parasitic replicators would correspond to the defective viruses that are observed for most types of viruses. Defective viruses are thus exactly the parasitic replicators of a functional virus, itself a parasitic replicator. These parasites of parasites are expected to have existed even under prebiotic conditions, as is described further below. In addition, computer-based modeling of replicator evolution, discussed below, also suggests a role for the parasitic replicators as well as the parasites of parasites.

Precellular RNA replicators: some dilemmas

As mentioned above, it is currently accepted that the prebiotic world may have been the domain of self-replicating RNA molecules, or ribozymes. However, as also mentioned, there are no surviving autonomous organisms or subcellular organelles that use RNA as their genetic material. Only some types of viruses (especially negative-strand single-stranded RNA viruses) and viroids may represent the sole remaining descendants of this RNA world, since they are the only biological entities, in addition to some naturally occurring ribozymes, that use only RNA as a genetic system of information. Furthermore, there are no known cellular analogues for RNA-dependent RNA polymerases. If this inference is correct, these families of viruses may retain some basic features present during the prebiotic period, such as frequent, conserved secondary structures in the RNA. In the cases of viroids and negative-strand RNA viruses, a DNA-based origin cannot be proposed, as there is no clear DNA-based process that might have led to the generation of these RNA systems. The viral replicase seems to be a very early invention in evolution that may therefore predate the evolution of DNA-based cellular life. Interestingly and consistent with this idea, it has also been proposed that the RNA-dependent RNA polymerases, which are central to the replication of both positive- and negative-strand single-stranded RNA viruses, may be ancestral to the DNA-dependent RNA polymerases, based on protein structure and phylogenetic considerations. Alternatively, these viral RNA polymerases may have evolved independently after the evolution of DNA. Scientists are currently unable to differentiate these two scenarios since there is no way to calibrate when the RNA viruses and DNA-based genomes might have emerged.

Currently, the most conserved features of many RNA viroids and RNA viruses are found in the secondary structures of the RNA molecules, especially the stem-loop structures associated with priming and replication as well as the RNA replicase gene in the case of RNA viruses. Mutation and reversion experiments confirm the importance of secondary structure independent from any coding potential for many RNA viruses. Frequently, there also exists a link between stem-loop structures and priming or RNA replication, as the replicase is usually covalently attached to the 3′ end of such stem-loop structures and is needed to prime RNA synthesis. This priming reaction is unique to viral systems, as no similar process is used by the host. Also, this priming step appears to define a distinct strategy to determine the molecular basis of virus identification and directs RNA replication to viral and not host RNA molecules. However, as alluded to above, protein-primed RNA replication poses a problem for current concepts of the RNA world, as it requires the simultaneous evolution of the template and the replicase without the coexistence of a translational system. This problem of simultaneous evolution of complementary functions is actually part of a much more general problem in evolutionary biology, the development of complex phenotype. The dilemma of complex phenotype is discussed further in chapter 3.

Adhering to the basic definition of a virus as a molecular genetic parasite, it can now be argued that even prebiotic replicators would be prone to generation of and infection by viruses. An autocatalytic, self-replicating RNA molecule not only would need to copy its own genetic information but also would need to replicate its catalytic activity in order to synthesize another self-replicating RNA molecule. In a sense, the utility as a template and the ability to perform the catalysis required for replication can be considered two separable functions. Because of this, a variant template might still function as a template but lose its ability to catalyze the synthesis of another template. In addition, for evolution to occur, some process of variation in the template that also results in variation in the catalytic properties of the RNA is needed to yield a more fit phenotype. If this process is considered from the perspective of parasites, it can be seen that it is possible for a parasite to separate the template or replicon function from the catalytic function of the RNA and thus drive the evolution of the system through competition with more efficient parasitic replicators. Therefore, the functional separation of template from catalysis renders the RNA replicator susceptible to molecular parasites and presents a situation that drives evolution to a higher efficiency.

In order for the molecular RNA parasites to initially emerge, all that is needed is a variant or defective of the catalytic RNA template that is able to be copied with greater relative efficiency than the original template and yet lacks catalytic activity. This defective RNA will therefore be continually copied without needing to invest the time to catalyze the synthesis of daughter molecules, thereby outcompeting its "host" catalytic RNA template. The resulting parasitic replicator is defective for catalysis but relatively efficient for replication. Such variations are expected to be frequent since they involve relatively simple loss-of-function mutations. However, such parasitic replicators would remain dependent on the occurrence of the replication-competent, ribozyme templates. Thus, parasitic replicators are more likely to initially occur under conditions where the catalytic replicators are prevalent. In addition, this parasitic dependence on the simultaneous or episodic presence of the catalytic replicator would establish selective conditions favoring the persistence of the parasitic replicator. If a parasitic replicator can persist in an inert state in the environment until a catalytic replicator is encountered, it will retain the capacity for competitive reproduction. Thus, defective replicators will also be selected for persistence in the environment. This process is, in fact, the well-established basis for the generation of defective viruses, noted above, and it can explain various biological phenomena associated with defective viruses. Defective viruses are infectious viruses with deletions that often have a replicative advantage over their infectious host genomes but may not contain any genes or catalytic activity themselves. The generation of defectives is conditional: they require the presence of an infectious helper virus, which will occur in infections having high virus frequency, such as during the high-multiplicity passage of virus stocks. Most virus systems (especially RNA and retroviruses) are prone to the generation of defectives, which can outnumber the infectious virus. Furthermore, defective viruses have been shown experimentally to mediate persistent infections under some conditions. Like these defective viruses, prebiotic replicators are expected to host parasitic versions of themselves. And we can expect these parasites to drive the evolution of the system.

Chemical Replicators

Another area of research, the study of chemical replicators, seeks to study the chemical principles of prebiotic replicators. In these systems, chemical substrates are presented to a chemical replicator molecule that is able to catalyze the assembly of the substrates into a copy of itself. The main problem with these systems is that the various chemical replicators able to catalyze their own synthesis have, for the most part, no clear relationship to biologically relevant molecules. These are mainly simple organic molecules that are able to stimulate the chemical bonding of two or more substrate molecules provided in the reaction medium. Thus, it is not clear whether the chemical replicator models are very informative about the more complicated biological catalysis needed for living systems. Still, it is hoped that some of the chemical characteristics of self-replicating systems can be applied to help elucidate the process of prebiotic replication. Some of these systems have interesting and complex topological behaviors and are able to form intriguing two-dimensional patterns of products, such as concentric circles and swirls, when reactions are conducted on flat surfaces. In some instances it has been reported that these reactions can terminate in circular patterns. It is interesting that at the boundaries of these patterns, there can at times be found defective versions of replicator molecules, which are assembled from substrates but fail to catalyze the assembly of daughter molecules. As mentioned above, the relationship of these chemical replicator models to prebiotic conditions has been questioned, since the molecules used often have little apparent chemical similarity to the molecules involved in living systems, such as RNA, amino acids, or proteins. In addition, chemical replicators frequently lack effective mechanisms to introduce "genetic" diversity, and they tend not to show the complex behaviors associated with living systems, such as increasing informational or chemical complexity. However, these replicators still retain the capacity to propagate information in future chemical generations. Thus, they do display the rather basic characteristic of information transmission. Nevertheless, these systems lack an essential element in that they do not link the production of the substrates needed for their synthesis to their replication.

Chemical replicators and hypercycles

Manfred Eigen has noted that RNA viruses reveal two principles of organization seen in all living systems, including prebiotic systems. These two principles are cyclic reaction pathways and compartment formation. In considering the dichotomy between genotype and phenotype, he has proposed that these two processes can be linked by cyclic feedback coupling. Thus, the replication of genetic information (genotype) can be considered one cycle and the catalysis and synthesis of the replication enzyme (phenotype) can be considered another cycle, and these two cycles are coupled (as a hypercycle). The coupling of these cycles requires compartmentalization. With respect to this coupling, Eigen wrote, "A mutation in the genotype that expresses itself in the phenotype brings about immediate evolutionary response" (see Eigen and Winkler [1992] in Recommended Reading). This leads to reaction cycles with a superimposed, but higher-order, coupling he called a hypercycle. The expected time-dependent behavior of these hypercycles could be expressed in a series of differential equations. A feedback loop would exist that connects the replication enzyme to its RNA template and requires that these two remain within each other's vicinity, i.e., that they be compartmentalized. The compartments would also allow the containment of template quasispecies. Hypercycles have the feature of limiting competition between different replicators through cyclic coupling, which allows complementing replicators to share any advantageous variations that might arise. Hypercycles may also allow a quasispecies to maintain information content over many generations. It was argued that the selective advantage of the entire hypercycle is strong due to the quadratic rate law that controls growth of the hypercycle. Thus, the hypercycle idea seems to account for some important behavior of biological systems and would be applicable to the early evolution of prebiotic chemical replicators. Eigen used the example of an RNA virus, such as influenza virus, to develop the concept of hypercycle. Figure 2.1A shows a hypercycle (n_1) as it would apply to an RNA virus that must replicate plus- and minus-strand templates (genotype). It also shows the reproduction of the replicase (E_1) and coupling to the

Figure 2.1 (A) Hypercycles (nonparasitic) as developed for an RNA virus by M. Eigen. Plus- and minus-strand templates (L_1) are indicated. Additional coupled hypercycles are not indicated. (B) Parasitic hypercycle. Shown is the emergence of a parasitic and defective template (L'_{1-}) with enhanced replication relative to that of the full (L_{1+}) template.

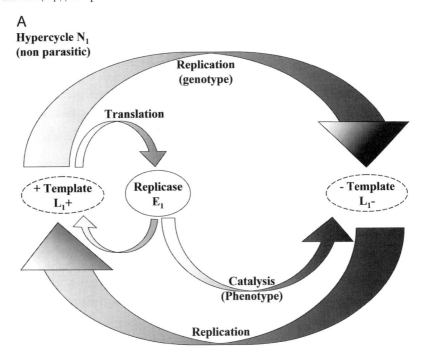

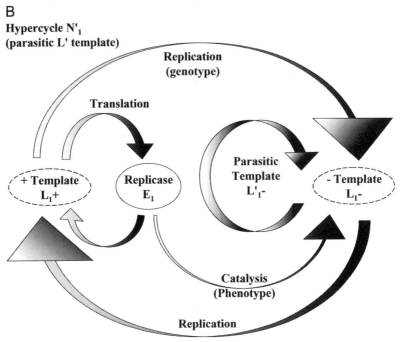

plus-strand template (L_{1+}) and the minus-strand template (L_{1-}). A cyclic linkage of distinct hypercycles (not shown) would be indicated by n_2, n_3, etc.

Although hypercycles unite the workings of several genes and can evolve as a unit, one of the curious features of hypercycles is that they tolerate no internal competition because the cycle tightly links phenotype (catalytic activity) to genotype. This intolerance and linkage do not seem biologically realistic to many. For example, the translational machinery of the cell translates all genes, not just a linked set of transcripts. In fact, this curious feature of hypercycles also limits the evolution of new species of replicators. This is because the hypercycle assumes that the replicase (expressed as E_i) has highest affinity for its own template molecule (I_i). If we consider the problem of n species generating a new species ($n + 1$), we see the dilemma that the new replicase and template must appear simultaneously (I_{n+1} and E_{n+1}), a scenario that is not accommodated by the hypercycle model. This is essentially the same problem as the acquisition of complex phenotype, noted above, and is also related to the possibility of punctuated evolutionary events. Thus, the evolution of new species poses a problem for the hypercycle model. Even more troubling, however, would be the occurrence of a parasitic replicator, with higher affinity for the replicase. If this were to happen, the hypercycle would collapse. Given the discussions above about the ease of generating parasitic replicators and also given the hypercycle requirement of compartmentalizing the replicase with the cognate template, these conditions can be expected to favor the emergence of a parasitic or defective replicator. The consequence of such a defective and parasitic template (L'_{1-}) is that it out-replicates the full template (L_{1-}) and inhibits the production of replicase (E_1). This will eventually shut down the entire hypercycle. This is shown diagrammatically in Fig. 2.1B. However, the hypercycle is not necessarily destroyed. It may simply be shut down in a state of persistence, awaiting conditions that allow replicase production. This is not always a bad thing. A replicating hypercycle needs to continually consume substrates to synthesize the replicator. But one can expect the environment to vary in its supply of substrate, occasionally being depleted of substrate. A parasitized hypercycle tends to survive these hard times intact. Eigen considered hypercycle selection to be a direct consequence of replication. However, selection can also be attained by persistence that is favored by parasites. Although parasitic networks that have interdependent elements with various half-lives and various links between phenotype and genotype might also be considered, they are not a component of the hypercycle model. Thus, the hypercycle model appears not to accommodate the likely effects and contributions of parasitic replicators.

Insights from Artificial-Life Simulations

Computer programs have also been used to try to model and understand the behavior of self-replicating genetic molecules subjected to Darwinian selection and evolution, including sexual exchange. These computer models are collectively known as artificial-life programs. The intent of these computer-based studies is to evaluate whether lifelike behaviors can emerge from man-made systems that utilize self-reproducing automata. As with the chemical replicators, the relevance of these systems to life has been questioned, since most computer-based programs do not need to consume physical substrates, such as food, to exist. Nonetheless, the behavior of an information system, such as those modeled with computers, is governed by many of the same basic issues as biological systems, including information content, copy fidelity, error rates, error thresholds, and complex, nondynamic behaviors. From the simple behaviors of these information systems, it is hoped (and has at times been observed) that the more complex behaviors of metabolism, reproduction, and evolution will emerge. Furthermore, many feel that these systems display nonlinear characteristics that make them worthy of continued investigation. As an aside, it is important to distinguish the field of artificial-life studies from that of artificial intelligence (AI). AI seeks to create programs that can solve problems from a top-down programming approach. The field of AI has often been met by hostility in the biological sciences, as it is accepted by biologists that nature is fundamentally parallel and that more complex properties emerge from the bottom up, not from the top down.

Artificial-life simulations generally depend on some type of self-reproduction process. The basic concepts of self-reproduction in a computer system were first put forward by John von Neumann. When von Neumann defined the simple, formal elements of self-reproduction, he noted that a self-replicating system would need to be able to copy the machine as well as the descriptions or instructions for the machine (see Levy [1992] in Recommended Reading). However, these descriptions would have both interpreted instructions, which are needed for self-construction, and uninterpreted instructions, which would be passive, unexpressed data needed to form the description of the offspring. Thus, inherent in self-reproducing automata is the requirement for both silent and active information. From these basic concepts, along with the addition of elements that are basic to evolution, such as heredity, variability, fecundity, and fitness, an array of computer-based simulations that attempt to model the behavior of living systems have been developed. Frequently, these programs are implemented as graphic simulations that compete for and occupy the computer screen. Some of these programs are of a more practical nature, such as genetic programs that are used to solve problems that may not initially be well defined. These programs emulate genetic processes, including sexual exchange, recombination, and crossover, to combine solutions into a maximized set.

Another fundamental characteristic of biological systems and their computer simulations is the ability to identify self from nonself and kin from nonkin. All biological systems, including viruses, differentiate their own and related lineages from others. In the simplest of genetic systems, this is often accomplished at the level of catalytic recognition of the template, such as polymerase recognition of an origin of replication. In more complex biological systems, such as cells and organisms, many other self-identification systems are known to exist. Thus, an automaton or program set from a computer simulation will also need to differentiate its own descendants or instruction sets from those of other, competing automata. Without a rigid kin definition, the rapid emergence of deceitful automata would be expected; these deceitful automata could elicit assistance from (complementing or "suckering" the other program) without contributing to propagation of the functional automaton. The deceitful automata would be helped in, but not obligated to help, either their own reproduction or that of other automata and would be allowed to utilize resources more efficiently than the functional automaton. Therefore, most artificial-life simulations also have some defined process by which kin identification is maintained.

Real and electronic viruses

If a biological virus is defined as a molecular genetic parasite that directs the host to maintain and copy the parasite, then the similarity to a computer virus is clear. A computer virus is basically a file or instruction set that instructs the host computer to maintain or to copy and transmit that instruction set. Thus, a computer virus is an electronic file able to parasitize a computer for the purpose of self-reproduction. Because both biological and computer information systems must copy information, a very basic question can be posed: is it possible to make a system that is able to copy information but is not also capable of being parasitized or making unauthorized file copies? Can it be made virusproof? In other words, is it possible to design a computer (or biological) information system that is able to prevent all disallowed or parasitic copies of information? Many would assume that it should indeed be possible to design such a system, given sufficiently elaborate safeguards to prevent unauthorized file copying or modification. Thus, there has ensued the endless effort to design computer operating systems and scanning software that prevent infection by computer viruses and unauthorized file copying. Still, it seems that some enterprising and creative computer hackers always manage to design viruses that get past all these protections, hence the need for endless upgrades to protection software involving the latest set of computer virus definitions. There seems to be no end to this process. And indeed, theory may tell us that this is to be expected. Interestingly, this question about preventing all strategies for unauthorized file copying can be and has been posed in mathematical terms and thus can be subjected to rigorous mathematical analysis. Clear but surprising results were obtained by an orthogonal proof which established that it will not be possible to design a self-reproducing information system that can prevent all forms of unauthorized file copying, although very restricted copying can be attained.

If self-reproducing information systems cannot be made virusproof, what does this imply for the evolution of biological viral and host systems? Will all biological systems inevitably become parasitized by the viruses that are allowable within that system? In terms of computers, computer viruses are often platform or software specific. Do we expect similar "platform-specific" (species-specific) biological viruses to develop? Are such infections inevitable? With computers, can we expect that all will eventually be exposed to some form of computer virus? Practically speaking, this certainly seems to be the case, as almost everyone in the world who works with a computer has had to deal with virus infections. And as predicted by theory, this situation appears to be unending. If so, what does this suggest for the future of computer system development, and what does this suggest for the evolution of biological information systems? Clearly, not all computer viral strategies are allowed on all computer systems, so some limitations are apparent. Do biological systems also show broad patterns of platform- or host-restricted virus parasitization? If so, how do these patterns of "allowed" information parasites affect the development or evolution of the system, especially with respect to the identity mechanisms of that system?

Artificial life and parasites

Simulated evolution has often been attempted on the two-dimensional world of computer screens. Programs such as Artificial BUGS (W. Packard) provide an artificial ecology in which simple graphic organisms that reside within a lattice seek and compete for "food" so that they can reproduce. Generally, these models predefine the nature of the replicator and also provide finite lifetimes during which they must succeed at reproduction. The basic features of Darwinian evolution (descent from common ancestors, genetic variation, sexual exchange, competition, survival of the fittest) are also programmed into the simulations. With these added features, the replicator programs are allowed to run their course of computer-simulated evolution. The results have often displayed complex behavior, sometime even "creepy" behavior that appears to emulate living things in their behaviors. However, the simulations have generally failed to develop the richness inherent in biological evolution, including the tendency to generate more complexity.

One particular model, however, did result in the evolution of greater complexity. This model was developed by T. S. Ray in 1992 and was called Tierra. In this model, the replicator is not specified ahead of time. In addition, time slicing of the central processing unit (CPU) gives each program limited time to execute the program, thereby simulating a basically parallel environment. The idea behind this model was to allow the replicator itself to evolve and compete, not to be predetermined. The time-sliced character of this model selects for speed of the replicator and removes the offspring of replicators with time. In addition, a built-in generator of diversity is provided. When the simulation was run, within a few million instructions parasitic species were seen to develop from single point mutations. These parasitic species were generally shorter replicators that lacked copy instructions and thus were copied more rapidly than parental replicators. The parasitic replicators were allowed use of other functional "organisms'" codes (complementation) and thus were very analogous to defective viruses. However, like defective viruses, these defective replicators needed these other similar, functional "organisms" to provide copy instructions. If the parasites became too successful, they consumed the available CPU space and did not copy themselves. This behavior is very reminiscent of the von Magnus phenomenon known to apply to many defective viruses. If a defective virus is too successful, it prevents the copying of the infectious virus, resulting in inhibition of replication for both the defective and the infectious viruses. Eventually, inhibition is so strong that only a few progeny are generated, and these will have to be nondefective infectious viruses in order to establish a subsequent infection. Then, the infectious virus is transiently free of defectives and replicates rapidly. But the abundance of infectious virus allows for the rapid generation of defectives, and the cycle starts again. This results in the cyclic, sinusoidal production of infectious virus with a phase-shifted sinusoidal production of interfering defective virus.

In the Tierra model, variants of these parasitic replicators then evolve and attempt to block or poison the preexisting parasites' use of the CPU. This results in the generation of a new, more complex order of parasites— the hyperparasites, which are parasitic to the existing parasites. Over a long period (two billion to three billion generations), yet even more complex parasites develop to be hyperhyperparasites. This evolutionary behavior can be punctuated, leading to "bust and sweeps" as new species of replicators become prominent. Overall, during this evolution it is seen that the initial "organisms," which started with relatively small instruction sets, eventually evolve to larger instruction sets (four to eight times) and require considerable time to take over the population of replicators.

Thus, starting from the premises of an undefined replicator and Darwinian evolution, these computer replicator programs spontaneously developed genetic parasites, which in turn drove the evolution of replicators to new and higher complexity. It seems possible that these results may identify some rather basic aspects of systems that replicate information. However, one would be hard-pressed to state that such results relate directly to the early evolution of biological replicators. We have little evidence to bring to bear on this issue. Faced with an inability both to reproduce early biological replicators and to reconstruct their history, computer simulations, however, remain our only system to explore some of these issues.

Persistence in Simulations

Computer-based models or simulations are only as good as the premises on which they are based. Many of the premises basic to our understanding of the Darwinian evolutionary process have thus been incorporated into the simulations. However, as discussed in chapter 1, not all of the basic issues of evolution, such as the fitness of persistence, have been defined sufficiently well to be implemented in computer programs. I have noted that the distinct acute and persistent life strategies of viruses have dissimilar features with respect to replicative success and fitness. Mathematical models have been developed that appear to accurately reflect the replication and population dynamics of acute viruses, using many of the accepted premises of the evolutionary process. However, the persistent life strategy, common to so many viruses, is not a well-developed topic, nor are there effective models of this life strategy. In considering the time-sliced Tierra model presented above, one quickly encounters a conundrum in the definition of fitness as it applies to persistence. The time-dependent character of the persistent life strategy and the manner in which this strategy is used to compete with rapid, acute replicators would seem not to fit most of the premises of the simulations that have been evaluated to date. As a biological example, consider the fitness of an *Escherichia coli* organism colonized by a lysogenic lambda phage. One hundred generations of the host can occur during which the virus replicates in direct concert with the host, yielding an amount of progeny virus equivalent to that from only a few rounds of replication of an acute virus. Yet even with this long period of viral silence, it would still be essential for the fitness of the lysogenic lambda to be able to move to a new host when host survival becomes problematic. For example, when the host sustains mortal damage from UV light irradiation, it is crucial for viral fitness for the virus to have maintained a high probability of reactivation, allowing for replication and transmission to another host. Thus, a temporal component, during which the phage is not replicating but is persistent, stable, and still capable of replication, is under positive selection. During the silent period, the virus is not under selective pressure to maximize virus replication. However, during this time, the virus may still need to compete with and exclude parasitic competitors (see chapter 3). Thus, the fitness profile of a persistent virus seems to operate on a different timescale than is normally considered in most acute models of virus replication. The temporal component for persistent virus selection may be even more extended than described above for lambda. For example, consider a phage that is lysogenic in a spore-forming, gram-positive bacterial host which has undergone sporulation. The lysogenized spore might sit idle for a very extended period, possibly thousands of years by some measurements. Yet viral fitness still depends on the phage's ability to survive this extended period in its dormant host, while retaining the ability to replicate progeny virus at the appropriate time and condition of spore reactivation. Clearly, such longevity can be a way to attain successful continuation of a viral lineage. There have been some attempts to model slow replicators and how they might compete with fast replicators (see Recommended Reading). However, these models still rely on relative replication rates for fitness definitions, and so our existing models of fitness and selection do not properly address persistence. For an extreme example of this issue, consider how fit an eternally living individual might be. With no progeny, our eternal life form would be unfit. Clearly, models must first develop more formal ways to define the issues of persistence before they can provide useful insights into the successful life strategies of a persistent virus.

Recommended Reading

Chemical Replicators

Gesteland, R. F., T. Cech, and J. F. Atkins. 1999. *The RNA World: the Nature of Modern RNA Suggests a Prebiotic RNA*, 2nd ed. Cold Spring Harbor Laboratory Press, Cold Spring Harbor, N.Y.

Hutton, T. J. 2002. Evolvable self-replicating molecules in an artificial chemistry. *Artif. Life* 8:341–356.

Shapiro, R. 2000. A replicator was not involved in the origin of life. *IUBMB Life* 49:173–176.

Szabo, P., I. Scheuring, T. Czaran, and E. Szathmary. 2002. In silico simulations reveal that replicators with limited dispersal evolve towards higher efficiency and fidelity. *Nature* 420:340–343.

Szathmary, E. 2000. The evolution of replicators. *Philos. Trans. R. Soc. Lond. B* 355:1669–1676.

Hypercycles

Cronhjort, M. B. 1995. Hypercycles versus parasites in the origin of life: model dependence in spatial hypercycle systems. *Origins Life Evol. Biosph.* 25:227–233.

Eigen, M., P. Schuster, K. Sigmund, and R. Wolff. 1980. Elementary step dynamics of catalytic hypercycles. *Biosystems* 13:1–22.

Eigen, M., and R. Winkler. 1992. *Steps towards Life: a Perspective on Evolution.* Oxford University Press, Oxford, England.

Szathmary, E. 1988. A hypercyclic illusion. *J. Theor. Biol.* 134:561–563.

Szathmary, E. 1992. Viral sex, levels of selection, and the origin of life. *J. Theor. Biol.* 159:99–109.

Artificial-Life and Evolution Simulations

Adami, C. 1998. *Introduction to Artificial Life*. Springer, New York, N.Y.

Gesteland, R. F., T. Cech, and J. F. Atkins. 1999. *The RNA World: the Nature of Modern RNA Suggests a Prebiotic RNA*, 2nd ed. Cold Spring Harbor Laboratory Press, Cold Spring Harbor, N.Y.

Huberman, B. A., and N. S. Glance. 1993. Evolutionary games and computer simulations. *Proc. Natl. Acad. Sci. USA* 90:7716–7718.

Hutton, T. J. 2002. Evolvable self-replicating molecules in an artificial chemistry. *Artif. Life* 8:341–356.

Langton, C. G. 1992. *Artificial Life II: Proceedings of the Workshop on Artificial Life Held February 1990 in Santa Fe, New Mexico*. Addison-Wesley, Redwood City, Calif.

Lenski, R. E., C. Ofria, R. T. Pennock, and C. Adami. 2003. The evolutionary origin of complex features. *Nature* 423:139–144.

Levy, S. 1992. *Artificial Life*. Pantheon, New York, N.Y.

Rowe, G. 1994. *Theoretical Models in Biology: the Origin of Life, the Immune System, and the Brain.* Clarendon Press, Oxford, England.

Szabo, P., I. Scheuring, T. Czaran, and E. Szathmary. 2002. In silico simulations reveal that replicators with limited dispersal evolve towards higher efficiency and fidelity. *Nature* **420**:340–343.

Szathmary, E. 1992. Viral sex, levels of selection, and the origin of life. *J. Theor. Biol.* **159**:99–109.

Ward, M. 2000. *Virtual Organisms: the Startling World of Artificial Life.* St. Martin's Press, New York, N.Y.

Wilke, C. O., J. L. Wang, C. Ofria, R. E. Lenski, and C. Adami. 2001. Evolution of digital organisms at high mutation rates leads to survival of the flattest. *Nature* **412**:331–333.

Yedid, G., and G. Bell. 2002. Macroevolution simulated with autonomously replicating computer programs. *Nature* **420**:810–812.

Viruses and Unicellular Organisms

History of Bacterial Viruses: a Conflict between the Concept of a Genetic Virus and an Acute Parasite

The viruses that infect unicellular organisms were among the very first viruses to be studied, and to this day, they remain the best understood of all viruses. In his 1899 paper, Martinus Beijerinck was first to clearly propose the idea of a virus (based on tobacco mosaic disease) as a parasite of the cell that replicates within its host by subverting cellular systems (see Brock [1999] in Recommended Reading). This idea was not initially widely accepted, and it was not until the study of bacterial viruses that existing views changed. F. W. Twort was the first to describe a filterable fluid (called a "glassy transformation") that would lyse bacterial cells, as discussed in his 1915 paper that reported his attempts to grow vaccinia virus as an autonomous agent on defined media (at odds with Beijerinck's idea of a virus and assuming that it was an autonomous organism). However, this initial report of a bacterial virus was generally ignored until a subsequent, more noted report introduced the concept of the bacteriophage. Early on, a close relationship between specific bacterial viruses and their specific bacterial hosts was recognized. It was observed that the very identity of some strains of bacteria could be best determined by the specific phage or virus type associated with it. This host identification became known as phage typing and is still used today. Current examples include phage typing of *Bacillus subtilis*, *Staphylococcus*, and *Mycobacterium*, all of which show virus-specific surface markers. For the initial phage studies, the investigated viruses were provided by the bacteria associated with both healthy and diseased human intestine (*Escherichia coli* or *Salmonella* and *Micrococcus*, respectively). However, as discussed below, phage from healthy and diseased human intestine can often show distinct biological characteristics.

In the now classical paper of Félix d'Herelle, it was reported that a filterable fluid could lyse bacteria from patients recovering from dysentery, and it was suggested that a virus of bacteria could explain the capacity of these

45

filtered fluids to kill bacteria. Due to its important medical implications, the idea that there existed a very small parasite of bacteria that was able to infect and kill specific bacteria gained much attention, especially prior to the discovery of antibiotics. In addition, it was realized early on that if such parasitic agents did indeed exist, then they might provide an ideal and simplified way to understand the nature of genes themselves, since the parasite seemed to be using host systems of genetic information. Almost immediately, however, a serious and essentially philosophical schism concerning the nature of phages and their relationships with their bacterial hosts developed among microbiologists. This schism was to last for 30 years.

The Bordet school: a genetic virus

In the early 1920s, Jules Bordet and M. Ciuca argued that phage-induced lysis was a normal characteristic of some bacteria. By various methods, they and others showed that the ability of a bacterial culture to produce phage and lyse susceptible cultures was a hereditary characteristic of the bacteria themselves. If a bacterial culture has the capacity to produce the "lytic principle," as suggested, then the hypothesis that the bacteriophage was the virulent parasite of bacteria was opposed. Bordet vigorously argued with the supporters of d'Herelle that the virus was a hereditary element of the bacteria.

The Delbrück school: a lytic bacterial virus

In the 1930s, Max Delbrück, at that time a quantum physicist, became interested in the study of phage as a way to understand the very elemental or molecular nature of genetic material and its reproduction. Using serologically related, T-even phages (T2 and T4, and others of the original phage isolates now named T1 to T7), Delbrück, and later Salvador Luria, conducted a series of elegant and precise experiments that established single-step viral growth curves and clearly showed that these phage were parasitic and unfailingly lytic viruses of their host bacteria. The Delbrück school of thought was thus formed and, armed with these clear experimental results, its adherents developed a violent disagreement with the followers of Bordet, who still maintained the hereditary nature of phage production and had coined the term *prophage* to describe this hereditary viral state.

Genetic- and lytic-virus schools reconciled

The "hereditary-phage" views of Bordet were supported by additional experimental results. Of particular note were studies by A. Lwoff, F. Jacob, and E. L. Wollman that carefully monitored individual bacterial offspring in microcultures and produced results clearly supporting the idea that ". . . the genetic material of the *E. coli* and the genetic material of the prophage have originated from the very same material." The clash of views between the Bordet and Delbrück schools continued for several decades. However, in the 1950s, great advances in the understanding of the molecular basis of bacteriophage led to the birth of the field of molecular biology. It became clear that both the idea of a virulent or lytic virus and the idea of a genetic or hereditary phage were correct. In 1953, the structure of DNA was discovered by James Watson and Francis Crick and could now be applied to understanding phages and viruses in general. In addition, the capacity of one hereditary virus to move into host genes and hence become one and the same as the cellular genetic material was clearly demonstrated by Joshua Lederberg and Esther Lederberg. Using the phage T4 as a model, Seymour Benzer and Francis Crick worked out the very nature of the gene and the genetic code. Thus, in the early 1950s, Luria was able to provide the currently accepted definition of a virus as a "molecular genetic parasite" that is dependent on host mechanisms for its replication. In the late 1950s, Lwoff published a more extensive definition of a virus that included both acute (lytic or virulent) viruses and persistent "proviruses," or hereditary viruses, which were called "lysogenic" or temperate phages. In 1962, the "prophage" model of A. Campbell finally provided the mechanistic details for the model of how this prophage worked, invoking integration of the viral chromosome into the bacterial chromosome. Thus, it was finally clear and accepted that there were two distinct life strategies, acute and temperate (or persistent), that applied to the viruses of bacteria and that both of these life strategies identified successful molecular genetic parasites of bacteria. Later, a third life strategy for virus replication, the continuous, nonlytic replication of the RNA and DNA miniphages (e.g., ϕX174 and M13), was discovered. This life strategy is distinguished by continued or chronic virus production and shedding, without the corresponding cellular lysis associated with lytic infections or the silent lysogeny associated with temperate phage.

As the above account reveals, the histories of virology and molecular biology have been filled with conceptual tension and confusion derived from the concurrent existence of viruses that are acute or lytic and viruses that can persist and/or colonize the host genome. When examined from a modern perspective, the cause of this long-standing disagreement seems to be differences in perception. These two processes, the lytic cycle and persistence, have been (and still are) perceived to be in direct opposition to and mutually exclusive of one another. In one sense, this absolute distinction seems correct. For example, T2 and T4 phages do not, with passage or in time, become hereditary, persistent viruses. They remain lytic and invariably lyse infected cells if their replication is successful. This acute behavior is a stable biological characteristic and is not compatible with the definition of a hereditary virus. This situation is also at odds with the prevalent views of some evolutionary biologists who feel that the relationships between lytic viruses and their hosts are "evolutionarily young" and tend to evolve into persistent, or benign, relationships given enough time for evolution to attain equilibrium. However, we know that lytic viruses like T4 do not evolve to become benign parasites of their hosts. Other similarly "lytic-only" viruses can also be found in many other host organisms, indicating that this life strategy is common. These acute viruses always harm or kill their hosts.

One source of confusion was the discovery that some specific viruses could be both lytic and persistent. Early on, it was clear that a specific viral agent (such as phage lambda) could have either an acute or persistent life strategy, depending on the specific bacterial host or the growth conditions. This capacity to switch between lytic and persistent life strategies is a characteristic of temperate viruses that tends to be highly host specific. The term *lysogenic* was originally coined to describe what happens when two strains of bacteria, one harboring virus and the other not, are mixed: the bacterial strain without the virus is lysed. One bacterial strain is thus lysed by the other "lysogenic" strain, while the lysogenic strain itself is protected from lysis. In a sense, the term lysogenic is confusing since the lysogenic strain does not itself lyse. We now know that the lysogenic strain is protected from lysis by the presence of the prophage but that this prophage can reactivate at a low rate to infect and lyse the second susceptible bacterial strain. In this instance, it appears obvious that colonization by the prophage provides a selective advantage to the lysogenic bacteria. The prophage provides protection from an otherwise lytic phage. Thus, in a competitive situation, where phage-colonized and non-phage-colonized bacterial cells might be found together, colonized bacteria have an immunity advantage. Furthermore, it was established early on that in addition to providing immunity against infection by the specific colonizing prophage, a prophage can often also confer immunity against similar and sometimes even dissimilar phage. Thus, phage colonization of the bacterial genome results in a clear and advantageous cellular phenotype. The colonized bacteria will have acquired a new, "virus-derived" molecular genetic identity, which is superimposed on the bacterial host identification system. Along with and inherent to this new identity, the host also acquires the ability to recognize and preclude other competitive genetic parasites. Furthermore, we now know that even defective versions of prophage lacking the ability to produce infectious virus can provide these immunity functions. Defective phage of various types can successfully colonize their hosts and preclude infection by other related parasites. Thus, the host phenotype that arises from prophage colonization is distinct and separable from those of both the noncolonized host and the acute virus.

Prokaryotes and Their Viruses: Lysis and Persistence

In chapter 1, I discussed the general issue of viral life strategy and the distinct characteristics of viral fitness that apply to acute and persistent viral life strategies. Here I argue that viruses of essentially all organisms tend to adopt one of these two life strategies. In some cases, as presented below, this pattern applies broadly to an entire family of virus or to a particular order of host. For example, fungi are frequently infected with persistent and inapparent versions of double-stranded RNA (dsRNA) viruses, whereas eukaryotic microalgae are susceptible to acute infections with large dsDNA-containing phycodnaviruses. In the case of prokaryotes, we now know that both acute lytic phages and persistent prophages are very common in essentially all microbiological communities. The most common of these ecologically abundant viruses resemble tailed phages like T4 and lambda. Many of these tailed phages are also known to be temperate (discussed below). This type of viral morphology is arguably the most abundant and dynamic life form on the entire planet, as it is highly abundant in the oceans and soil (discussed below). How can these abundant phages affect the evolution of life? How do phages contribute to the origin and evolution of prokaryotic host genomes? Historical accounts in evolutionary biology generally have not posed these questions, both because phages were usually thought of as simply destructive entities and because disagreement about the basic nature of phages continued. What was missing from the historical disagreement concerning lytic and hereditary viruses, however, was the idea that there exists a dynamic but enduring tension between these two states. Acute and persistent viruses (or their defectives) exert an enduring tension on each other and their hosts that is stable on the evolutionary timescale. This means that each of these types of virus must adapt not only to the host it parasitizes but also to the prevailing acute and persistent viruses that inevitably seek to occupy its ecological habitat and its host. This represents a previously unrecognized adaptation that molds the entire prokaryotic world. A striking example of this "acute-persistent" virus dynamic can be found in the first molecular genetic element to be identified as a "gene" during the foundation of molecular biology. This is the *rII* gene of phage T4, which is considered nonessential because it is not needed for growth in most *E. coli* hosts. However, *rII* is essential for a T4 phage that infects an *E. coli* host harboring a lambda prophage or its defective lambda phage. The *rII* gene represents a class of "accessory" genes that are well conserved in clinical isolates of T4 from *E. coli*. Such genes are not unique to lambda and represent a general trait of viruses, as is discussed below. Such accessory phage genes, whose role it is to counter the effects of other phages, are commonly observed and such an exclusion also applies to the function of various other T4 genes. Given the high abundance of phage in natural environments, we can expect the existence of related genes in essentially all microbiological communities. The study of phage-phage gene function is especially well developed in the very large microbiological populations of the dairy fermentation industry, as is discussed below.

Terminology for persistent, lysogenic, and temperate infections

The various terms used here to describe the temperate lambda lifestyle should be clarified. In chapter 1, where I present the distinct life strategies of viruses and the fitness associated with them, I use the term *persistence* to describe the capacity of a virus that has individually infected an organism to produce progeny virus at a later time. This is a more general use of the term than is typical of the scientific literature. In order to avoid confusion, this general use needs to be emphasized, and it should also be noted that this use is inclusive of the temperate life strategy. The terms *temperate* and *lysogenic* are often used interchangeably, although they can sometimes be differentiated. As defined by Lwoff, a lysogenic infection is one in which the infected bacteria have the hereditary capacity to make lytic virus at a later time. Not all temperate infections result in production of lytic virus. However, the ability of an infected individual host cell to subsequently make virus requires the persistence of viral genetic information. Thus, these infections require the stable and persistent maintenance of the viral genome within the host. In the case of lambda and P1 phages, this highly stable relationship can involve an "epigenetic" type of stability, capable of being maintained for hundreds of generations of bacterial cells. As noted above, the term lysogenic was derived from the ability of such persistently infected bacteria to induce hereditary virus production and thereby lytically infect a mixed culture containing a susceptible (nonlysogenized) bacterial host. However, the term *temperate* describes a more restricted process in which a phage infection, rather than leading to lysis, establishes a nonlytic, lysogenic state that does not kill the host. In the case of lambda, the lysogenic state usually involves the integration of lambda DNA into specific regions of the host chromosome. When in this integrated state, the phage is called a "prophage." However, nonintegrated persistence is also known and involves episomes (exogenous genetic elements that resemble plasmids and replicate independently from the host genome). Some temperate phages, such as P1, normally persist as episomes and seldom integrate. Thus, DNA integration is not required for temperate phages. Another related term that is sometimes used is *vegetative phage replication*. This refers to the production of lytic virus and death of host cells that are associated with either the reactivation of a temperate virus or the lytic infection of a susceptible host.

Overall patterns of prokaryotic viruses: high prevalence of tailed phages

Some might consider the preceding historical account as biologically misleading because the cited studies focus overwhelmingly on enteric bacteria and the phages associated with them. Table 3.1 summarizes the major types of viruses known to infect bacteria, and many of these virus types relate to commonly studied bacteria. How representative might these systems be of other bacteria and their viruses? There is good reason to worry about this issue. Historically, *Bacteria* and their corresponding viral ecology were not as well studied as many virologists might suppose. Thus, this issue is more difficult to address than might be expected. However, in the last 10 years the topic of phage ecology has received more attention, and it seems clear that both lytic and temperate viruses are common to many bacterial species in various habitats. In addition, we recognize two distinct and ancient do-

Table 3.1 Signature genes and genome size range for the proposed phage groups[a]

| Phage group | Genetic material | Subgroup | No. of: | | Genome size range (kbp) |
			Genomes in subgroup	Signature genes	
Leviphage	ssRNA	None	10	2	3.5–4.3
Inophage	ssDNA	None	13	0	5.8–8.8
Plectrophage	ssDNA	None	2	6	6.8–8.3
Microphage	ssDNA	φX174-like	5	10	5.4–6.1
		Chp1-like	6	5	4.4–4.8
Podophage	dsDNA	PZA-like	6	1	11.7–21.1
		T7-like	3	7	39.6–39.9
Siphophage	dsDNA	Lambda-like	11	0	36.5–61.7
		D29-like	3	53	49.1–52.3
		SK1-like	4	15	22.2–31.8
		TP901-like	7	9	37.7–49.7
		SFI21-like	13	0	14.5–52.2[b]
Myophage	dsDNA	P2-like	6	2	30.6–35.6

[a]Signature genes were described as loci found in all of the genomes within the proposed group and that have a BLASTP E value <0.1 between all members. Details about the signature genes, links to alignments, sequences, and more information can be found at http://salmonella.utmem.edu/phage/tree/signature.html.
[b]*Leuconostoc oenos* φL5 was not included in this analysis.

mains of prokaryotic unicellular life, *Bacteria* and *Archaea,* which can be further subdivided into distinct orders of prokaryotic organisms. The ecology of archaeal phages is even less well understood than that of bacterial phages, but our knowledge is developing. If we limit our consideration to the viruses that infect both of these prokaryotic domains (*Bacteria* and *Archaea*), then we observe distinct patterns and relationships of virus to specific host. The great majority of prokaryotic viruses are dsDNA viruses with moderate to large (20 to 180 kbp) genomes of either linear of circular morphology that are packaged into icosahedral capsids (or filamentous forms in the case of viruses specific to hyperthermophiles). Currently, about 96% of known bacterial phages are tailed; the remaining 4% are isometric. The abundance of archaeal phages differs significantly from this distribution, with tailed phages amounting to only 5%. Of the tailed phages, those with linear and circular dsDNA genomes are classified as *Myoviridae* (contractile tails), *Siphoviridae* (noncontractile long tails and linear DNA) and *Podoviridae* (short tails and linear DNA). Both lytic and temperate (episomal and integrated) life strategies are found in all these groups, although this characteristic tends to be associated with the specific genus of phage and host within each of these families. In addition, acute viruses (T4, T7, and PDR1) show various other characteristics, such as a strong tendency to code for their own replication, recombination, and repair proteins (DNA polymerase, ligase, etc.), as opposed to most temperate phages, which tend to lack such genes and instead utilize the cellular replication systems.

Membrane-bound or nucleoprotein-associated dsDNA viruses (like animal herpesviruses or algal phycodnaviruses) are rarely found in *Bacteria* (but are found in some *Archaea*). The next most common virus type is the single-stranded DNA (ssDNA) virus of smaller genome size that has icosahedral and filamentous capsids, such as φX174 and M13 or coliphage f1, respectively. Unlike the other ssDNA viruses, such as the plant geminiviruses that replicate using rolling circular replicons (RCR), these bacterial RCR viruses are not segmented. dsRNA viruses are uncommon in *Bacteria,* but a few are known and well studied, such as φ6. φ6 is clearly related to the dsRNA viruses of animals. ssRNA viruses of positive polarity are also much less common in prokaryotes than eukaryotes, but examples are known for *Bacteria,* such as Qβ. These viruses are currently less abundant or not observed in *Archaea*. ssRNA viruses of negative polarity are essentially unknown in prokaryotes, as are authentic retroviruses such as the autonomous retroviruses of animals or the pararetroviruses of plants. However, retroposons (defective retroviral elements) are found in some *Archaea* and in *Micrococcus* spp. As is discussed below, these virus-host relationships are distinct among various archaeal and bacterial orders and specific even to subdivisions of host species within these orders. Possible reasons for and evolutionary implications of these associations are presented.

Bacterial cells as a viral habitat

Bacteria present viruses with a specific cellular and molecular habitat to which these phages must adapt. Generally, *Bacteria* are unicellular organisms with rigid cell walls able to withstand high osmotic pressures. The composition of the cell wall varies substantially among types of bacteria and thereby presents a diverse chemical surface to the virus for recognition, attachment, and penetration. To accomplish penetration, the peptidoglycans that make up the bacterial cell walls need to be mechanically breached. The presence of this physical barrier probably accounts for the fact that the great majority of prokaryotic viruses do not physically enter their host cells but rather attach to the surface and inject their genomes (usually DNA) into the host cell, often by active mechanisms such as contractile tails or "pilot" proteins. The rigid cell wall also presents a problem for the exit of progeny virus because this barrier must again be breached. Virus release is often achieved by host cell burst, which results from the actions of virus-induced lytic enzymes. In addition to the cell wall, the internal workings of a bacterial cell present specific molecular situations that the virus needs to deal with. Bacterial cells have DNA genomes that are not highly organized into topological superstructures and are not tightly packaged into chromatin. Although there do exist some bacterial DNA-associated proteins, they do not interact strongly with the chromatin to produce highly stable structures, which are observed in eukaryotes. Therefore, the free bacterial DNA molecules are more likely to interact directly with the cellular replication and transcription machinery. Bacterial DNA is generally circular, with a unique origin of DNA replication. In contrast, most viral dsDNA genomes are packaged as linear DNA that contains short regions of terminal repeats. These repeats facilitate replication via circular theta forms and RCR intermediates. Since *Bacteria* have no nuclei, or cytoplasm- or endoplasmic reticulum-specific transport systems, bacterial viruses do not need methods to move through these cellular systems or to breach a nuclear membrane or nuclear pore complex, abilities required for some eukaryotic viruses. In addition, bacterial transcripts are not capped, spliced, or polyadenylated and are not transported from a nucleus. In conclusion, because viruses of *Bacteria* are expected to have host-specific molecular adaptations to all these situations, they differ in many fundamental ways from viruses of eukaryotes.

The bacterial system for DNA replication is also distinct from that of the eukaryote. As mentioned above, most bacterial chromosomes are circular, sometimes occur as multiples per cell, and initiate DNA synthesis from a single bidirectional origin of replication, which is often able to reinitiate replication prior to the completion of cell division. Thus, bacteria lack the basic components of the cell cycle seen in eukaryotes. Viruses (such as many temperate or episomal phage) that persistently parasitize cellular replication systems need to have mechanisms that ensure that their DNA replication is coordinated with that of the host. Conversely, lytic viruses that replicate using their own virus-encoded replication proteins will need to bypass existing host controls on extrachromosomal DNA replication. Another common feature of the bacterial cell habitat is the occurrence of restriction/modification systems. These systems consist of two matched enzymes: a DNA endonuclease that degrades unmodified DNA, usually by cutting at a specific palindromic sequence, and a matched DNA modification enzyme (usually a methylase) that typically covalently modifies DNA at specific bases during replication, thereby protecting it from the matching endonuclease. Restriction/modification systems are highly diverse systems found throughout *Bacteria* and *Archaea*. Statistical analysis of the occurrence of palindromic sequences in prokaryotic genomes strongly suggests that most prokaryotes have been under intense phage selection by restriction/modification systems: their genomes are highly underpopulated by potential restriction sites. As discussed below, many restriction/modification systems are themselves coded for by both temperate and lytic viruses, and these systems can be virus specific. Other systems of virus restriction are also known, such as small interfering RNAs. These virus restriction systems are common to all bacteria and, among prokaryotes, are essentially invariant. To conclude, it is expected that bacterial viruses will not be able to evolve replication mechanisms outside of the molecular habitat described above and that the barriers presented by this habitat are fundamental. Curiously, some bacterial species, such as mycoplasmas and likely descendants of *Bacteria* (including eukaryotic mitochondria), have genomes that lack the palindromic restriction sequence bias noted above, suggesting that they are not under phage selection.

It seems possible that by becoming a parasite within another, eukaryotic cell, these degenerate bacteria may have developed a way to escape exposure to and selection by viruses and thus no longer need to maintain the avoidance of nucleotide palindromes that otherwise prevails.

Bacterial population structure and ecology of the viral habitat: prevalent viruses and host fitness

In addition to the cellular and intracellular habitat provided by bacteria for viral replication, the population structure and ecology of bacteria also present specific circumstances for virus adaptation and replication. Bacteria are haploid and generally do not require sexual exchange for reproduction. This might suggest that bacteria tend to have genetically uniform populations, since a successful individual bacterium would be expected to rapidly generate a large clonal population of descendants. However, due to very rapid growth rates and large populations, most bacteria have the capacity to rapidly select for rare mutants. This, along with the activity of various insertion sequences that rearrange DNA, can explain why bacteria are generally able to generate a lot of genetic variation and can quickly select for genetic variants with greater fitness. In nature, bacteria are the most genetically adaptable organisms known and have been observed to adapt even to multifactorial changes in their habitats, including intense heat and even intense radiation that will normally break DNA into small fragments.

However, the very high rates of adaptation of bacteria to changes in their environment do not simply stem from their high reproductive rates and associated ability to select for rare mutants. Often, *Bacteria* have shown an ability to acquire DNA segments bearing genes in complete and complex sets from external sources in their environment. Because these gene sets were not present in any direct cellular predecessors, they are not a basic component of the genetic lineage of that bacterial population. This type of gene acquisition generally occurs through an infectious or horizontal process by one of several known mechanisms. Some bacteria possess specific systems for DNA transfer, such as conjugative plasmids and fertility pili (sex systems). However, these transfer systems are not uniformly found and are absent even from some highly adaptable bacteria. Furthermore, these sex systems have clear associations with bacterial viruses. For example, the origins of these sex systems may be viral in nature, as discussed below. Also, it is often observed that these transfer systems will make the host cells carrying them susceptible to infection with various viruses. For example, PDR1, the multivalent lytic phage able to infect a broad range of gram-negative bacteria, was originally isolated from bacteria that were multiply drug resistant, a phenotype produced by an acquired and complex set of genes. However, it was not the multidrug resistance that allowed PDR1 to infect these cells but the pili associated with DNA transfer: PDR1 infects via pili. Thus, it is the presence of these pili that allows PDR1 entry into its bacterial host.

The overriding mechanisms of bacterial genetic exchange and adaptation are basically infectious in nature. However, such infections must result in persistent genetic adaptation if they are to affect host evolution. The acquisitions of multiple drug resistance and virulence factors are probably the most-studied and best-known examples of this type of rapid but persisting bacterial adaptation that clearly involves the acquisition of complex multifactorial gene function. Phage exclusion is another well-studied example of a complex phenotype that can be similarly acquired by many bacteria. Furthermore, the capacity to exclude phage identifies how various competing phage can themselves select for the adaptation of a complex phenotype. Phage-phage interactions can directly contribute to the acquisition of a complex phenotype, such as phage immunity. A bacterial virus that is well adapted to its bacterial host is, in a sense, in direct competition with other temperate and lytic viruses (and their genetic derivatives) that attempt to parasitize or colonize the same host. Thus, the ability of one phage to compete with and exclude others provides the host with a new source of complex genetic information that can result in resistance for that host. This has a large effect on the survivability and hence fitness of the parasitized bacterial host. However, the main point is that such enhanced host fitness is an attribute of the successful genetic colonizer (the phage), and for success, this colonizer must successfully exclude prevalent competing colonizers in the environment. Thus, these fitness genes tend to be derived from a virus, not another host lineage. Therefore, in order to understand host and viral fitness and evolution, the phage (viral) ecology and the interaction of prevalent genetic parasites must also be considered.

Host Habitats and Associated Viruses

Oceans: a viral soup

In nature, there are several environments that represent large bacterial populations and ecologies from which virus-host and virus-virus interactions can be considered. Probably the largest of these is the oceans. Aquatic systems typically contain up to a million bacteria per ml. The unicellular cyanobacteria and the filamentous cyanobacteria as well as heterotrophic marine bacteria constitute the major prokaryotic components of the oceans. In addition, eukaryotic algae along with their phycodnaviruses are important oceanic organisms, although less abundant than the prokaryotes. Viruses are abundant in the oceans and are generally observed by electron microscopy in at least a 10-fold numeric excess over bacteria. Both acute and temperate phages are known for most of these bacteria. Physical counts of viruses in the oceans from concentrated ocean water, using electron microscopy, indicate that, in total, the world's oceans harbor about 10^{31} viral particles. The majority of these particles appear to be tailed phages (large DNA-containing virions). Measurements of viability suggest that the half-life of these phages is less than 1 day. Also, these phages are highly diverse, as determined by mass cloning and sequencing of the mixed oceanic phage, and most types are not represented in the current genetic database of known phage families and host genes. In cyanobacteria, the major classes of phage are similar to the dsDNA phage of *E. coli* (*Myoviridae*, *Siphoviridae*, and *Podoviridae*). There are also many large, poorly characterized DNA viruses of eukaryotic algae and amoebae, making this oceanic viral soup by far the most abundant and diverse collection of life forms on the planet.

Measurements of the phage and gene transfer rates in the oceans have demonstrated high rates of transduction. For example, it is estimated that 10^{14} transduction events per year occur in Tampa Bay, Fla., alone. This suggests that the tremendous viral genetic diversity of the oceans has an established pathway by which some of these genes can become parts of the host genomes. Given the age and the volume of the ocean, and its central role in the early evolution of multicellular life, the bacterial and viral genetic activity within the oceans must be considered the main candidate for the creative genetic process that ultimately resulted in higher, more complex life forms. The infectious nature of bacterial evolution and the tendency for phages, the main transmissible genetic parasites, to evolve by shuffling subgene cassettes (described below) suggest that this ongoing process may have also played a central role in the origination of the higher life forms that emerged from the oceans (discussed in chapter 4).

Soil: a viral slurry

Soil is another natural environment that maintains large populations of bacteria and hence would also be expected to have large populations of viruses. However, soil virus ecology has not been as well studied as ocean virus ecology because it has been very difficult technically to measure populations of phage found in soil. Soil phage estimates have ranged from 10^2 to 10^7 PFU per g of dry topsoil. Due to poor energy sources, soil bacteria tend to be concentrated at and frequently associated with plant root regions. Thus, the evolution of root systems must have had a significant influence on the emergence of soil bacteria and their phages. In addition, many soil bacteria generate spores, in which infection by temperate viruses is not apparent unless spore germination and virus reactivation can be monitored. *Streptomyces* spp. are examples of soil host bacteria from which both lytic and temperate DNA phages have frequently been isolated. Some of these phages have shown a high degree of polyvalency with respect to the broad range of bacterial host genera they infect, reminiscent of PDR1 in coliforms (gram-negative, non-spore-forming bacteria). For example, phage F22 infects a very broad range of *Streptomyces* species. This phage is known to form lysogens in *Streptomyces ambofaciens*. Similar observations have been made for thermophilic *Bacillus* species. In one study, 19 strains of *Bacillus megaterium* were examined, and all were shown to harbor temperate phages (most harbored several phages, but some also harbored defective phages) that could be induced with mitomycin C. It thus appears that, like ocean bacteria, soil bacteria have high rates of lysogenization. In comparison with aquatic organisms, however, much less is known concerning the possible rates of gene transduction in soil organisms or the accuracy of counts of soil phage. Although the size of the soil bacterial population and their phages must be immense, few specific measurements are available. Still, given the similarity to the enormous bacterial populations present in the oceans and the known presence of both lytic and temperate phages in soil environments, we can conclude that soil bacterial fitness, adaptation, and evolution are likely to be strongly influenced by the ecology and activity of these viral genetic parasites.

The enteric bacterium phage habitat of animals

In contrast to soil, with its relatively low bacterial counts, the gastrointestinal, or enteric, tract of animals has exceedingly high counts of bacteria. Bacteria in fecal animal waste are so concentrated that they can constitute up to 80% of the fecal dry mass. The quantity of bacteria growing in the enteric habitat is thus sufficiently large enough to affect environmental measurements of bacteria, such as in water runoff from soil. The evolution of the animal gut and its ability to support very high concentrations of bacteria, which for the most part turn over every day with waste excretion, have created a significant habitat for bacterial growth. With so many enteric bacteria, we can predict that this is an excellent habitat for viruses or phage. As mentioned in chapter 2, the very discovery of lytic and lysogenic phages is due to studies of the human enteric flora. The best-studied phage in this context is T4 of *E. coli*. T4 has frequently been isolated from clinical specimens. Surprisingly, the large majority of clinical isolates of *E. coli* (the normal host for T4) do not support the replication of T4, for unknown reasons. Relatively few of these clinical isolates were lysogenic for lambda; thus, exclusion by lambda is not common in a clinical setting. In one study, T4 was able to replicate in only 38 of 200 clinical *E. coli* isolates. Clearly, non-lambda-mediated restriction of T4 infection is common in clinical settings. As the P1 temperate phage is also a potent restrictor of T4 infection, and as temperate phage and plasmid colonization is also common in clinical *E. coli* isolates, it seems likely that other genetic symbionts, such as P1, are responsible for restricting T4 permissivity in these *E. coli* isolates. This issue, however, has not been well studied and needs to be further evaluated. The isolation of T4-related phage from human feces or raw sewage is affected by the health of the individual human subject. For example, 209 of 607 healthy human subjects were reported to harbor phage that could be isolated. These phages were mostly temperate (36% related to F80, 27% related to lambda, and 17% related to F28). However, when diarrheal patients were studied, 98 of 140 (70%) were found to carry isolatable phage at high concentrations. These phages were mainly virulent (T4, T5, and

TU23 related; see the book by Goyal et al. in Recommended Reading). This indicates that lytic phages are more often associated with disturbed, possibly rapidly growing bacterial populations in enterically ill humans.

A healthy human maintains a growing population of enteric bacteria (including *E. coli*) that regenerates daily in association with the digestion of food and passage of stool. Therefore, it might be concluded that the intestinal tract harbors a large population of rapidly and continuously dividing bacteria. However, we know that the bacteria throughout most of the intestinal tract do not actively divide; bacterial division occurs rapidly in a spatially limited portion of the gut after the transition from the small to the large intestine. In patients with diarrhea, it is likely that this bacterial growth pattern is disturbed. The relationship between lytic and temperate phage isolation with respect to bacterial growth and human intestinal health may explain early disagreements in the study of phages presented at the beginning of this chapter. d'Herelle, working with dysentery patients, believed that lytic phage was the norm, whereas Twort, working with a natural micrococcal contaminant, viewed temperate phage as more typical. As discussed below, with both acute lytic and persistent lysogenic phages, virus reproduction is affected by bacterial growth, but lysogenic establishment is more frequently associated with nondividing states. This issue of host cell growth in relation to phage growth has also been well studied in the context of the dairy industry, which deals with very large and rapidly growing populations of bacteria (discussed below). Clearly, virulent phages are common in numerous bacterial populations in nature. And these lytic phages appear to be stable in their natural ecologies. Therefore, T4 appears to be quite representative of a prevailing natural phage strategy. The populational and evolutionary stabilities of a strictly lytic agent have also been considered from a theoretical perspective, and it has been concluded that virulent phages can exist in stable, dynamic (albeit sometimes chaotic) relationships with their host bacteria.

The Lytic Phages of Bacteria: T4 as the Paradigm

The history of viruses of bacteria is presented above from the perspective of acute and persistent (temperate) viruses. Although this focus will continue to be used in subsequent chapters, emphasis will also be placed on the best-studied experimental models for the various topics discussed. The first model presented is the lytic phage T4, which is probably the most thoroughly studied lytic virus in all of virology. T4 phage and the serologically related T-even (T2 and T6) viruses contain large, linear dsDNA genomes (about 170,000 bp) that contain about 140 genes. Other phages with related acute life strategies include T5, T7, and SPO1 (specific for the gram-positive *Bacillus*). The DNA termini of the T-even phages contain repetitions of 400 to 800 bp and are involved in DNA replication via circular forms. The template for T4 DNA replication and recombination is not naked DNA but a DNA-protein complex containing numerous copies of gene 32 protein, an ssDNA binding protein able to remove hairpins from ss-DNA. The DNA is packaged into an icosahedral head, and the tails are contractile. Interestingly, T4 packaging accepts DNA longer than the genome length and imports up to 20% of the genome length in additional sequence into the phage head. Thus, the virus is always partially diploid and may also carry host and other resident DNA sequences. The virus attaches to the bacterial wall via a baseplate at the end of the tail, which provides cell-specific binding. Although the T-even phages have morphologically complex virion structures, these virions tend to be highly efficient structures. Phage T4 (and many other phages, including lambda) can have a particle-to-PFU ratio of 1, indicating that essentially every phage is biologically active. This is in sharp contrast to most animal viruses, which tend to have a particle-to-PFU ratio in the hundreds.

Lytic virus, autonomous replicators, and marked DNA

All acute phages lyse their susceptible host bacteria. Generally, these acute phages code for virus-specific DNA replication proteins (such as DNA polymerase, DNA ligase, and thymidylate synthetase), thereby establishing a replication strategy that is relatively autonomous from that of the host cell. In addition, this group of viruses codes for virus-specific DNA repair proteins, some of which have no homologues in the corresponding host cells. These proteins support the high rates of DNA repair (such as occurs following UV damage) that are also a characteristic of T-even phages (compared to the much greater UV sensitivity of temperate phages). Another intriguing feature of these lytic phages is the presence of one of the few prokaryotic examples of group I self-splicing introns. The occurrence of such introns in these well-conserved viruses of bacteria argues that introns may have evolved in viruses before the evolution of eukaryotes. Also, many of these phages code for a set of tRNA molecules, which can be deleted without prohibiting phage replication in laboratory strains of *E. coli*. These tRNA genes are conserved in natural populations, suggesting some type of accessory function. One very distinguishing feature in the DNA of these lytic phages is the high frequency of modified nucleotides, which include hydroxymethylcytosine (HMC) in place of cytidine in the case of T4. The synthesis of modified phage DNA serves to mark the molecular identity of the phage genome distinctly from the host DNA and to protect the phage DNA from phage-encoded restriction endonucleases (II and IV) that degrade unmodified host DNA. However, HMC DNA modification renders it sensitive to restriction by Mcr endonucleases (discussed below). The glycosylation at HMC residues of T4 DNA serves to prevent this Mcr restriction. The modification of T4 DNA makes it difficult to study with restriction enzymes. However, modification of T4 DNA is not essential, and deletions of the dCTPase and endonuclease IV, along with other alterations, can be used to make T4 with unmodified DNA when grown in nonrestricting *E. coli*. SPO1 modifies DNA using hydroxymethyluracil in place of thymidine, rather than HMC, but does not degrade host DNA. SPO1 also modifies phage DNA by methylation. It thus seems to be a common principle that lytic phages can mark or modify their DNA, distinguishing it from host DNA.

T4 and the definition of a viral species

T4 has been considered above as an example of an acute, virulent bacterial virus. Can T4 be considered to belong to a species of virus? In contemplating this question, the nature of any "species" as it applies to a virus deserves additional consideration. An acceptable definition of a biological species (proposed by Ernst Mayr) is a population of interbreeding individual organisms. If this definition is applied to T4, one might also conclude that all the T-even phages may represent one species of virus, since there is evidence of genetic exchange within this phage group. All the T-even phages have 85% sequence homology. In addition, all T-even phages show high rates of genetic recombination among the different types. Thus, many of the differences between T2 and T4, for example, could be considered as the normal variation within one population of species. However, this variation would seem to represent more heterogeneity than is usually associated with the genetic variation within one species. Also, distinct gene sets (sometimes called accessory genes) that distinguish the T-even phages can be identified. Thus, grouping all T-even phages as a single species seems to allow for genetic differences beyond the usual species variation. In terms of similarity, all T-even phages have conserved a set of genes that includes both essential and, curiously, nonessential genes. For example, phage T4 contains 140 genes. Genetic analysis indicates that only 69 of these genes are essential for replication in laboratory (nonlysogenic) strains of *E. coli*. These genes code for structural proteins (capsids, baseplate, etc.) and basic replicative proteins (polymerase, ligase, etc.), which are thought to be essential for virus replication. However, these essential genes are only a fraction of the T-even conserved genes. The high degree of conservation of the other 50 or so nonessential genes in different T-even phages suggests that they also have essential and selectively conserved functions in natural settings. These nonessential conserved genes include the tRNA cluster and the *rII* genes. In general, these conserved genes are not related to sequences found in *E. coli*, so they appear to be derived from virus, not host, genomes. For example, the conserved DNA methylation and repair enzymes found in T4 are distinct and have no host counterparts. Curiously, neither the highly conserved HMC content of T4 DNA nor its methylation is essential; they have been called accessory functions. Yet the hydroxymethylation of cytosine glucosylation, and methylation of T-even DNA, is characteristic of and used to identify all the T-even phages, which argues against an accessory role in vivo.

T4 gene conservation thus shows some very general characteristics that can be seen in most, but, interestingly, not all, other viral families (such as lambdoid phages). Those general characteristics include the existence of a "core" subset of genes, as well as a conserved genetic map or gene order (common in most viral families). However, an even more basic aspect of all viral families appears to be the maintenance of a particular molecular strategy of viral replication (discussed below). For example, T-even phages have distinct, non-host-like DNA replicases (discussed further in chapter 4). Viruses often display and maintain specific (non-host-like) systems for their replication, which include the replicase and corresponding *cis*-restricted origins of replication or regulatory nucleic acid sequences. In addition, most virus families also maintain a set of accessory genes that are relatively unique to the particular lineage of virus but also generally dissimilar from host analogue genes.

Conservation of accessory genes and interaction with prophage: the need to compete with other viruses

The only known biological role for T4 DNA methylation of adenine is to protect T4 against degradation by the restriction/modification genes (or addiction module) of the lysogenic P1 prophage. *E. coli* organisms free of P1 prophage do not have a restriction endonuclease that degrades methylated T4 DNA. The *rII*A and -B genes (conserved in T2, T4, and T6), which played such an essential role in the history of molecular biology (i.e., the molecular concept of the cistron and the genetic code), function only to allow T4 replication in a host colonized with prophage lambda. *rII* has no function in an *E. coli* organism free of a lambda prophage (or defective lambda) expressing *rex*A and *rex*B genes. Even more surprising is that HMC incorporation into T4 DNA is also not essential: HMC-free T4 that contains additional mutations (such as in the T4 restriction endonuclease gene) can be grown in nonrestricted *E. coli*. The presence of HMC modifications, however, renders the T4 DNA susceptible to degradation by McrA and McrB restriction endonucleases. Glycosylation at HMC residues counteracts this Mcr restriction sensitivity. Although *McrA* is often thought of as a host gene, it actually resides within the e14 genetic element, a cryptic prophage found only in some strains of *E. coli*. The e14 element also codes for proteins that inhibit T4 translation. Thus, both T4 methylation and T4 accessory genes appear to be aimed at countering the effects of other genetic parasites of *E. coli*.

Phylogeny of the T-even phages

T-even phages appear to have a monophyletic origin. Various characteristics, such as common DNA modification, capsid morphogenesis, the nature and order of genes, and the mechanism of replication, all support a common lineage. Yet T4-like phages (lytic with contractile tails) can be found to infect many diverse types of prokaryotes. T4 also shows some clear similarities to other polyvalent lytic tailed phages, such as T7 and RD114. In addition, a broad distribution of nucleotide bias patterns is observed in the lytic phages (including T4) but not temperate phages (discussed below), suggesting evolutionary isolation of these two viral life strategies. The existence of broadly conserved genetic patterns in the lytic phages suggests an evolutionary connection among the lytic tailed phages that is more distant than can currently be determined by sequence analysis.

Another group of viruses with clear similarity to T4 is the pseudo-T-even bacteriophages. This diverse group of viruses can cross-hybridize, even under stringent conditions, with T4 and RB49 DNA. The pseudo-T-even sequences most like T4 correspond to the head and contractile tail regions of these phage but also include some early genes (those for DNA topoisomerase, DNA ligase, and ribonucleotide reductase), consistent with the maintenance of a T-even replication strategy. Yet the DNA polymerase is not in this conserved set. Furthermore, more than one-third of the pseudo-T-even phage DNA has no homology to T4. And this DNA is not modified with HMC. Phylogenetic analysis of all known phages based on the similarity of 105 phage proteins places T4 at the unresolved root of the tree that includes podophage and other phage families. This analysis is shown in Fig. 3.1. From this, we can see that T4 shows evolutionary connections to a broad array of phages. Furthermore, T4 also shows clear similarity to eukaryotic DNA viruses, such as herpesvirus (see chapter 4), in terms of viral morphogenesis as well as sequence similarity of some replication proteins. Thus, T4 appears to represent a very ancient viral system whose relationship to other viruses that infect distantly related hosts is still apparent. However, T4 also has conserved specific similarities to eukaryotic cells themselves and not only their viruses. These similarities are presented mainly in chapter 4. As an example, T2 and T-even-like RB3, as well as SPO1 and its relatives (SP82, Φe, 2C) and numerous phage of *Streptococcus thermophilus*, all have group I self-splicing introns. Group I introns are also found in fungal mitochondria, protist nuclei, plant chloroplasts, and mitochondria but not in most prokaryotes. Interestingly, group I introns do occur in some species of purple bacteria and cyanobacteria (discussed below), but these are polyphyletic lineages that seem to have acquired the introns by horizontal transmission during relatively recent evolution. T4 DNA polymerase, lysozyme, and several other phage proteins are also clearly similar to eukaryotic proteins but not the homologues of their prokaryotic hosts. Yet these genes of the lytic viruses do not seem to evolve by the same process(es) as host genes. The host genes are more stable (as entire genes), and their ancestral relationships are easier to discern. The T-even genomes, unlike the cellular genomes, clearly appear to have evolved by modular, subgene evolution from a network of both closely and distantly related genomes. The one evolutionary force that binds all these lytic phages together appears to be their common strategies

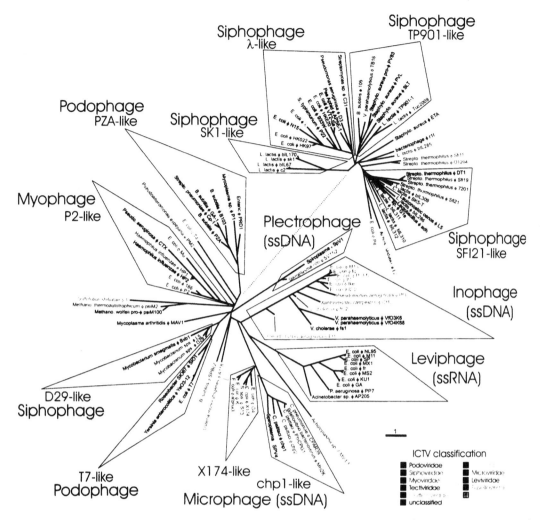

Figure 3.1 Dendrogram of DNA phage proteomic tree. The tree was constructed from 105 completely sequenced phage genomes and was generated from length-corrected protein distance scores with a penalty of 10 for missing proteins. Each phage genome is colored according to its International Committee on Taxonomy of Viruses classification, as shown in the key. To make the figure easier to read, the large group of siphophage has been manually shifted away from the other groups. Reprinted from F. Rohwer and B. Edwards, *J. Bacteriol.* **184:**4529–4535, 2002, with permission.

of replication and morphogenesis. Thus, the most reasonable view is that these phages constitute a common, compatible, and ancient network or pool of genetically interchanging replicators—not a single lineage. Thus, phage nomenclature, such as the "T-even" designation, does have strong evolutionary significance (compared to host species nomenclature), as the phage nomenclature represents at best a viral mosaic which we call a virus family but which cannot be represented as a traditional evolutionary family tree.

Persistent Temperate Phages: the Phage Lambda Paradigm

Phage T4 is considered the best-studied example of an acute lytic bacterial virus. I now turn to the *E. coli* phage lambda to examine the characteristics of a persistent or temperate life strategy and compare this model to temperate viruses of other prokaryotes. In the lambda model for prokaryotic virus persistence and integration, a number of broad patterns can be discerned. For example, the existence of a large number of similar dsDNA viruses that infect a broad array (possibly all types) of prokaryotic cells seems peculiar to prokaryotes, including *Bacteria*, *Archaea*, purple bacteria, and cyanobacteria. Although dsDNA viruses are prevalent in many eukaryotic organisms, such as insects and all vertebrates, few if any of these viruses persist by integrating their DNA into a host chromosome. For example, herpesviruses often persist in their animal host, but as episomal forms. Prokaryotes seem especially prone to persistent infection by larger dsDNA proviruses. Given the vast ecological habitats occupied by prokaryotes, we can only imagine the enormous impact that these parasites have had on their host genomes.

Characteristics of temperate phages and role of immunity

Lambda represents a rather large family of related tailed phages (*Siphoviridae*) whose genome is approximately 40 kbp of dsDNA, with around 40 genes. However, in contrast to analysis of T-even phages and many other viral families, recent phylogenetic analysis of 105 phage genes failed to identify any core genes that were common to the entire family of lambda-like siphophages. Nonetheless, the lambdoid phages, like the T-even phages, still represent a viral family with a common replication strategy that allows genetic exchange and maintains both the virus-specific replication strategy and morphology. The best-studied temperate phages with the closest relationships to lambda are P2 and P22 of salmonellae. These phages all show some serological cross-reactivity to each other. At the nucleotide level, P22 is most similar to lambda and shares 22% sequence homology, whereas P2 shares only 10% homology with lambda. These three phages can recombine with each other, and thus they fit the previously discussed view of a viral species. As in the case of the virulent T-even phages discussed above, the regions of conserved sequence similarity are distributed throughout the genome in a patchy manner often involving subgene regions. The lambdoid phages differ from the lytic T-even phages in that, within the lambda family, there are also more distant members that show little similarity within their genes. Thus, the lambdoid phages are even more heterogeneous than the T-even phages. This raises many questions about common mechanisms of virus evolution. How can we account for this broad difference? Why are the core genes of this lambda virus family not conserved? Do these unrelated members represent an independent lineage of phages? Are the temperate phage replicators so dependent on host cell replication systems that they do not need to maintain any core or distinctive viral replication genes? Can the viral persistent life strategy be involved in this difference, and do temperate phages exist in a gene pool distinct from that of their hosts?

Because lambda is a persistent prophage, its expression is controlled by a bistable genetic switch that allows the prophage to express only the genes associated with immunity (*cI* and *rex*). Stable protein-protein interactions with promoter sequences are used to achieve this epigenetic stability. The switch is sensitive to small changes in concentrations of the Cro repressor, which switches expression and leads to the induction of lytic virus replication. Due to the expression of one of these genes (*cI*), a lambda lysogen is immune to superinfection by phages related to lambda. In addition, as noted above, a lambda lysogen is resistant to superinfection with T4 (via Rex proteins). However, the mechanism of lambda immunity is specific to the individual lambdoid phage: other related temperate phages (e.g., P2 and P22) differ in the mechanism of and genes associated with immunity. This variation represents one of the most diverse characteristics of the lambda-like phages. For example, P22 and lambda are very similar, but they differ completely in the immunity region. Yet it also seems clear that these phages are part of the same family, as they conserve their relative gene order. In essentially all cases, lysogenic bacteria are immune to similar and sometimes dissimilar phage types. This situation is also called "lysogenic conversion" (although this frequently refers to surface receptor modification).

In the case of P22, three immunity genes (*mnt*, *sic*A, and α1) are expressed from the prophage, and all affect the replication of other phages (via phage immunity, phage exclusion, and altered surface proteins affecting phage attachment). This means that 12% of the P22 genome is dedicated to preventing the growth of competing viruses. P2 is similar in this respect, but in its case the genes *old*, *tin*, and *fun* are involved; these genes resist lambda infection, block T-even phages, and inhibit T5 phage, respectively. Interestingly, this region of the P2 genome has a high A+T content that distinguishes it as a more recently acquired sequence. It is also very interesting that P2, unlike lambda, is a noninducible prophage that appears to be locked into its host. Thus, P2 is not lysogenic in the usual sense. From these facts, it would appear that a major selective pressure on a persistent phage is to resist competition and superinfection by other phages. However, this exclusion has clear limits, and it is still possible to establish multiple prophage infections with as many as eight distinct prophages under laboratory conditions.

A less autonomous replicator that senses host physiology

The establishment of lysogeny is affected by the physiology of the host cell at the time of infection. Cells that are starved for media or in cold environments tend to establish a lysogenic infection rather than a lytic infection. Since bacteria seldom grow logarithmically in natural ecological situations, lysogeny is expected to prevail. In *E. coli*, stationary phase is also associated with hypermutability due to high rates of recombination. In addition, although some temperate phages (such as P2) can infect multiple host cell types, most temperate phages are highly specific to their host bacteria. This is in contrast to the lytic phages, which tend to be more host polyvalent.

The induction of lytic phage from lysogeny can be highly efficient under some conditions. In rich media, lambda prophage is induced from *E. coli* lysogens by irradiation with UV light in essentially every cell, resulting in mass lysis of a culture. Many other temperate phages are also induced by UV irradiation. Thus, this is a common assay for the presence of prophage in bacteria. It seems that disturbances of DNA replication might be responsible for this induction, since thymidine starvation or treatment with mitomycin C also frequently induces prophage. By using such an assay, it has been determined that 20% of the bacteria in ruminant intestines harbor prophage. However, the biological relevance of this UV induction for enteric bacteria is debatable, and it is clearly not always the case that prophage is induced with UV or any other treatment. Defective or episomal versions of lambda, for example, are not induced by exposure to UV light, although they still provide immunity and inhibit T4 *rII* mutants. More important is the example of P2. P2 is a member of a distinct, very large, and widely dispersed family of temperate phages and can infect *E. coli* and *Shigella, Serratia, Klebsiella,* and *Yersinia* species. Yet P2 is not induced under UV light and is essentially noninducible. P2 propagates by rolling-circle replication, but this replication can itself be parasitized and induced by the satellite virus P4 (discussed below). In this case, one silent virus needs another for induction.

Another process of induction can also be observed for lambda prophage: zygotic induction. When a lambda prophage has been transduced as a part of fertility factor (F^+)-mediated chromosome transduction to an F^- and nonlysogenic recipient, lambda will be induced in the zygotic recipient to produce lytic phage due to the absence of immunity function in the zygote. In a sense, lambda prophage itself behaves like an addiction module and is toxic (lytic) to *E. coli* lacking lambda, but it is ineffectual in lysogenized *E. coli* due to the persistence of lambda prophage. More importantly, this situation also suggests how the presence of a persistent virus can lead to the reproductive isolation of its host, since the infected and noninfected hosts are no longer compatible sex partners. This issue has major general implications for evolutionary biology and is discussed in subsequent chapters. It should be noted that not all prophages (e.g., P2) undergo zygotic induction. Furthermore, versions of stable persistence that use both harmful genes and genes that prevent harm, or "addiction strategies," are commonly used by various persistent genetic parasites (discussed below).

Persistence without integration or induction: P1, P2, and defective lysogens

It has been noted that P2 prophage protects *E. coli* against lambda infection by expressing the *old* gene and inhibiting lambda-specific DNA replication. Thus, even among temperate phage there is competition and exclusion. It seems likely that this viral capacity to successfully compete in order to colonize the host must be under positive selection and may explain the evolutionary conservation of so many viral accessory genes. P2 is clinically prevalent and can be isolated from about one-quarter of clinical human isolates of *E. coli*. Thus, P2 lysogens are much more common and successful colonizers than lambda lysogens. Like various other prophage, P2 integrates in a site-specific manner adjacent to various tRNA sites (the 7-bp anticodon loop). It is interesting to consider the life strategy and fitness of P2 because it is essentially a noninducible prophage. How does P2 survive and transmit if it cannot be reactivated? P2 has clearly conserved the ability to make virus and thus does not appear to be defective. P2 fitness thus appears to require the capacity to produce infectious virions (presumably for transmission to other hosts) at some time other than during extended persistence. The SOS-inducible nature for both P2 and the related phages 186 and HP1 is consistent with this view. Interestingly, 186 also does not interfere with exogenous phage infection nearly as efficiently as P2 does. How does one rationalize the persistence of P2, a noninducible prophage? A commonly expressed view is that a bacterial host is under some positive selection for the P2 prophage to persist in it because P2 provides protection against environmental stress and damage. This relationship is considered symbiotic. If this is the case, then we are still left with trying to explain how the virus moves to new hosts and why it conserves virion structural proteins. In response to such issues, it has been reasoned that if the infected host is damaged, then it is in the best interest of the virus to induce lytic phage production and seek other hosts ("jump ship"). However, P2 and various other persistent phages (including episomal versions of lambda and RP4), which are clearly prevalent and fit for persistence, do not induce phage production even following lethal host genome damage. It is possible that phage persistence itself has a fitness advantage (such as preventing other viruses and squelching competition). In this case, the life strategy of P2 and various other defective prophages might be more like a "king of the hill" game that is won simply by successful prevention of host infection by competing viruses. Thus, the maintenance of persistence, not reactivation, is the main target of selective pressure, and this can explain the success of defective lambda or 186 prophage. Any such defective genetic parasites that improve the competitive fitness of persistence will be under positive selection. In this regard, the presence of a retron (reverse transcriptase-coding element) within P2 and similar defective prophages (ϕR67 and ϕR86) may provide a parasitic element that can disrupt the genes of competing phages that have colonized host genomes. By interrupting the competitors' genomes (resulting in the loss of the resident immunity-addiction modules), one persistent phage can defeat the already resident colonized prophage. Thus, invasive parasitic elements within P2 can enhance the persistence function of P2, thereby improving the fitness of P2 relative to its competitors. If this is in fact the function of these retrons, then it suggests the existence of significant competitive interactions among

A

Addiction Module: restriction/modification

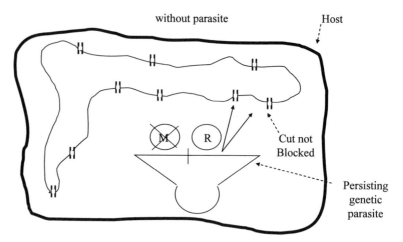

Antitoxic: unstable beneficial/protective agent: M = Modification enzyme
Toxic: stable harmful/destructive agent: R = Restriction enzyme

B

Addiction Modules: Toxic pore

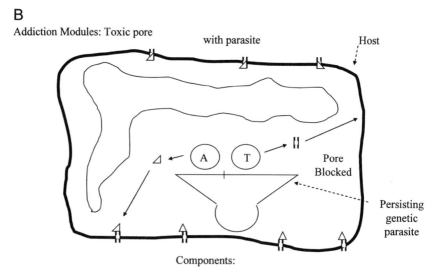

Components:
Antitoxic: unstable beneficial/protective agent: A = anti-pore toxin
Toxic: stable harmful/destructive agent: T = toxic pore

Figure 3.2 Nature of addiction modules. (A) A phage P1-like addiction module based on restriction/modification enzymes. (B) An addiction module based on toxic pore proteins and antitoxins to the pore; shown is the state when the parasite is present. (C) Same as panel B but with the absence of the parasite. (D) An addiction model based on a prophage providing immunity and protection from either mating or infection with exogenous phage.

C

Addiction Module: toxic pore

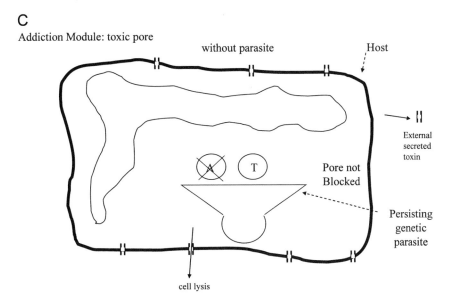

without parasite

D

Addiction Modules: Virus addiction

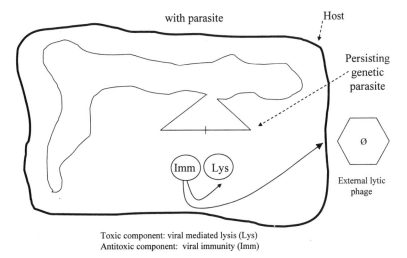

with parasite

Toxic component: viral mediated lysis (Lys)
Antitoxic component: viral immunity (Imm)

Figure 3.2 (*continued*)

various persistent genetic parasites which have distinct fitness profiles. Clearly, P2 does interact with other genetic parasites. And, as discussed below, P2 is efficiently mobilized to produce infectious virions only after infection with an associated, but defective, member of the satellite phage P4 family.

Episomal persistence

The temperate phage P1 (related to P7) represents another family of temperate phages that was first identified due to its ability to exclude growth of lambda. P1, however, is more complex than most other temperate phages and contains a genome of about 100 kb, containing about 100 genes. P1 displays a variety of biological characteristics that make it interesting to consider from the perspective of persistence. Of particular interest, P1 appears to have several addiction modules that ensure the maintenance of the virus in the host. An addiction module can be defined as a set of functions or gene products (which are generally stable) that are toxic or harmful to the host in association with a matching set of functions or gene products (which are generally unstable) that counteract, inhibit, or provide immunity to the same toxin agents. The two are necessary as a set to prevent harm to the host and maintain the parasitic agent. P1 is generally maintained as an episome and thus seems similar to F factors. Although it is often observed that the persistent plasmid versions of phages, such as lambda, are less stable than integrated prophages, P1 has evolved a rather elaborate addiction module that maintains plasmid stability in daughter host bacteria (loss rate, 10^{-5}/generation). This system involves both the coordination of cellular and viral DNA replication and the partitioning of daughter plasmids into daughter bacteria, thereby precluding the coexistence of more than one plasmid per daughter cell. In addition, P1 has one of the most complicated systems for expressing immunity of the temperate phages, in that it uses three distinct immunity regions. Of particular interest is the expression of a P1-encoded restriction/modification system that is involved in the exclusion of other phages, as well as the maintenance of the P1 plasmid. P1 was the first system in which host restriction of this type was observed. P1 (as well as other parasitic plasmids) encodes fast-acting modification enzymes (e.g., EcoP1) that act on replicating DNA, along with a more stable, slow-acting type III restriction endonuclease that cleaves unmodified DNA. Together, these enzymes can function as an addiction module, with an unstable modification enzyme and a stable restriction endonuclease. In daughter cells that have lost the P1 episome, the restriction endonuclease causes postsegregation killing of these uninfected daughter cells, since they do not express the corresponding modification enzyme. Figure 3.2 outlines the essential characteristics of an addiction module. Figure 3.2A demonstrates how the P1 restriction/modification addiction module functions. Figure 3.2B and C show how addiction modules use toxic pores and antitoxin to the pore functions. Figure 3.2D shows how a prophage, which provides an immunity module, can by itself function as an addiction module by preventing infection and lysis by prevalent phage, as discussed above, during zygotic induction or phage exclusion by lambda.

Mechanisms of persistence: molecular identity markers and addiction modules

In 1993, Michael Yarmolinsky first coined the term *addiction module* to describe the killing action of a serine protease of phage P1 in order to explain postsegregation killing. This killing is a type of programmed cell death that occurs via action of the *Phd-Doc* (<u>d</u>eath <u>on</u> <u>c</u>uring) addiction module after the curing of the P1 plasmid. This system, along with others, is designed to compel the infected cell to retain the viral genome. It results in a very stable but persistently infected cell lineage. The stability is such that only one cell in 10^5 cell generations spontaneously loses the P1 plasmid. To accomplish this, P1 codes for a second addiction module consisting of a toxin-antitoxin gene set. The toxin gene (*Doc*) is stable, while the antitoxin gene (*Phd*) is unstable; thus, like the restriction endonuclease noted above, *Doc* kills host cells (or daughters) that lack the P1 episome. These two examples of mechanisms (or phenotypes) that compel viral persistence appear to be crucial for the persistent lifestyle. However, as a consequence of the genes of P1 forcing viral persistence onto its host, two very important effects can be observed. First, P1 lysogens are immune to superinfection with unmodified phage. Thus, upon infection, the bacterial host acquires a new P1 identity that prevents the addition of further viral identities, including multiple occurrences of P1 itself. Second, P1-lysogenized bacteria are reproductively isolated in that uninfected *E. coli* organisms which mate via F transduction with P1-lysogenized *E. coli* are killed, as are daughter cells that have lost their P1 plasmids. In addition, only phage grown in a P1 lysogen will be properly modified and able to subsequently infect P1 lysogenic *E. coli,* so even permissive phage susceptibility is host restricted via P1. This type of persistence appears to be an infectious molecular identity process that could lead to reproductive isolation of the host organism.

There are several other interesting characteristics of P1 lysogens that may affect virus-host evolution. P1 contains the insertion sequence (IS) element IS*1,* which is involved in the occasional chromosomal integration of P1. In addition, P1 undergoes switching of a genetic module that controls host restriction. Using a site-specific recombination system, P1, similar to phage Mu, can invert a coding sequence allowing for expression of one of two sets of tail fibers. These two sets of tail fibers have distinct host specificities. Another rather surprising characteristic of P1 is its ability to provide the host chromosome with a second, functional origin of replication. P1 can replicate both as a plasmid (*oriL*) and as an integrated prophage (*oriR*). When integrated, P1 (and P7) allows *E. coli* bearing an impaired *Dna*A gene to replicate the bacterial chromosome from the integrated P1 prophage origin. Thus, the phage has the capacity to superimpose a new replication system on its host in a way that replaces the host ori function. This situation has important implications for the evolution of new host replication systems, as is discussed in chapter 4. Figure 3.2 shows three types of addiction modules (restriction/modification, toxic pores/antitoxin, and prophage addiction).

General implications of persistent phage and fitness

Several general conclusions from the considerations of phage P1 and P2 can be drawn. First, a temperate phage (which is a persisting genomic parasite) can be highly successful, and hence fit, yet lack the ability to induce the prophage to produce lytic virus, even when colonizing a dying host cell. Thus, it would be difficult to define this persistent fitness in terms of the reproductive ratio. In addition, the defective and noninducible variants of an inducible phage such as lambda can similarly be fit, as evidenced by their biological and competitive stability. This fitness and the capacity of these variants to later produce phage have a basic temporal component that allows for some subsequent, infrequent, but dependable event (such as infection with a helper virus) to propagate the variants with high probability and success. Second, temperate phages can be very successful and stably maintained as episomes that do not usually integrate into the host chromosome. In this situation, the phage must express genes that ensure coordinated replication of the two replicons, the virus DNA and the host DNA, as well as stable partitioning of viral chromosomes into daughter cells. These two systems, which are needed for persistence of both normal and defective viruses, are elaborate and in sharp contrast to the concept of selfish DNA, which has no phenotype in the host. Viral persistence seems to always require a phenotype or strategy that compels the host to maintain the viral genome yet maintain the capacity to recognize the distinct molecular genetic identity of that virus. In many bacteria, these persistent phenotypes usually involve immunity functions or addiction modules, but sometimes no gene products are involved and persistence simply provides a genetic system that is efficiently parasitic to and competitive with other prevalent genetic parasites. Phage genomes contribute a significant portion of their coding capacity to these interparasite functions. And these genes compose an especially dynamic component of phage (viral) DNA. It must again be emphasized that persistent phage fitness (and hence host fitness) requires the capacity to compete, preclude, or be parasitic to other prevalent viruses, both persistent and acute. If we attempt to evaluate the fitness of persistence in a host cell without such competing parasites, we will fail to see the essential contribution of these genes and think of them as accessory in function.

Although colonization by persistent phage is forced onto the host by addiction modules and other molecular strategies, it often appears more benign, as a mutualistic, or symbiotic, relationship between virus and host. These stable persistent states often involve integration into the host chromosome, but they can also be episomal. However, the existence of episomal phage persistence diminishes the distinctions between a persistent phage and a parasitic episome, especially considering that many parasitic episomes appear to be defective prophages. This blurred distinction between virus and plasmid is discussed further below. Finally, several examples of phages that have themselves become parasitized by other genetic parasites (hyperparasites) have been presented. In some cases, these secondary genetic parasites are specific to a virus and are not otherwise present in the host, such as the invasive introns of T-even phages. In other cases, the parasites are interdependent. As is discussed below, both temperate phages and lytic phages appear to become parasitized by either defective viral or subviral agents. In the case of temperate phages, such hyperparasites can result in the destruction of the immunity module that controls persistence, leading to the induction of a lytic phage. Thus, there is also good evidence that nested interactions among genetic parasites exist in natural settings and that these interactions affect the virus-virus and virus-host relationships, some of which appear to be mutualistic. It is worth noting the similarity of these hyperparasites to those observed in the Tierra simulated-life program described in chapter 2.

Distinct gene pools of persistent and acute viruses

The view that the viruses of bacteria can be considered to have at least two distinct life strategies—acute and persistent—has been presented previously. It was also shown that these situations involve distinct fitness and relationships between virus and host. Beyond the noted examples with gram-negative bacterial hosts, there also appear to be broad evolutionary distinctions between acute and persistent viruses and their relationships to their hosts. It is now well established that all organisms, including viruses, have rather distinct frequencies of occurrence of the four nucleotide bases (such as A+T or G+C content) as well as distinct patterns of nucleotide "words" (di-, tri-, and tetranucleotides) and nucleotide palindromes. It has been observed that acute bacterial phage have nucleotide word frequencies that are fully dissimilar from those of their hosts. However, temperate phages (and parasitic episomes) have word frequencies that are the same as those of their hosts. This observation confirms the distinction between acute and persistent viral life strategies in bacteria and also indicates that temperate phages are in the same gene pool as their hosts. Furthermore, this distinction between the word frequencies of acute and persistent viruses is not unique to bacterial viruses: it can be seen in essentially all viruses. In the case of bacteria, persistence provides a pathway by which virally derived genes can contribute to the evolution of the bacterial host. Nonetheless, temperate phages, for the most part, tend to acquire new genes (and sometime complex gene sets) from recombinational processes involving other temperate phages or genetic parasites.

The P2-P4 satellite phage system, a parasite-of-parasites paradigm: survival of the more defective

Lysogenic phages are highly successful in many natural bacterial populations. But their high rates of colonization and success can themselves lead to additional opportunities for viral parasitization. In chapters 1 and 2, I discussed the general tendency of viruses to generate defective versions that are parasitic to nondefective helper viruses. In this chapter, I have also discussed defective versions of lambda and P2 and the relationship of P2 to P1. However, besides affecting P1, P2 is a satellite virus of P4. A satellite virus is defective for autonomous replication and requires a helper virus. Thus, it resembles a defective virus. However, unlike defectives, a satellite virus is not directly derived from the helper virus. Nonetheless, for a satellite virus to be prevalent, a helper must also be prevalent. The best-studied satellite-helper virus system of bacteria is the P2-P4 phage system of *E. coli*. Satellite viruses are also very common in plants, as discussed in chapter 7.

As discussed above, P2 is a member of a large family of temperate phages related to lambda that are able to exclude T4 and other phage by several mechanisms. P2 was originally isolated from an *E. coli* strain that also harbored P1 and P3. When integrated into *E. coli*, P2 becomes a molecular "defective," in that integration interrupts P2 transcription. Furthermore, P2 is not readily induced from lysogeny and is not activated by UV irradiation or zygotic induction. It persists as a seemingly stable defective parasite expressing only immunity functions. Yet P2 and all of its relatives can function as helper viruses for the smaller P4 and can be induced by P4 to produce phage at low levels. P4, which has no P2-related genes, is a 12-kb linear dsDNA phage that packages and replicates via circular DNA using a bidirectional origin of replication (as opposed to rolling-circle DNA replication, used by P2). P4 replication is dependent on a P2 helper: although P4 produces a distinct and smaller version of a capsid protein, it derives all the other numerous structural proteins, including tail proteins, from the P2 helper. Infection with P4 has several possible outcomes. If the host is not colonized by a P2-like helper, P4 either integrates at a unique tRNA site to become a prophage or, more rarely, establishes a multicopy (30 to 50) episomal state. In both of these situations, P4 always expresses immunity functions, which are mediated by rather distinct mechanisms involving expression of a small stem-loop-like RNA and transcriptional termination. These immunity functions must be suppressed for lytic replication. Thus, in the absence of P2, persistence and immunity constitute the default mode of P4 infection. P4 also carries genes for various addiction modules, such as the *gop* killer toxin and the β gene (which prevents *gop*-mediated cell killing), as well as several other potentially toxic genes and numerous small genes. If a P4 lysogen is subsequently infected with P2, the P4 prophage will suppress P2 replication and produce P4 instead. When P4 infects a host colonized by P2, it can establish a lysogenic state by either integrating or becoming an episome. A frequent outcome, however, is the induction of P2 and the efficient lytic replication (using P2 lysis genes) of P4. During this induction, some P2 is also produced, thereby ensuring the propagation of P2.

In its episomal form, P4 (like P2, described above) closely resembles an array of plasmids that resemble cryptic phage elements. P4, however, is a highly successful parasite in nature, and hence it is clearly fit in a natural habitat. Yet, one in four clinical isolates of *E. coli* is observed to harbor P4 or a defective derivative of P4. Furthermore, like P2, P4 can be parasitized by a retron. An apparent product of this parasitization is phage φR73. This phage is nearly identical to P4 but contains a retron element (coding for reverse transcriptase) that has integrated to the right of the *att* site and provides a different tRNA target site, relative to that of P4, for prophage integration. These two P4-like phages display cross-immunity to each other. Retrons are generally rare in prokaryotes and are found mainly in various phage and other genetic parasites, including the viruses that infect cyanobacteria (discussed below).

Host fitness in the context of successful defective parasites

The P4 satellite virus example brings up several important and general points that should now be emphasized. A defective and generally persistent virus can be highly fit and adapted to its host, yet its own transmission may be parasitic to other prevalent and persistent (and possibly also defective) genetic parasites. This almost sounds like an oxymoron. How can two unrelated defective viruses have evolved to work together? Can two wrongs make a right? Is this an example of group selection? Yet the presence of addiction modules and other strategies makes this persistent life strategy highly successful. It is the viral genes and virus-specific phenotypes (such as viral immunity, addiction, and toxins) that are generally involved in successful persistence. In the case of P4, the best-studied example of a parasite of parasites (hyperparasite), we must also consider the fact that an *E. coli* host multicolonized by P4 and other phages will be strongly affected by the superimposition of these viral phenotypes. This host will have a different adaptive and fitness profile and a distinct evolutionary trajectory (relative to a singly colonized *E. coli* host) as a consequence of multiple parasite colonization. Especially affected will be the way in which this colonized host interacts with other genetic parasites. In the case of φR73, there is yet another level of parasitization that can be further considered—the retron, a hyperhyperparasite. More complex situations are also possible. Whereas the example given here simply considers the interaction of two persistent parasites (P2 and P4), the original P4 clinical isolate also contained other lytic phages, including T4, P1, and P3, all of which could have effects on the host. For example, these lytic phages are also known to be important to the outcome of P4 colonization and host survival. What we see in these interactions is the existence of a highly sophisticated and complex web of parasites in the context of their natural habitat and host. This cauldron of nested sets of competing and interacting and (often seemingly defective) genetic parasites is indeed reminiscent of the observations in chapter 2 on computer-simulated evolution, in which parasites of parasites bring about the evolution of much more complex and higher-order systems. In the case of P2, the phage is unable to be reactive from a lysogenic state unless P4 is also present. Thus, P2 may be considered to depend on its own genetic parasite, the satellite P4, in order to undergo reactivation. Once liberated by low-level P4-mediated reactivation, P2 can undergo a much more efficient lytic replication in a susceptible *E. coli* host. Yet P2 will become host trapped if it again undergoes lysogenization. Thus, P4 appears to be necessary to mobilize P2 at a low but successful rate from its colonized host. This mobilization explains why P2 is under selection to retain all the gene functions of a virus. It can now been clearly seen why the fitness of a persistent virus and its host is exceedingly difficult to measure in the absence of the other collaborating or competing genetic parasites.

The hyperparasites, such as the P4 retron, must also maintain mechanisms of persistence in their parasite hosts. To do this, they can utilize addiction modules, as described below, in which genes or functions that are both harmful and beneficial to the host (which can be a virus) are stably maintained in the hyperparasite. Examples of this include the numerous homing endonuclease genes (HEGs) which code for a harmful endonuclease within an intron or intein (a self-splicing protein; see the work by Gimble in Recommended Reading). The utilization of self-splicing introns and inteins suggests an interesting molecular strategy in which persistence can be attained by hiding as an intervening element. The presence of this mobile element, however, prevents HEG cleavage and thus protects the host genome. Such an intron-borne HEG is conserved in the *B. subtilis* phage SP82. Furthermore, the presence of this mobile HEG intron imparts to SP82 a competitive advantage, as the related phage SPO1 is cleaved but not colonized by the SP82 intron. It is interesting that the colicin plasmids of members of the family *Enterobacteriaceae*, which produce homing endonucleases during stress, cleave rRNA and chromosomal DNA of competing bacteria and kill them.

Phages That Infect via Host Pili

Acute infections

As mentioned above, enteric bacteria often carry sex plasmids that confer the capacity for conjugational transfer of DNA via a pilus structure and code for an integrase (and often also drug resistance factors). The conjugational transfer factors encoded on these plasmids include F (fertility) factors and N pili. The likely origins of pili are discussed below, but their similarity to the capsid proteins of filamentous phages suggests that these sex plasmids are derived from persistent viral parasites. The presence of pili, external appendages that transport DNA, makes conjugated bacterial cells susceptible to infection with various types of virus. In fact, the specificity of virus infection often depends more on the pilus than on the bacterial host. These pilus-infecting viruses are distinct from the larger dsDNA tailed phage considered earlier. Some of the pilus-infecting viruses are acute agents and do not establish either provirus or persisting infections. φ6 is one of the better-studied examples of a pilus-restricted acute virus. φ6 is a small, strictly lytic virus with a dsRNA genome containing 14 genes, including a

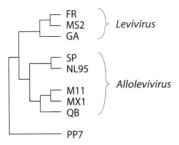

Figure 3.3 Dendrogram showing the proposed evolutionary relationships of RNA bacteriophage of the genera *Levivirus* and *Allolevivirus* which show F-plasmid specificity. PP7 is proposed to represent the basal phage which also has the simplest genetic map. Redrawn from J. P. Bollback and J. P. Huelsenbeck, *J. Mol. Evol.* 52:117–128, 2001, with permission.

gene for an RNA-dependent RNA replicase. It is most interesting that, for unknown reasons, ϕ6 infection is restricted to *Pseudomonas* spp. and is frequently found in association with degrading plant material. Land plants are known to support symbiotic filamentous fungi, essentially all of which are infected with various types of dsRNA viruses (see chapter 4). The ϕ6 family of viruses is clearly related to the dsRNA viruses found in animals, such as reovirus (see chapter 8), and thus represents a well-established viral lineage. In contrast to the capsids of tailed phages, however, the ϕ6 viral capsid also contains an essential lipid envelope. A dendrogram of RNA phage genomic evolution is shown in Fig. 3.3, which attempts to also relate evolution of F plasmid specificity.

Another lytic phage that infects via a pilus is Qβ. Qβ belongs to a distinct family of phages that have small icosahedral capsids and genomes of plus polarity ssRNA. This family of viruses is organized into several groups which can infect a wide range of bacteria and are clearly related to one another (such as M52 and f2). These phages are strictly lytic, and selection for resistant bacteria results in the loss of the pilus. Phages of this family are mostly found in association with animal feces, where phage counts can be as high as 10^7 PFU/ml. Curiously, F-specific coliphages are rare in human and cattle feces but common in feces of pigs and birds. The RNA-dependent replicase of Qβ has been especially well studied and has provided considerable insight into the biochemical evolution of a replicase and its intact or defective substrate (see chapter 2). Early on, Sol Spiegelman showed that when highly purified, this polymerase could spontaneously assemble substrate nucleotide triphosphates into Qβ origin-containing templates, which would then be amplified in vitro to very high levels. This process is called "monster formation" and resembles spontaneous biogenesis. Although the Qβ replicase has one of the highest polymerase error rates ever measured, natural and field isolates of this family of phages show very little sequence variation. However, variants of Qβ can easily be isolated in lab settings, where they grow well. Only when these variants must compete with wild-type Qβ is the relative weakness of the variant's fitness observed, as the variant is lost from the passed culture.

Nonlytic pilus infections

Filamentous phages, such as Ff and M13, also infect their hosts via pilus structures. However, in contrast to infections with the RNA phages described above, infections with these phages do not result in lysis. Instead, a chronic ongoing, but nonlytic, production of virus is established. The cells do not need to lyse because virus-encoded functions facilitate continuous virus extrusion through the cell membrane and cell wall. This is a variation of viral persistence, since virus production can be continuous. Whereas some of these filamentous phages become prophages by integrating into the host genome, others remain as episomal or plasmid-like prophages and establish a "pseudolysogenic" relationship with the host, during which phages do not integrate but do continuously produce progeny. Chronic virus production, very much like production of these filamentous phages of bacteria, is exceedingly common in archaeal prokaryotes and is further discussed below. Filamentous phages of bacteria have circular ssDNA genomes that are packaged into rodlike protein structures that have very high alpha-helical content and are extruded from the infected host without lysis. There is a clear structural relationship between capsid genes of filamentous phages and the gene for bacterial pili. Furthermore, it has been observed that M13 coat protein (which is so useful for phage display of cloned protein epitopes) is tolerant of a surprisingly high level of variation. The amino acid sequence of M13 can be inverted from its natural polarity yet still result in a highly efficient, totally novel capsid protein. Filamentous phages replicate as RCR, involving a virus-encoded, site-specific endonuclease that leads to the covalent attachment of nucleotides to a viral protein, which then functions as a primer for the replication of viral DNA. This process of protein-primed replication is found in various families of DNA and RNA viruses but is absent from host genomes. This family of filamentous phages also carries addiction module genes and toxin genes. Of specific interest is the CTXphi phage of *Vibrio cholerae*, which carries the cholera toxin genes (discussed below).

It seems clear that there is a strong relationship between the pilus-mediated sexual system of bacteria and various phages that produce either acute or nonlytic infections. It seems likely that the pilus structure, along with the integrase and the transmembrane transport of DNA used for transduction, has itself originated and evolved from an ancient virus infection (discussed below). However, it is also clear that the presence of a conjugative sex plasmid has an important impact on the relationship the host has with the various prevalent viral agents. Therefore, viruses can be expected to have an important and perhaps central impact on the evolutionary potential and consequence of this sexual process. However, there are very few laboratory or direct measurements of these interactions, so one would be hard-pressed to make any definitive evaluation of this issue. For example, if a bacterium that harbors an F factor is also infected with M13, what consequence does this have with respect to colonization by other temperate phages or infection by lytic phages? How would the expected milieu of virus-virus inter-actions affect sexual exchange and evolution? We currently lack answers to such questions.

Other types of acute RCR phage

The filamentous phages are very similar in their replication strategy to an-other well-studied *E. coli* phage, φX174. However, unlike the filamentous phages, φX174 is a lytic virus that is not pilus dependent. Infection with φX174 excludes virus reinfection. The φX174 genome codes for capsid and replication primer proteins as well as numerous other small proteins, sev-eral of which are not essential for replication in culture and have unknown functions. Although φX174 has a circular ssDNA genome, it is packaged into an icosahedral capsid, not a filamentous rod. φX174 is the best-studied phage that replicates by an RCR. It also uses a phage-specific and sequence-specific endonuclease and covalently attaches a primer protein to the viral DNA to prime rolling circular replication of viral DNA using host DNA polymerase III. φX174 is structurally more closely related to small ssDNA viruses found in plants and animals than to other phages, so it is considered a better model for an ancestor for these eukaryotic viruses. The φX174 pro-tein primer (gpA) is thus a basal member of a very large family of RCR primer proteins and has similarities to the viruses of plants and animals. Rolling-circle replication is specific to numerous viruses (including P2, P1, and lambda) but is not used to replicate any host chromosome. Thus, the presence of a rolling-circle replication strategy is a reliable marker for an ancient virus-specific replication system.

Relationship of Persistent Phages to Plasmids and Sex in Prokaryotes

How phages resemble plasmids

I have previously mentioned that in several cases, the distinction between plasmids and episomal persistent phages can be very blurred. Yet plasmids and fertility factors are often thought of as distinct entities from bacterio-phage since they lack many of the structural genes characteristic of viruses. Entire books that address the issues of plasmids have been written from this perspective. However, persistent episomal phages closely resemble plasmids in various ways and may themselves lack the genes coding for virion struc-tural proteins. As noted above, the P4 satellite virus and other defective viruses are often defective for most, if not all, of the viral structural genes. Yet these persistent phages retain an essentially virus-dependent replication strategy, requiring the help of another viral agent for their mobilization and transmission. In some of the previous examples, the persistent virus was a hyperparasite. In order for such a persistent and hyperparasitic virus to be fit, it must retain some phenotype or strategy that compels persistence by providing either a competitive advantage to the host or a new molecular identity system that recognizes and precludes other genetic parasites. Plasmids are often identical to persistent defective phages in these characteristics.

Another similarity between prophages and plasmids is observed in their respective coding functions. Both phages and plasmids can code for a specific integrase that directs DNA integration at specific chromosomal sites associated with specific tRNA genes. In some cases, the integrase itself can function as a phage-specific virulence factor. For example, the prophage-encoded integrase of *Dichelobacter nodosus* (the cause of ovine foot rot) is one such virulence factor. Bacterial virulence itself also establishes the similarity between plasmids and phages. Bacterial virulence is probably the best-studied example of an important complex host phenotype that is acquired in one genetic event. It has always been clear that a prophage can confer on its host bacterium the rather complex phenotypes associated with the acquisition of virulence factors. For the most part, these factors are phage-borne toxin genes, such as those for diphtheria toxin, erythrogenic toxins, staphylokinase, enterotoxin A, Shiga-like toxin, and botulinum toxin. In addition, alterations in the bacterial cell surface, such as the O antigen of *Shigella,* can be due to phage-encoded virulence factors. These virulence-associated genes are, for the most part, virus-derived genes and generally have no host counterpart. Also, they tend to reproductively isolate their hosts from uncolonized hosts. Finally, it should be noted that by acquiring one of these virulence-associated prophage, the host has acquired a complex new phenotype that can include the acquisition of over 100 new genes, all in one genetic event.

How plasmids resemble phages

All of the above-mentioned characteristics of phage persistence are also seen in plasmids. Let us now consider how some well-studied plasmids closely resemble phage. Bacteriocins are plasmid-encoded bactericidal particles that are highly specific to and active against other bacterial strains that lack the plasmid. Bacteriocins exist in two categories, large particles and small molecules. The large bacteriocin particles are clearly related to phage particles (forming both icosahedral and filamentous forms), although such particles often lack DNA. Thus, they closely resemble phage virion structural proteins able to forms holes in and kill susceptible cells. The small molecules are toxins that specifically kill host strains that lack the plasmid-encoded antitoxin. In this regard, the bacteriocins clearly resemble the addiction modules found in numerous persistent phage. Colicin K is a plasmid-encoded cell wall antigen that confers virulence and also clearly resembles a converted phage. Thus, it seems likely that these bacteriocin plasmids and other plasmids have originated from persistent cryptic phages and retained the associated persistent phenotype. It is also well established that plasmids can code for restriction/modification systems that also function as addiction modules and are widespread in both *Bacteria* and *Archaea*. In the case of *Lactobacillus* spp., there is a plasmid-based example of both the restriction and modification activities residing within one peptide that still confers resistance to phage infection and represents the simplest known version of a restriction/modification system. Clearly, these plasmids are essentially identical to persisting cryptic phages.

Multidrug resistance and virulence plasmids are also very well studied and are known to carry large numbers of genes. In some cases, very large virulence plasmids (e.g., *Bacillus anthracis* pXO1, which is more than 180 kbp long) that, in addition to being virulence determinants, also appear to be sites for the acquisition of other plasmids, various addiction and immunity modules, and transposable elements have been observed. However, in some cases these plasmids are so large that they also appear to function as second bacterial chromosomes. In a sense, these large plasmids function as sinks or traps for other plasmids and prophages, making the host more able to adapt to complex environmental changes. A good example of this situation is the genome of *V. cholerae,* which has a large plasmid that can be considered a second chromosome. As a rule, it is often stated that the presence of these plasmids (especially large ones) presents a fitness burden for the host bacteria in the absence of clear selective pressure, such as the presence of a drug in the growth media. However, it has been experimentally observed that following selection for plasmid persistence, plasmids can confer a competitive advantage on parasitized hosts in the absence of an obvious selective pressure. These plasmids also resemble phages in several other respects, in that they are stable, can have alternative replication strategies, and can have base compositions (GC versus AT) distinct from that of the cellular chromosome. As discussed below, mobile plasmids often code for integrases that are clearly related to phage-encoded integrases and use the same tRNA integration sites. Although many plasmids are related to one another, plasmids do not generally have the same degree of phylogenetic conservation as do phage families. It therefore seems very likely that, for the most part, plasmids have evolved from the polythetic lines of cryptic persistent phage.

PAIs: origins and phages

Pathogenicity islands (PAIs) constitute a well-studied, plasmid-mediated genetic system which has received much attention due to its obvious medical importance. Usually these sequences are integrated, not episomal. Studies of PAIs make it clear that in one genetic event, a bacterial cell can acquire a very large and complex set of interacting genes that confers on the host bacteria the ability to colonize human hosts, affect immune recognition, alter or regulate cell physiology, and cause disease. Thus, PAIs provide a clear example of the acquisition of a complex phenotype by the host. Seventy-five percent of these PAIs are associated with tRNAs at a sequence junction at the site of integration. This observation suggests phage involvement, since phage (not host) integrases target tRNA DNA sequences. Consistent with this idea, it is known that the PAI integrase, which is essential for host colonization, is related to the integrases of P4 and φ73. As noted above, these phage integrases are encoded by retrons present in the phage. PAI integrases are frequently defective, indicating that they are inactive. It thus seems likely that PAI elements need a helper phage to mobilize and then colonize additional hosts. Some of these PAIs have in fact been demonstrated to be excised and transmitted by helper phages. For example, the *Salmonella* SaP1 island is mobilized by phage φ13 or φ80. Thus, the PAI colonization process appears to essentially be an infectious event involving defective replicator elements and phage. The distinction between this process and the defective prophage relationships outlined above seems minimal. In a sense the term PAI is really a misnomer; a better term is "fitness island." These islands clearly introduce new phenotypes into their host and alter host fitness. Thus, PAIs can also be thought of as persistent genetic colonizers that are defective for mobilization. It is clear that PAIs cannot be considered selfish genetic elements, since they clearly bestow important and complex phenotypes on their hosts. However, PAIs are often thought of as having moved genes from one host to another, thereby effecting lateral transfer of gene sets. The problem with this view, however, is that it fails to address the origin of these complex gene sets, which are very often unlike other host genes. These extended regions of acquired DNA constitute the most dynamic portion of the bacterial genome. Since it is clear that PAI acquisition results from an infectious colonization event, it is also logical to propose that an infectious agent itself, such as a persistent phage, might have provided the original genetic material from which were assembled these fitness islands.

Sex factors and transposable elements: relationships to phages

Of special interest are the mobile plasmids and sex factors of bacteria, both because they are thought to be of major importance to bacterial evolution and because they have a close resemblance to viruses. However, F-factor distribution in natural populations is not as uniform as phage distribution. I have already discussed the clear physical similarity between the pilus structures of F factors and the capsids of filamentous phages, as well as the relationship of the sex plasmids' integrases and tRNA *att* sites to those of phages and the likelihood that phages are progenitors of sex plasmids. However, low sequence similarity between these plasmids and phages prevents the general conclusion that phages were indeed the predecessors of sex factors. In the case of phage, the integrase can also be part of the primer-replication system of the phage genome, and the particular *att* sites used define the specific phage lineages. Except for that of Mu, all known phage integrases mediate site-specific recombination (via tRNA sites) and their genes belong to the same gene family, the lambda-type integrase family. Thus, it has been argued that all phage integrases appear to be monophyletic and to have evolved from a common ancestor. It has also been noted how pili make bacteria susceptible to various acute phage infections and are thus directly subjected to phage-based selection. In addition, transduction requires the transmembrane movement of DNA. The plasmid protein responsible for this DNA movement clearly resembles the DNA ring helicase, which is characteristic of various phage replicators. Thus, for the most part, these essential F-plasmid genes have clear viral counterparts but do not have host counterparts. Furthermore, in many cases, these factors can affect the outcome of phage infections. These effects include F factor-mediated phage resistance, induction of prophage production, and F-factor invasion of silent prophage. The invasion of silent prophage can lead to the loss of the prophage immunity module, suggesting that F factors are in competition with some prophage. In addition, F-factor mutation or reactivation of lytic phage, along with F-factor invasion of the prophage, can result in the mobilization and hitchhiking of the F factor within the phage during transmission to another host. All of these characteristics suggest that sex plasmids are derived from phages and are a component of the continuum of genetic parasites that we call viruses, perhaps being most involved in a defective persistent life strategy that often depends on acute viruses for transmission to other hosts.

F factors are efficient transposable elements. However, the Mu phage of *E. coli* is by far the most highly adapted, complex, and efficient transposable element. Mu is a temperate linear dsDNA tailed phage (similar to T-even phages) that is able to transpose at rates 100 to 1,000 times greater than those of nonviral transposable elements. In addition, Mu can transpose to almost any site in the *E. coli* chromosome, hence the name Mu for "mutagenic." Like other temperate phages, Mu codes for a DNA-modifying enzyme. Moreover, Mu is resistant to P1 phage-mediated restriction. Most often, Mu infections are lytic, as lysogenic establishment is not efficient. In addition, unlike lambda, Mu prophage is not induced following UV irradiation or mitomycin C treatment. However, unlike that of all other phage types presented so far, Mu lytic replication is coupled to transposition. Thus, the transpositional activity of Mu proteins (which provide *att* site recognition or nick DNA to prime integration) is essential for Mu lytic replication. About 100 transposition events occur in each lytically infected cell. Furthermore, it is interesting that Mu can integrate into and inactivate a lambda prophage, suggesting that its high-level transposition might allow Mu to compete successfully with other prophages that have colonized the same host cell. One thing is clear, however: viruses are by far the most efficient transposable elements known, and their unmatched rates of genetic adaptation and evolution make it likely that they were the progenitors of both sex plasmids and other bacterial transposons.

Phage Variation and Evolution in Bacterial Populations

In 1981, it was suggested by D. Botstein that virus genomes undergo "modular evolution" in which new viruses originate from a combination of genes or gene clusters derived from multiple sources, including chromosomes, defective viruses, plasmids, transposable elements, etc. The observations that have accumulated in the ensuing years have, for the most part, been consistent with this view of phage evolution. However, it now also seems clear that the origins of most viral lineages were other predecessor viruses (possibly networks of related viruses) and not escaped cellular replicons, as was frequently suggested in the early literature. In addition, the sources of most new individual viral genes generally cannot be traced to host chromosomal (nonprophage) genes. For example, phylogenetic analysis has generally shown that the acquired genes that distinguish diverged lineages of phage have few, if any, counterparts in their hosts. Phage genes are mostly unique to the specific phage lineage. In some viral lineages, very few or none of the viral genes show similarity to any host genes. One characteristic that appears to distinguish phage genes from chromosomal genes is that phage (and most viruses) have an overrepresented level of small, single-domain genes (100 amino acids or less). The most recognized of these genes are the small genes of human immunodeficiency virus 1 and human papillomavirus (Tat, Rev, E6, E7, etc.). Similar small regulatory genes are found in the genomes of essentially all viruses, including phages. Gene loss has also been observed in specific viral lineages, but it appears to be much less common than gene acquisition, which is mostly characteristic of new viral lineages. The phycodnaviruses of algae seem especially prone to contain such "simple" examples of proteins, as discussed in chapter 4.

The dairy industry and phage variation

The best-studied system for phage variation in large bacterial populations is found in the dairy industry. Due to the large economic impact of lytic phages, bacterial fermentation of milk lactose into lactic acid by *Streptococcus lactis* in yogurt and cheese has been carefully monitored for over 30 years. The enormous culture volumes involved (up to 50,000 liters per day) would seem to be ideal for observing the dynamics of lytic phage adaptation in large populations. Although for many lytic and temperate phages, the genomes have been examined and interactions have been observed, it does not appear that the dairy industry provides an environment that puts evolution in fast-forward mode and creates new phage types. Instead, most of the new lytic variants appear to enter these cultures from preexisting outside (natural) sources, such as raw milk. In addition, specific lytic variants can be stable or fit for extended periods, suggesting that they neither are selected to become temperate or nonlytic nor possess high rates of variation.

The dairy industry observations, however, do confirm the patchy or modular nature of phage evolution, which is likely due to high-level recombination. Comparisons of 60 isolated and related lytic phage types, such as Sfi19 of *S. lactis,* show extensive cross-hybridization and patchy sequence similarity within gene-sized and sub-gene-sized regions, corresponding to sub-gene domains. The pattern of cross-hybridization is often distinct for the different genes; that is, individual genes cross-hybridize with distinct sets of phage. Sequence analysis indicates that these genes are mainly derived from other viruses, including cryptic proviruses, but seldom from genes of host chromosomes. This is especially true for regions coding for phage tails and base plates, as well as the diverse immunity regions. These genes seem to be assembled from numerous other phage sources rather than one single progenitor virus. In keeping with this idea, the phylogenetic analysis of the integrase gene has established that it is clearly virus specific and not congruent with the phylogenetic patterns of other phage genes. Thus, it seems that a gene-specific network of various viral lineages contributes to many phage genes, although more basic genes, such as those for helicases, are better conserved within one viral lineage. Of specific interest was the observation that the strictly lytic virus Sfi19 is highly related to a temperate phage, Sfi21, differing by only 10% in its sequence. The differences between the two phages also show the invasion of an intron into the lysin gene, again suggesting that a temperate phage can lose immunity function and be induced following colonization by a transposon to generate a strictly lytic variant. This result suggests a clear strategy by which the intron parasitizes a temperate phage for its mobilization. Interestingly, *Lactobacillus* phage LL-H contains a group II intron, which has not been observed to occur within its host genome, although group II introns (with reverse transcriptase domains) are present in the pRS01 conjugational element. Thus, there may exist a dynamic relationship between a persistent or temperate phage and the lytic variants that were derived from a successful persistent state and escaped virus-specific immunity.

Practical observations of phage control: use of plasmids and defectives

The dairy industry has been mainly interested in understanding how to make starter cultures resistant to an array of lytic phage. Thus, starter cultures provide much practical insight into the genetic factors that affect phage-host interactions. It has been observed that colonization of the starter culture by various plasmids, such as those that code for restriction/modification addiction modules, provides some of the most robust protection against lytic phages. This includes plasmid W10, which codes for one protein providing both restriction and modification function. However, the most impressive resistance to a broad array of phages (23 of 25 phages evaluated) was accomplished with defective plasmids of the phages themselves. By using two distinct types of phage-based origin sequences, synthetic recombinant replicons that would amplify following complementing phage infection and interfere with phage replication were constructed. This confirmed that defective or cryptic phage-based plasmids can be highly fit for persistence if subjected to acute, phage-based selection. This observation also suggests a genetic pathway by which persistence might evolve from an acute infection.

Bacterial differentiation and phage production

Among the prokaryotes, there also exist examples of *Bacteria* that can undergo cellular differentiation. As this is a characteristic associated with higher organisms, it is interesting to examine what is known about the relationship of such *Bacteria* with their viruses. During the differentiation-sporulation life cycle of *Thermoactinomyces vulgaris*, there is a clear linkage between the replication of the virulent bacteriophage Ta1 and cellular differentiation. In this case, the primary mycelium arising from spores is the only bacterial stage permissive for phage replication. Curiously, infection of mycelium or of late sporulation stages results in a loss of phage. And, if phage is added at the beginning of spore formation, the phage genome becomes integrated in the developing spores rather than producing a lytic infection. Subsequent outgrowth of these prophage-carrying spores leads to reactivated lytic phage production. This linkage raises the question of whether the phage life cycle contributed to the evolution of this cellular differentiation life cycle.

Comparative bacterial genomics, evolution, and dynamic genomes

With the completion of the genomic sequences of numerous bacterial species, we can now examine the specific types of global changes associated with separation between pairs of bacterial species. The first of these comparative genomic analyses has already been completed for *E. coli* and *B. subtilis*. Comparing the genomes of *E. coli* and *B. subtilis* shows that bacterial speciation occurs in a patchy manner involving gene sets: there are about 230 regions of distinct dissimilarity between these genomes. The great majority of these regions of difference are flanked by tRNA sequences, which mark the integration events associated with these gene sets. As presented above, tRNA-primed integration is characteristic of viral integrases, and some plasmids, but is neither typical of nor essential for host gene function. Thus, these flanking tRNA sequences clearly indicate that an infectious process involving the colonization of host genomes by genetic parasites was primarily responsible for most of the genetic events that led to the speciation of *E. coli* from *B. subtilis*. It has often been proposed that such types of insertional events are most likely be mediated by IS elements. However, *B. subtilis* DNA contains no IS elements or transposons. However, *B. subtilis* is known to currently contain 10 proviral genomes (including cryptic defective phage) in its chromosome. Thus, it seems clear that IS elements are not always involved in the alteration or adaptation of bacterial genomes. Interestingly, during the early bacterial genomic sequencing projects, it was observed that about one-third of *B. subtilis* sequences cannot be cloned in (i.e., are toxic to) *E. coli,* preventing the construction of an exhaustive phage library for *B. subtilis* proteins. Thus, there seem to exist clear limits on the degree of species compatibility for many bacterial genes, which could also suggest that some horizontal gene movement may not be tolerated between even closely related bacterial genomes. This observation also raises a question concerning the origin of novel bacterium-specific sequences. Rather than agreeing that new genes tend to come from other lineages of bacterial cells, I suggest that viral lineages may be a more likely source for the origination of most new genes. Bacterial evolution is now established to be essentially infectious, but the resulting changes seen must be persistent in the lineage in order to result in a new species. Bacteria are the most adaptable organisms we know and also have the most dynamic genomes.

In terms of naturally dynamic genomes, natural isolates of *E. coli* strains are known to vary in DNA content from 4.5 to 5.5 Mbp. Thus, at least 20% of the *E. coli* genome is normally dynamic. These strains differ significantly in gene number. This variation includes some iconic operons of *E. coli,* such as the lactose operon, which are not found in all natural *E. coli* isolates. Most of these variable genes (755 genes, of which 515 are in 62 sets) are now proposed to be associated with mobile accessory elements such as IS elements or prophages (including some large defective prophages). The prophage-associated genes comprise the largest set and include over 120 genes. The most variable and thus most dynamic of these genes include those that code for restriction enzymes as well as those coding for surface lipopolysaccharides. If a possible phage-based origin is considered for such genes, in keeping with the discussions above, such changes clearly seem to be related to phage colonization events, which often involve acquisition of addiction modules, and are consistent with the view that stable host colonization is a major force in sculpting the bacterial genome. I have already argued that restriction/modification systems represent examples of such addiction modules. Prior to the recent comparative genomic analyses, restriction enzymes had already been established to be highly variable and mobile between bacterial strains, indicating a strong association with genetic colonization.

The concept of the stable bacterial genome versus the unstable genome and the role of plasmids and phages

Currently, most of the dynamic portion of the *E. coli* genome is believed to have been generated by phage colonization rather than IS activity. In contrast, a major portion of the remaining *E. coli* genome is considered stable: it has been maintained during evolution and does not appear to have resulted from phage colonization. Evolutionists often consider this to be the true core genome and consider the dynamic portion to be an accessory part of the bacterial genome. It is interesting that many DNA viruses also show the same characteristic of core and accessory, or dynamic, genome regions. Yet some virus families, such as the lambdoid phages, show no conserved core sequences but maintain sequence signatures and other characteristics, including compatible recombination, which clearly suggest that they are in a common gene pool with their hosts. We might therefore ask, What constitutes an evolutionary (or phylogenetically) stable genome? What genes go on to "persist" in evolution, and why? Can these core genes also originate from virus? We have seen examples in which a persistent phage (e.g., P1) can replace the most basic elements of the host replication machinery: the origin of DNA replication and the corresponding origin recognition proteins. Thus, it is clear that viruses can create and superimpose on the most basic core host replicative functions. Yet such basic functions would seem to define the host itself. It would thus follow that essentially any host function might similarly have been derived from or replaced by a stable, persisting provirus. Given the preceding discussion, might it now be concluded that infectious genetic colonization must indeed be a major driving force in prokaryotic evolution? The genomic data that are now available, at least for the prokaryotes, appear to support such a claim for a viral role. However, this raises a conundrum. If the infectious colonizing genetic mechanism was so important for the creation of genetic novelty and evolution in *Bacteria*, why was such a process not maintained in more complex organisms such as eukaryotes? DNA viruses (prophages) and their defective derivatives generally do not colonize in and excise from the genomes of any eukaryotes. However, other colonizing genetic parasites may yet play a major role in eukaryotic evolution. As is presented in the following chapters, persistent germ line genetic parasites are exceedingly common in eukaryotes and are specific to their host species. These genomic parasites, however, are seldom derived from DNA viruses but instead are mostly related to retroviruses. Such stable genetic parasites may provide an answer to the dilemma of the apparent absence of an infectious mechanism involved in eukaryotic evolution.

Plasmids as chromosomes and origination of multiple chromosomes

One significant and general distinction between prokaryotic and eukaryotic chromosomes is that eukaryotic chromosomes are multiple and linear, not circular as in prokaryotes. However, some prokaryotes do harbor multiple chromosomes, and these warrant examination. *V. cholerae* is one example of a prokaryote that has a second chromosome of significant size. However, unlike the primary chromosome, the second chromosome has many nonessential accessory genes, including sequences from plasmids as well as genes for various addiction and immunity modules. This striking occurrence of so many accessory sequences has led some to propose that the second chromosome constitutes a plasmid capture system that facilitates the acquisition of new genes useful for adaptation. In general, the existence of multiple chromosomes in some prokaryotes raises several questions concerning the mechanisms that allow stable maintenance of multiple chromosomes in relation to those in eukaryotes. Could these mechanisms be derived from a common ancestor? Chromosome coordination would seem to require highly coordinated control of DNA replication and segregation. As noted above, some phage-derived plasmids, such as P1, are persistent and extrachromosomal. P1 achieves plasmid stability by tightly coordinating plasmid replication with host chromosome replication. Thus, it seems worth considering whether such phage-related strategies might have led to the origination of multiple chromosomes as seen in *V. cholerae*.

Addiction and multigenome stability

For the P1 episome, the coordinated replication of plasmid and host chromosome and the resulting stability are accomplished with the help of addiction modules. As noted above, Yarmolinsky first coined the term *addiction module* and applied it to the serine protease of phage P1 to explain postsegregation killing, or programmed cell death. This system, along with others, makes retention of the viral genome very stable. I have reasoned that addiction modules, involving stable toxins and unstable antitoxins or the restriction/modification systems, are one of the general strategies that allow an infectious genetic parasite to successfully attain stable persistence. Often, phage and plasmid addiction modules use various types of toxins and antitoxins for this purpose. I have also argued that persistent phages can themselves function as an addiction module in that they can kill uninfected members of the same or related bacterial species. In this case, the addiction module is a general genetic strategy since there are no specific toxin or antitoxin genes. The lytic action of the reactivated vegetative phage provides the harm (toxin), whereas the protective action of the viral immunity module is the antitoxin. Bacterial sex is also affected by such addiction strategies. For example, the loss of an F plasmid can kill its host due to the inhibitory effect of the small toxic *ccd* gene product on host gyrase A. Interestingly, although host gyrase resistance to *ccd* killing can be selected for, such mutations are unstable and recessive to the wild-type gyrase. Thus, natural selective pressures favor the maintenance of host sensitivity to the actions of *ccd* protein. Such toxin genes are often small proteins (less than 100 amino acids composing a single protein domain, often with an active site). Also, and like *ccd* protein, these toxins frequently target the most basic host machinery (such as cellular gyrase) or can create holes or pores in the target cell. In some cases, the antitoxin can be an antisense RNA (e.g., the hok/suk family). Often one system has several independent addiction modules, suggesting the major importance of such strategies to virus fitness. Clearly, such a persistent virus-host relationship is under strong selection. An unexpected example of the importance of phage biology to various addiction modules and to phage survival can be found in lambda. During lysogeny, lambda expresses only *rex*A and *rex*B (T4 *rII* exclusion), along with the *cII* repressor gene. Although this immunity function controls lytic lambda replication, it is also directed at excluding unrelated phage. However, the lambda immunity genes can also exclude the addiction modules of other persistent phage, such as P1, that might occupy a lambda host. RexB protein prevents degradation of lambda O protein, which is involved in DNA replication. However, by affecting targets of ClpP protease, RexB protein also inhibits the degradation of antitoxin proteins Phd (of P1) and Maze (of the *rel* operon), thus stabilizing these antitoxin proteins to prevent postsegregation cell killing. With respect to the Maze protein of the *rel* operon, RexB protein prevents the starvation-induced killing that would otherwise occur. Thus, lambda RexB can be an "antideath" or "antiaddiction" protein by stabilizing and extending the life span of the P1-addicted host. In this way, *rex*B provides a competitive advantage to lambda relative to other potential colonizers, thereby allowing lambda-colonized *E. coli* to exclude other persistent parasites. The occurrence of a large number of similar types of addiction modules within the second chromosome of *V. cholerae* might well suggest that the second chromosome did indeed originate from a persistent phage and has attained stability.

The Remainder of the Prokaryotic World: the Domain *Archaea* and Its Viruses

In the above discussion, I have examined viruses that infect *Bacteria* and considered their relationship to bacterial hosts. For the most part, viruses that infect *E. coli* have been the best-examined models for both lytic and persistent bacterial infections. However, we know that the prokaryotic world comprises two of the three distinct domains of life: *Bacteria* and *Archaea*. The *Archaea* are further divided into two kingdoms, the *Euryarchaeota*, which include methanogens and extreme halophiles, and the *Crenarchaeota*, which include thermophiles and sulfur-metabolizing organisms. In keeping with the general observation that all life forms have their own particular types of viruses, it can be seen that *Archaea* also have relationships with their viruses that differ from those of *Bacteria*. As for *Bacteria*, the large majority of the viruses so far characterized for *Archaea* are also dsDNA viruses. However, there are significant overall differences between the viruses that infect *Archaea* and *Bacteria*.

Euryarchaeota

One of the best-studied archaeal phages is the halophage φH, which infects halophiles (*Euryarchaeota*). φH is a tailed phage strikingly similar to T-even phages in morphology, replication, and transcription strategy. It contains a 59-kbp dsDNA linear genome. Hs1 and HF1/HF2 are similar tailed phages of halophiles, but the Ja1 halophage is notable for having a very large genome

(230 kbp) and lacking modified nucleotides. In contrast to the T-even phages, φH is a temperate phage that is induced by stationary-phase growth. In addition (and like P1), φH persists as a stable episome, or autonomous plasmid, that confers immunity on its host. Like T4, φH packages almost 20% more DNA than is contained in a single copy of the phage genome. Typically the φH genome contains tandem copies of the ISH 1.8 insertion element, which leads to genetic instability of the phage genome. The HF1/HF2 phages are acute versions of tailed halophages and replicate only in a lytic mode. They are resistant to type II restriction enzymes. ΨM1 is the best-studied phage of the methanogens, which constitute one of the largest groups of *Archaea*. Like φH, ΨM1 is a temperate phage, and like T4, it has a linear dsDNA genome that occupies less than a headful of DNA. Multiple copies of the pME2001 cryptic plasmid provide additional DNA to fill the phage head. Thus, in the *Euryarchaeota* (which is the prokaryotic kingdom most closely related to the *Eukarya*), the types of phage present and their relationships to their hosts are similar to but distinct from those seen in *Bacteria*.

Crenarchaeota

In the kingdom *Crenarchaeota,* the types of viruses found and their relationships with their hosts are unique. The overall pattern of phage found in these hosts differs strikingly from the pattern found in *Bacteria* and *Euryarchaeota*. Ninety-five percent of phages characterized for *Bacteria* have tailed dsDNA-containing capsids, whereas only 5% of the phages currently characterized for the *Crenarchaeota* are of this tailed type. Ninety-five percent of these phages have other morphologies, with filamentous forms being the most abundant. No RNA viruses have yet been reported for the *Archaea*. However, nine distinct morphotypes of various DNA viruses have been observed in these hosts, most of which have unique physical structures. The most striking difference, however, concerns the prevailing life strategy of these viruses. All of the nine types of viruses are nonlytic and are produced by continuous extrusion and not by a cell burst process, as is common in *Bacteria*. One of these viruses has a completely unique double-tailed morphology. Thus, most or all infections involve some type of persistent carrier state, rather than being lytic. As stated above, the *Crenarchaeota* include hyperthermophiles and sulfur-metabolizing bacteria. The viruses that infect hyperthermophiles have attracted the most attention because they present a potentially rich source of proteins that have high thermal stability and might have much commercial value. Recent thermophile phage surveys and sequencing projects by D. Prangishvili, W. Zillig, and others have begun to give us a better picture of these remarkable viruses and their hosts. The extrachromosomal carrier-infected states are exceedingly common in thermal habitats. One of these viruses is also known to integrate into the host chromosome. All thermophilic cells isolated so far appear to host some and often multiple phage infections. Strikingly, in one study, one isolated hyperthermophile hosted all nine phage morphotypes. Thus, mixed persistent carrier infections are very common. Several genomes of these viruses have recently been sequenced. One sequenced genome, that of *Pyrobaculum* spherical virus, was shown to have a linear dsDNA in which all open reading frames are on one DNA strand. Most remarkably, however, not one of these open reading frames showed any recognizable similarity to any gene in the GenBank database! This includes viral core and replication genes, which often show clear similarity to other viral genes. Furthermore, initial screening of other phage clones suggests that the low degree of similarity (less than 5%) of these phage genes to those in the database might be a general characteristic of most of the uncharacterized genomes. The implication is that a vast repertoire of unique genes exists in these hyperthermophilic phage. Another phage, *Acidianus* filamentous virus 1, that was also sequenced did show some gene similarity and also had some highly interesting properties. For example, *Acidianus* filamentous virus 1 uses eukaryote-like TATA promoters, which are not used by the hyperthermophile hosts. Of high relevance to chapter 4 (which discusses the possible origin of the nucleus), this phage has a linear dsDNA with ends composed of GC-rich 11-mer repeats, which clearly resemble the telomeres of eukaryotic chromosomes. Furthermore, sequence analysis suggests that this family of virus is basal to and resembles the *Chlorella* viruses, the poxviruses, and *African swine fever virus*—all large DNA viruses of eukaryotes. No other prokaryotic DNA virus had been shown to occupy this basal phylogenetic position.

For the sulfur-metabolizing *Crenarchaeota*, the best-studied virus is *Sulfolobus virus 1*, which has a novel lemon shape with a very short tail not seen in any other type of virus. *Sulfolobus virus 1* is distinct from bacterial viruses in having a closed circular dsDNA genome of about 15.5 kbp. Intriguingly, and possibly unique in the biological world, this DNA is packaged as a positively supercoiled topoisomer and requires a reverse gyrase for replication. In addition, this virus is not lytic; it is lysogenic (with a tRNA^{Arg} integration site) in its host and does not reactivate lytic virus production. Instead, it appears to spread in a unique fashion, from lysogenic host to uninfected host by direct cell-cell contacts, involving only low-level, nonlytic virus production. This virus therefore has a highly inapparent and persistent life strategy (more so than most viruses of bacteria) and seems to be adapted to essentially never make large quantities of virus. The *Sulfolobus neozealandicus* droplet-shaped virus is another virus with a closed circular DNA genome (20 kbp), so it appears that circular viral genomes are the norm in these sulfur-metabolizing hosts, unlike for hyperthermophilic *Archaea*, which host mostly linear DNA viruses. There also exist unique filamentous forms of DNA viruses for the *Crenarchaeota*, such as *Thermoproteus tenax* virus 1 (TTV-1), TTV-2, TTV-3, and TTV-4. Unlike the filamentous viruses of *Bacteria*, these viruses have linear dsDNA genomes and are highly heat stable. Also in distinction from viruses of *Bacteria*, the DNA of these viruses is stoichiometrically bound by one or several highly basic, virus-encoded proteins. Thus, there is a chromatin-like structure for the viral genome, more like the dsDNA viruses of eukaryotes. In addition, these virions have lipid envelopes that are either internal or external to the capsids. Both temperate (TTV1) and lytic (TTV4) life strategies are found in these filamentous viruses. In *Sulfolobus* spp., six unique virus particle morphologies have been observed, three of which were completely novel. Essentially every currently known virus of the *Crenarchaeota* is of a unique type not found in either *Bacteria* or *Eucarya*.

Plasmids are also known to occur in *Archaea*. Besides the ISH elements present in halophage genomes, noted above, conjugative plasmids have been observed. As for *Bacteria*, many archaeal viruses appear to bind to pili. In addition, a widespread plasmid, pDL10, that is found in *Sulfolobus* allows alternative oxidative or reductive metabolism of sulfur. Its copy number is amplified in response to energy metabolism. Another plasmid, pTIK4, has an addiction module and is able to induce killing of uncolonized cells via a cell-cell contact-associated process. However, plasmid-virus interactions have not been well characterized for *Archaea*, although the existence of plasmid-encoded restriction/modification addiction modules and the mobilization of plasmids by parasitizing infectious phage suggest that considerable interaction must occur.

Overall, *Archaea* support virus infections that are clearly similar in some respects to those of *Bacteria* (lytic and temperate dsDNA viruses), which can have restriction/modification- and toxin-based recognition systems. Yet the two kingdoms of *Archaea* are distinct from each other and from *Bacteria* in the types of viruses they support. Of special interest is the presence of linear dsDNA viruses that have chromatin-bound DNA as well as lipid envelopes. These are also characteristics of eukaryotic chromosomes, as is discussed in chapter 4. Given the tendency for viruses to distinguish their genomes from those of their hosts by various covalent modifications, it can be inferred that the tight association of viral DNA with protein seen in archaeal viruses also functions as a molecular system that differentiates viral from host chromosomes while also providing protection against sequence-specific host recognition systems, such as restriction/modification or DNA integration.

T-even-like phages predating the *Bacteria-Archaea* host divergence

Although it was noted above that archaeal and bacterial phages have numerous distinctions, the striking similarities between the halophage-like ϕH and bacterial T-even phages in terms of structure (including contractile tails), DNA replication strategy (including concatemers and headful packaging), and transcriptional organization (including back-to-back promoters controlling early to late transcription) are all hallmarks found in related viruses. This makes it likely that these phages originated from a common ancestor, and it has been argued that all tailed phages have a monophyletic origin. Given the highly host-dependent nature of these phages and the major differences in the lifestyles and physiologies of their hosts, it also seems likely that the host for the common ancestor of the tailed phages would be the ancestral cell progenitor to both *Archaea* and *Bacteria* (an undifferentiated prokaryote). If so, this argues that at least this lineage of phage has been present in prokaryotes since before the divergence of *Archaea* from *Bacteria*. It also argues against the idea that these phages evolved later (or frequently) from escaped host replicons, as has often been suggested. However, the mosaic and network nature of phage evolution makes it very difficult to trace ancestry. It is interesting that sequence analysis of 105 phage genes places T4 at the unresolved center of the major phage tree. Such placement would be consistent with a very early origin of tailed T-even phages.

The Cyanobacteria and Their Viruses: Steps toward Eukaryotic Evolution

The marine environment is of special interest to evolutionary biology, as it is the birthplace of so many lineages of life. The marine bacterial environment accounts for about 70 to 90% of marine organic matter. Thus, the ocean is a very large bacterial and viral habitat. Cyanobacteria and their viruses (phages) are of special interest due to the more developed nature of these photosynthetic and nitrogen-fixing bacteria. Cyanobacteria are thought to have diverged from *Bacteria* and *Archaea* about 3.5 billion years before the present and thus are one of the first and oldest living domains to have diverged from prokaryotes. Cyanobacteria are themselves prokaryotes and exist in five groups or orders. These groups show both asexual and sexual reproduction as well as morphological differentiation. Two groups are unicellular and divide by fission. Two groups are filamentous; one divides by fission, whereas the other forms sexual heterocysts (e.g., *Anabaena* spp.). The fifth group is morphologically complex and undergoes differentiation, giving rise to filamentous, branching, and multicellular morphology, and heterocyst formation. Curiously, lytic phages of cyanobacteria are well known for the first four groups but not known for the more complex fifth group (*Stigonematales*). Cyanobacteria are more complex than most prokaryotes in that they have photosynthetic, chlorophyll-containing, multimembraned structures (thylakoids) that are internal to the cell wall and function in absorbing light and CO_2 and emitting O_2. Cyanobacterial cell walls resemble those of gram-negative *Bacteria*, from which they appear to have evolved. Figure 3.4 is a dendrogram for a currently accepted view of cyanobacterial evolution. The less differentiated forms are at the base of the dendrogram. Interestingly, the most recently evolved forms appear to correlate to adaptation to low and high light levels. The phages that infect cyanobacteria are mainly large, tailed dsDNA phages that closely resemble T7. Mostly, these ubiquitous and lytic phages are specific to their host orders (e.g., unicellular or filamentous). Interestingly, some of these phages infect and replicate in the thylakoid structures, whereas others infect the nucleoplasm. LDP-1 is an example of a *Plectonema* cyanophage that infects thylakoids, displaces the

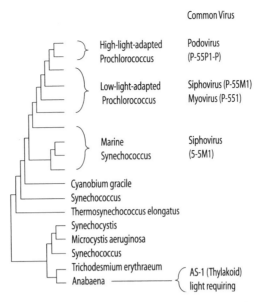

Figure 3.4 Evolution and common phage of cyanobacteria. The dendrogram is based on ribosomal DNA data reproduced from M. B. Sullivan, J. B. Waterbury, and S. W. Chisholm, *Nature* **424**:1047–1051, 2003, with permission.

photosynthetic lamellae, and stops CO_2 photoassimilation. Thus, some of these phages can clearly alter and regulate the bacterial photosynthetic system, as discussed in chapter 4, and can even code for the core photosynthetic enzymes. The prominent phage types infecting the higher orders of cyanobacteria are also indicated in Fig. 3.4. Interestingly, the high-light-adapted genus *Prochlorococcus* shows a strong basis for supporting podovirus (P-SSP1-8) infection.

The cyanophages for unicellular cyanobacteria are mainly lytic (such as SM-1, AS-1, and S2-L), and until recently no lysogenic phage had been described. Three genetically distinct phages of unicellular phycoerythrin-containing cyanobacteria are known, and they conserve much of the sequence encoding a capsid assembly protein found in T4 (gp20). Cyanophage S-PM2 has conserved a genetic module that includes g18 to g23, including the major virion components. However, in filamentous cyanobacteria of the LPP group (the genera *Lyngbya*, *Plectonema*, and *Phormidium*), lysogenic phages (such as LPP-1) are common and have long been observed. It is possible that the paucity of lysogenic phages from unicellular cyanobacteria is due to a lack of screening and reliable methods for phage induction. LDP-1D, a variant of LDP-1, is temperate in *Plectonema* and is not induced following UV irradiation, but it can be induced with mitomycin C.

CO_2 fixation and phage production

AS-1 is a cyanophage of *Anabaena* that also infects thylakoids, but in this case phage replication is obligate to photosynthesis, needing light. Viruses infecting subcellular structures (precursor "organelles") are thus common in this genus. There is also a relationship between phage infection and heterocyst formation. For example, phage A4 of *Anabaena* infects vegetative but not heterocyst cells. Curiously, selection for vegetative cells that resist A4 infection can result in mutation of the HU (histone-like) gene that is required for both A4 replication and heterocyst formation, suggesting a link between virus replication and sexual reproduction.

As was observed for both *Archaea* and their viruses, a strong association exists for cyanobacteria and their phages between the host order and the nature of viruses it supports. As mentioned above, lytic phages are most common in unicellular cyanobacteria, whereas lysogenic phages are more common in filamentous cyanobacteria. In addition, the linkage of virus replication to organelle function could suggest the involvement of viruses in the origins of these organelles. Generally, the accepted view is that the more autonomous organelles of eukaryotes (e.g., chloroplasts and mitochondria) evolved from endosymbiotic relationships between prokaryotic hosts and degenerate symbiotic cells. This view is supported by the presence of distinct 16S and 23S rRNA genes in these organelles because these genes are characteristic of cells but not viruses or plasmids. Recently, however, it has been observed that in red algae, there is a large (150-kbp) plasmid (from host strain RK-1) that contains these rRNA genes but is an autonomous replicon containing inverted repeat regions, characteristic of viral genomes. Thus, I suggest that the possibility that phages were involved in the origins of the eukaryotic organelles remains open. The possible viral origins of organelles is discussed further in the context of fungal mitochondria in chapter 4.

Recommended Reading

Phage History, Classification, and Host Association

Ackermann, H. W. 2001. Frequency of morphological phage descriptions in the year 2000. *Arch. Virol.* **146:**843–857.

Ackermann, H. W. 1998. Tailed bacteriophages: the order Caudovirales. *Adv. Virus Res.* **51:**135–201.

Brock, T. D. (ed.). 1999. *Milestones in Microbiology: 1546 to 1940.* ASM Press, Washington, D.C.

Brussow, H. 2001. Phages of dairy bacteria. *Annu. Rev. Microbiol.* **55:**283–303.

Cairns, J., and M. Delbrück. 1966. *Phage and the Origins of Molecular Biology.* Cold Spring Harbor Laboratory Press, Cold Spring Harbor, N.Y.

Davis, R. H. 2003. *The Microbial Models of Molecular Biology: from Genes to Genomes.* Oxford University Press, New York, N.Y.

Delbrück, M. 1950. *Viruses 1950. Proceedings of a Conference on the Similarities and Dissimilarities between Viruses Attacking Animals, Plants, and Bacteria, Respectively.* Division of Biology, California Institute of Technology, Pasadena.

Goyal, S. M., C. P. Gerba, and G. Bitton. 1987. *Phage Ecology.* Wiley, New York, N.Y.

Tetart, F., C. Desplats, M. Kutateladze, C. Monod, H. W. Ackermann, and H. M. Krisch. 2001. Phylogeny of the major head and tail genes of the wide-ranging T4-type bacteriophages. *J. Bacteriol.* **183:**358–366.

Phage and Plasmid Evolution

Bernstein, H., and C. Bernstein. 1989. Bacteriophage T4 genetic homologies with bacteria and eucaryotes. *J. Bacteriol.* **171:**2265–2270.

Blaisdell, B. E., A. M. Campbell, and S. Karlin. 1996. Similarities and dissimilarities of phage genomes. *Proc. Natl. Acad. Sci. USA* **93:**5854–5859.

Bollback, J. P., and J. P. Huelsenbeck. 2001. Phylogeny, genome evolution, and host specificity of single-stranded RNA bacteriophage (family Leviviridae). *J. Mol. Evol.* **52:**117–128.

Botstein, D. 1981. A modular theory of virus evolution, p. 363–384. *In* B. N. Fields, R. Jaenisch, and C. F. Fox (ed.), *Animal Virus Genetics.* Academic Press, New York, N.Y.

Brussow, H., and F. Desiere. 2001. Comparative phage genomics and the evolution of Siphoviridae: insights from dairy phages. *Mol. Microbiol.* **39:**213–222.

Gimble, G. S. 2000. Invasion of a multitude of genetic niches by mobile endonuclease genes. *FEMS Microbiol. Lett.* **185:**99–107.

Hendrix, R. W. 2002. Bacteriophages: evolution of the majority. *Theor. Popul. Biol.* **61:**471–480.

Hendrix, R. W. 1999. Evolution: the long evolutionary reach of viruses. *Curr. Biol.* **9:**R914–R917.

Oshima, K., S. Kakizawa, H. Nishigawa, T. Kuboyama, S. Miyata, M. Ugaki, and S. Namba. 2001. A plasmid of phytoplasma encodes a unique replication protein having both plasmid- and virus-like domains: clue to viral ancestry or result of virus/plasmid recombination? *Virology* **285:**270–277.

Rohwer, F., and R. Edwards. 2002. The phage proteomic tree: a genome-based taxonomy for phage. *J. Bacteriol.* **184:**4529–4535.

Tetart, F., C. Desplats, M. Kutateladze, C. Monod, H. W. Ackermann, and H. M. Krisch. 2001. Phylogeny of the major head and tail genes of the wide-ranging T4-type bacteriophages. *J. Bacteriol.* **183:**358–366.

Lytic Phages

Karam, J. D., et al. (ed.). 1994. *Molecular Biology of Bacteriophage T4.* American Society for Microbiology, Washington, D.C.

Mathews, C. K., E. M. Kutter, G. Mosig, and P. B. Berget (ed.). 1983. *Bacteriophage T4.* American Society for Microbiology, Washington, D.C.

Tetart, F., C. Desplats, M. Kutateladze, C. Monod, H. W. Ackermann, and H. M. Krisch. 2001. Phylogeny of the major head and tail genes of the wide-ranging T4-type bacteriophages. *J. Bacteriol.* **183:**358–366.

Lysogenic Episomal Phages

Gordon, G. S., and A. Wright. 2000. DNA segregation in bacteria. *Annu. Rev. Microbiol.* **54:**681–708.

Hendrix, R. W. 1983. *Lambda II.* Cold Spring Harbor Laboratory, Cold Spring Harbor, N.Y.

Hershey, A. D. 1971. *The Bacteriophage Lambda.* Cold Spring Harbor Laboratory Press, Cold Spring Harbor, N.Y.

Phage Mu

Morgan, G. J., G. F. Hatfull, S. Casjens, and R. W. Hendrix. 2002. Bacteriophage Mu genome sequence: analysis and comparison with Mu-like prophages in *Haemophilus, Neisseria* and *Deinococcus. J. Mol. Biol.* **317:**337–359.

Symonds, N. 1987. *Phage Mu*. Cold Spring Harbor Laboratory Press, Cold Spring Harbor, N.Y.

Viruses of *Archaea*

Bettstetter, M., X. Peng, R. A. Garrett, and D. Prangishvili. 2003. AFV1, a novel virus infecting hyperthermophilic Archaea of the genus Acidianus. *Virology* **315**:68–79.

Prangishvili, D. 2003. Evolutionary insights from studies on viruses of hyperthermophilic *Archaea*. *Res. Microbiol*. **154**:289–294.

Rachel, R., M. Bettstetter, B. P. Hedlund, M. Haring, A. Kessler, K. O. Stetter, and D. Prangishvili. 2002. Remarkable morphological diversity of viruses and virus-like particles in hot terrestrial environments. *Arch. Virol*. **147**:2419–2429.

Rice, G., K. Stedman, J. Snyder, B. Wiedenheft, D. Willits, S. Brumfield, T. McDermott, and M. J. Young. 2001. Viruses from extreme thermal environments. *Proc. Natl. Acad. Sci. USA* **98**:13341–13345.

Zillig, W., D. Prangishvili, C. Schleper, M. Elferink, I. Holz, S. Albers, D. Janekovic, and D. Gotz. 1996. Viruses, plasmids and other genetic elements of thermophilic and hyperthermophilic Archaea. *FEMS Microbiol. Rev*. **18**:225–236.

Oceanic and Soil Phages

Ackermann, H. W. 2001. Frequency of morphological phage descriptions in the year 2000. *Arch. Virol*. **146**:843–857.

Ackermann, H. W. 1998. Tailed bacteriophages: the order Caudovirales. *Adv. Virus. Res*. **51**:135–201.

Bernstein, H., and C. Bernstein. 1989. Bacteriophage T4 genetic homologies with bacteria and eucaryotes. *J. Bacteriol*. **171**:2265–2270.

Bettstetter, M., X. Peng, R. A. Garrett, and D. Prangishvili. 2003. AFV1, a novel virus infecting hyperthermophilic Archaea of the genus Acidianus. *Virology* **315**:68–79.

Blaisdell, B. E., A. M. Campbell, and S. Karlin. 1996. Similarities and dissimilarities of phage genomes. *Proc. Natl. Acad. Sci. USA* **93**:5854–5859.

Bollback, J. P., and J. P. Huelsenbeck. 2001. Phylogeny, genome evolution, and host specificity of single-stranded RNA bacteriophage (family Leviviridae). *J. Mol. Evol*. **52**:117–128.

Botstein, D. 1981. A modular theory of virus evolution, p. 363–384. *In* B. N. Fields, R. Jaenisch, and C. F. Fox (ed.), *Animal Virus Genetics*. Academic Press, New York, N.Y.

Brussow, H. 2001. Phages of dairy bacteria. *Annu. Rev. Microbiol*. **55**:283–303.

Brussow, H., and F. Desiere. 2001. Comparative phage genomics and the evolution of Siphoviridae: insights from dairy phages. *Mol. Microbiol*. **39**:213–222.

Cairns, J., and M. Delbrück. 1966. *Phage and the Origins of Molecular Biology*. Cold Spring Harbor Laboratory Press, Cold Spring Harbor, N.Y.

Delbrück, M. 1950. *Viruses 1950. Proceedings of a Conference on the Similarities and Dissimilarities between Viruses Attacking Animals, Plants, and Bacteria, Respectively*. Division of Biology, California Institute of Technology, Pasadena.

Fuller, N. J., W. H. Wilson, I. R. Joint, and N. H. Mann. 1998. Occurrence of a sequence in marine cyanophages similar to that of T4 g20 and its application to PCR-based detection and quantification techniques. *Appl. Environ. Microbiol*. **64**:2051–2060.

Goyal, S. M., C. P. Gerba, and G. Bitton. 1987. *Phage Ecology*. Wiley, New York, N.Y.

Hacker, J., and J. B. Kaper. 2000. Pathogenicity islands and the evolution of microbes. *Annu. Rev. Microbiol.* **54:**641–679.

Hambly, E., F. Tetart, C. Desplats, W. H. Wilson, H. M. Krisch, and N. H. Mann. 2001. A conserved genetic module that encodes the major virion components in both the coliphage T4 and the marine cyanophage S-PM2. *Proc. Natl. Acad. Sci. USA* **98:**11411–11416.

Hendrix, R. W. 2002. Bacteriophages: evolution of the majority. *Theor. Popul. Biol.* **61:**471–480.

Hendrix, R. W. 1999. Evolution: the long evolutionary reach of viruses. *Curr. Biol.* **9:**R914–R917.

Hendrix, R. W. 1983. *Lambda II*. Cold Spring Harbor Laboratory, Cold Spring Harbor, N.Y.

Hershey, A. D. 1971. *The Bacteriophage Lambda*. Cold Spring Harbor Laboratory, Cold Spring Harbor, N.Y.

Hurst, C. J. 2000. *Viral Ecology*. Academic Press, San Diego, Calif.

Karam, J. D., et al. (ed.). 1994. *Molecular Biology of Bacteriophage T4*. American Society for Microbiology, Washington, D.C.

Lindqvist, B. H., G. Deho, and R. Calendar. 1993. Mechanisms of genome propagation and helper exploitation by satellite phage P4. *Microbiol. Rev.* **57:**683–702.

Mathews, C. K., E. M. Kutter, G. Mosig, and P. B. Berget (ed.). 1983. *Bacteriophage T4*. American Society for Microbiology, Washington, D.C.

Morgan, G. J., G. F. Hatfull, S. Casjens, and R. W. Hendrix. 2002. Bacteriophage Mu genome sequence: analysis and comparison with Mu-like prophages in *Haemophilus*, *Neisseria* and *Deinococcus*. *J. Mol. Biol.* **317:**337–359.

Perova, E. V., A. S. Tikhonenko, T. I. Bulatova, and B. N. Il'iashenko. 1977. Bacteriophage induction in cultures of C1. botulinum type A. *Zh. Mikrobiol. Epidemiol. Immunobiol.* **11:**125–128. (In Russian.)

Prangishvili, D. 2003. Evolutionary insights from studies on viruses of hyperthermophilic Archaea. *Res. Microbiol.* **154:**289–294.

Rachel, R., M. Bettstetter, B. P. Hedlund, M. Haring, A. Kessler, K. O. Stetter, and D. Prangishvili. 2002. Remarkable morphological diversity of viruses and virus-like particles in hot terrestrial environments. *Arch. Virol.* **147:**2419–2429.

Rice, G., K. Stedman, J. Snyder, B. Wiedenheft, D. Willits, S. Brumfield, T. McDermott, and M. J. Young. 2001. Viruses from extreme thermal environments. *Proc. Natl. Acad. Sci. USA* **98:**13341–13345.

Rohwer, F., and R. Edwards. 2002. The phage proteomic tree: a genome-based taxonomy for phage. *J. Bacteriol.* **184:**4529–4535.

Symonds, N. 1987. *Phage Mu*. Cold Spring Harbor Laboratory, Cold Spring Harbor, N.Y.

Tetart, F., C. Desplats, M. Kutateladze, C. Monod, H. W. Ackermann, and H. M. Krisch. 2001. Phylogeny of the major head and tail genes of the wide-ranging T4-type bacteriophages. *J. Bacteriol.* **183:**358–366.

Tikhonenko, A. S., N. N. Belyaeva, and G. Ivanovics. 1975. Electron microscopy of phages liberated by megacin A producing lysogenic *Bacillus megaterium* strains. *Acta. Microbiol. Acad. Sci. Hung.* **22:**58–59.

Zillig, W., D. Prangishvili, C. Schleper, M. Elferink, I. Holz, S. Albers, D. Janekovic, and D. Gotz. 1996. Viruses, plasmids and other genetic elements of thermophilic and hyperthermophilic Archaea. *FEMS Microbiol. Rev.* **18**:225–236.

The P2-P4 Helper Phage System

Bertani, G., and G. Deho. 2001. Bacteriophage P2: recombination in the super-infection preprophage state and under replication control by phage P4. *Mol. Genet. Genomics* **266**:406–416.

Lindqvist, B. H., G. Deho, and R. Calendar. 1993. Mechanisms of genome propagation and helper exploitation by satellite phage P4. *Microbiol. Rev.* **57**:683–702.

Addiction Modules and Persistence

Engelberg-Kulka, H., and G. Glaser. 1999. Addiction modules and programmed cell death and antideath in bacterial cultures. *Annu. Rev. Microbiol.* **53**:43–70.

PAIs

Cheetham, B. F., and M. E. Katz. 1995. A role for bacteriophages in the evolution and transfer of bacterial virulence determinants. *Mol. Microbiol.* **18**:201–208.

Finlay, B. B., and S. Falkow. 1997. Common themes in microbial pathogenicity revisited. *Microbiol. Mol. Biol. Rev.* **61**:136–169.

Hacker, J., and J. B. Kaper. 2000. Pathogenicity islands and the evolution of microbes. *Annu. Rev. Microbiol.* **54**:641–679.

Oshima, K., S. Kakizawa, H. Nishigawa, T. Kuboyama, S. Miyata, M. Ugaki, and S. Namba. 2001. A plasmid of phytoplasma encodes a unique replication protein having both plasmid- and virus-like domains: clue to viral ancestry or result of virus/plasmid recombination? *Virology* **285**:270–277.

Bacterial Genomes

Davis, R. H. 2003. *The Microbial Models of Molecular Biology: from Genes to Genomes.* Oxford University Press, New York, N.Y.

Gelfand, M. S., and E. V. Koonin. 1997. Avoidance of palindromic words in bacterial and archaeal genomes: a close connection with restriction enzymes. *Nucleic Acids Res.* **25**:2430–2439.

Gordon, G. S., and A. Wright. 2000. DNA segregation in bacteria. *Annu. Rev. Microbiol.* **54**:681–708.

Karlin, S., and C. Burge. 1995. Dinucleotide relative abundance extremes: a genomic signature. *Trends Genet.* **11**:283–290.

Karlin, S., A. M. Campbell, and J. Mrazek. 1998. Comparative DNA analysis across diverse genomes. *Annu. Rev. Genet.* **32**:185–225.

Mrazek, J., and S. Karlin. 1999. Detecting alien genes in bacterial genomes. *Ann. N.Y. Acad. Sci.* **870**:314–329.

Riley, M., and M. H. Serres. 2000. Interim report on genomics of *Escherichia coli.* *Annu. Rev. Microbiol.* **54**:341–411.

The Dilemma of the Big Transition in Evolution: the Eukaryotes

Introduction

The evolution of the eukaryotic nucleus and the eukaryotic cell was the largest discontinuity in the evolution of all life. In addition to the dilemma this presents to evolutionary biologists, the origination of the eukaryote also raises important issues for virologists, as this produced a new and distinctly different habitat for viruses. As mentioned in chapter 1, geological evidence of fossilized cellular structures suggests that prokaryotes have been present for at least 3.5 billion years. Fossil evidence also suggests that cyanobacteria had developed by 2.7 billion to 2.8 billion years before present. Cyanobacteria appear to have originated prior to the evolution of the first eukaryotes. The earliest eukaryotes for which there exist clear geological fossil data are the microalgae (similar to red and brown algae), which first appeared at the pre-Cambrian boundary. The microalgal fossils date to about 2.0 billion to 2.2 billion years before present. On a geological timescale, the Cambrian radiation occurred relatively soon after the appearance of the first eukaryotes and resulted in the generation of diverse eukaryotic phyla and species, including organisms with shell and bone structures. Thus, according to the fossil record, prokaryotes existed for over 1 billion years before the first unicellular eukaryotic cell emerged and radiated into diverse eukaryotic organisms. Why was a prokaryotic world so stable, and why did this change so rapidly? Although it appears that the origination of eukaryotic green algae is associated with the symbiotic acquisition of the chloroplast, it seems probable that the eukaryotic nucleus, along with various other cytoplasmic characteristics (such as cytoplasmic motility and the endoplasmic reticulum [ER]), was acquired even before this event. *Giardia* species, for example, constitute one of the most primitive forms of eukaryotes: they lack mitochondria but still have nuclei. Recent phylogenetic analysis suggests that the eukaryotic nucleus evolved prior to the symbiotic acquisition of plastids (both mitochondria and chloroplasts). If so, then the acquisition of the nucleus represents the most basal event in the evolution of eukaryotes. In this chapter, I present the possible role of viruses in acquisition of the nucleus, as well as other evolutionary discontinuities. For the most part, evolutionary biology has not considered the possible role of viruses in host biogenesis.

Current view: symbiosis between several prokaryotes

A current and prominent view concerning the origination of eukaryotes is that they represent a fusion of two or more symbiotic progenitor cells. This kind of symbiosis is also thought to have resulted in generation of chloroplasts and mitochondria in addition to the nuclear structures. Each of these structures is thought to have been derived from a distinct prokaryotic cellular predecessor. For example, the symbiosis that produced the nucleus is thought to have resulted from the engulfment of a nuclear prokarotyic progenitor (that lacks a cell wall) by another cell wall-lacking prokaryotic predecessor. The resulting unicellular eukaryote cell would have resembled a primitive alga-like cell. Cyanobacteria are believed by many to be the likely ancestors of both mitochondria and chloroplasts. Numerous similarities have been noted to support this view, such as the similarity of chloroplast and mitochondrial 16S rRNAs to those of cyanobacteria and purple bacteria. The evolution of these bacteria prior to eukaryotic evolution is also thought to have resulted in the change in oxidation state of the early atmosphere, which substantially altered the world's ecology and set the stage for the emergence of eukaryotes with their mitochondrion-based oxidative metabolism. Photosynthesis is thought to have been a central participant in the origination of eukaryotes because it allows unicellular organisms to live without a dependence on chemical energy, instead using photosynthetic phosphorylation to provide energy. Photosynthesis, however, needs to "pump away" excess energy to prevent photooxidation of chlorophyll. Therefore, some system for this purpose (such as carotenoids) would also need to have been created in the early eukaryote. More problematic for a eukaryotic cell, however, is the incompatibility between photosynthesis and the free oxygen needed for mitochondrial oxidative respiration. These two features do not appear to be chemically compatible within an open system and would require a strict chemical separation. Unicellular algae have clearly solved this problem by compartmentalization of the two plastids, chloroplasts and mitochondria. Thus, cyanobacteria or purple bacteria which could provide the basis for both chloroplasts and mitochondria seem likely possible participants in these plastid acquisitions and symbiosis.

However, the origin of the nucleus, the most basal distinction of all eukaryotes from prokaryotes, presents the biggest challenge for theories based on a symbiotic origin. As is discussed below, most molecular and structural features of the nucleus pose a problem for having originated from prokaryotes. A widely accepted view, proposed by T. Cavalier-Smith in 1987, is that the early eukaryotic cell must have lacked a cell wall, allowing motility and cytoplasmic engulfment of food. Mycoplasmas are bacteria that lack cell walls and thus have been proposed as the likely source of this progenitor cell. However, there do not now exist any known prokaryotic organisms (including mycoplasmas) that feed by engulfment (phagocytosis) of food, as do eukaryotes, which suggests that the phagocytic character of the eukaryote did not exist in the prokaryotic progenitor. As mentioned above, recent phylogenetic analysis has suggested that the acquisition of the nucleus in early eukaryotes may predate the acquisition of both the mitochondria and chloroplasts. Thus, a conundrum seems to exist concerning the origin of the nucleus, and, as discussed below, no existing prokaryotic cell stands out as being the likely progenitor of the nucleus.

Evolution of the Nucleus

The nucleus contains numerous basic and distinguishing features of the eukaryotic cell, including all the highly coordinated proteins involved in genome replication. The eukaryotic replication proteins and apparatus, although functionally homologous to the replication proteins and apparatus of prokaryotes, are very distinct. Eukaryotic replication proteins have amino acid sequence compositions that differ almost completely from those of prokaryotes. This sequence difference is so large that prokaryotic proteins do not appear able to have been the progenitors of most of these functionally homologous eukaryotic proteins. However, the prokaryotes of the domain *Archaea* do have some replication proteins with notably greater sequence similarity to those of eukaryotes than do *Bacteria*. This observation has led some to suggest that *Archaea* nuclear proteins were the likely symbiotic progenitors of the eukaryotic nucleus. There are, however, major problems with this scenario. In particular, this leaves unexplained the origin of too many other features of the eukaryotic nucleus. In 2001, A. Poole and D. Penny reviewed the evidence for the archaeal origin of the nucleus and concluded that existing data argue against archaeal origins. This conclusion is also consistent with the observation that *Archaea* are much more like *Bacteria* than they are like eukaryotes: *Archaea* have four times more *Bacteria*-like proteins than eukaryote-like proteins. Thus, *Archaea* are significantly more related to *Bacteria* than they are to eukaryotes. This dilemma has led J. Maynard Smith and E. Szathmary to conclude that the evidence of the symbiotic origin of the eukaryotic nucleus is presently weak and that we still lack a sensible scenario for the origin of the nucleus. There currently exists no living cell that has all or even many of the characteristics needed to have become the nucleus. Below, I list specific examples of nuclear characteristics that lack a sensible explanation based on having originated from a prokaryotic cell. Each of these characteristics alone raises a dilemma for explaining the origin of the nucleus. Yet all are considered unique cellular and molecular characteristics of all eukaryotes.

The workings of the nucleus

The existence within the nucleus of numerous molecular differences from prokaryotes raises several issues. Each of these differences requires an explanation according to the theory of prokaryotic symbiosis as has been presented by L. Margulis and D. Sagan. In addition, these molecular differences of the nucleus have major implications for the function of eukaryotic RNA and especially for DNA viruses. Not only does the nucleus segregate the process of transcription and DNA replication from that of translation, but also it provides a very distinctive molecular and chromosomal environment for both DNA replication and transcription. Essentially all currently characterized prokaryotic organisms have circular DNA genomes with unique origins of replication that attach to the cell membrane and facilitate daughter chromosome segregation. A few examples of linear large plasmids or accessory chromosomes are also now known to occur in some bacteria (e.g., *Agrobacterium tumefaciens*). However, as mentioned in chapter 3, the genes within these uncommon linear plasmids are usually associated with the accessory functions, such as tumor/transforming DNA transfer to host plant cells or the presence of addiction modules, pathogenicity islands, and transposable elements. The core replication and biosynthetic genes are generally located in the circular chromosome, suggesting that the linear DNAs represent remnants of a past colonizer (molecular genetic parasite). Another very interesting exception to circular chromosomes of prokaryotes is that of the parasitic *Borrelia* spirochetes. They can contain sets with both circular and linear double-stranded DNA (dsDNA) chromosomes, the latter being associated with pathogenicity and having covalently closed snapback ends. As is described below, covalently closed snapback ends of DNA are molecular characteristics that clearly resemble the genomes of some DNA viruses, such as the poxviruses. Another example of a prokaryote that has additional chromosomes is *Vibrio cholerae*. As mentioned previously, the additional chromosome of *V. cholerae* has also been called a gene capture system, which contains addiction genes and toxin genes and is associated with prophages. In fact, it is the prophage genes within these chromosomes that include the gene for the cholera toxin. In contrast to the central role of the circular chromosomes in prokaryotes, eukaryotes have only multiple linear chromosomes bearing some type of repeated telomeric sequences at their ends. In eukaryotes, circular chromosomes or chromosomes with only one origin are essentially nonexistent. The circular genomes that are found in eukaryotes either are all viral episomes or result from differentiation-linked DNA amplification (endoreduplication) of specific replicons, such as ribosomal DNA in diplomonads.

Another major molecular distinction between prokaryotes and eukaryotes is the control of DNA packaging and replication. Unlike eukaryotic DNA, prokaryotic DNA is not tightly associated with stoichiometrically bound, basic chromatin proteins, such as histones. This distinction appears to have affected viral strategies in prokaryotes. For example, the great majority of prokaryotic viruses inject naked DNA into their host cells, and integration of viral DNA into host chromosomes is a common viral strategy, especially during lysogenic persistent states. In contrast, eukaryotic chromosomal DNA is always tightly associated with small basic DNA binding proteins, usually histone-like, with the interesting exception of the naked DNA present in a gamete just after sperm penetration and uncoating. In keeping with this host characteristic, eukaryotic DNA viruses all use some type of basic polymer- or protein-bound chromatin structure that is used to package virion DNA and to infect host cells. Eukaryotic viruses appear to avoid naked DNA in their replication strategy. Another major distinction from prokaryotes is that eukaryotic DNA replication initiates in numerous (generally thousands) of sites, which can have either a loosely defined origin sequence (corresponding to regions of initiation) or a specific origin sequence (such as the amplified ribosomal DNA origins of replication). In stark contrast to the case for prokaryotes, the reinitiation of eukaryotic DNA replication is very stringently regulated within a complex cell cycle control system (except for the diplomonad macronucleus discussed below) and shows exceedingly low overreplication error rates (generally less than 1 in 10^7). Daughter eukaryotic DNA molecules segregate via attachment to tubular proteins that make up the spindles, not by membrane attachment as in prokaryotes. Furthermore, a complex set of more than 10 proteins is involved in the control of the initiation and extension of DNA replication in eukaryotes. Although functionally analogous, all these proteins are distinct in sequence from those of prokaryotes (see the work by Forterre in Recommended Reading). In general, the replication control proteins in eukaryotes lack close prokaryotic homologues and are not part of the universally conserved set of proteins found in all domains of cellular life. However, essentially all of these eukaryotic replication proteins can be found to exist as identifiable homologues within various DNA phages and eukaryotic DNA viruses. The most distinctive of these replication proteins are those involved in the very tight control of the initiation of DNA replication (the origin recognition complex [ORC]). Eukaryotic ORC proteins have no direct prokaryotic homologues. Interestingly, the only prokaryotic ORC system that clearly resembles that of eukaryotes includes the ORC proteins of lysogenic prophages such as lambda (whereas the lytic phages, such as T4, lack these homologues).

Eukaryotic viruses appear to adhere tightly to the same basic molecular characteristics and strategies that are used by the eukaryotic nucleus, except for the notable need of DNA viruses to overreplicate acute viral genomes, which must occur independently from host cell cycle control. Interestingly, latent eukaryotic DNA viruses often replicate their DNA in coordination with host cell cycle control, which is also typically linked to cellular differentiation. Except for the DNA viruses of algae, eukaryotic DNA viruses do not inject naked DNA into host cells as do bacteriophages. Eukaryotic DNA viruses generally use either virally or cellularly encoded, DNA-associated basic polymers or histone-like proteins to stoichiometrically coat and condense their chromosomes. In addition, all eukaryotic nuclear viruses appear to have specific mechanisms for nuclear entry, often involving viral structural proteins. Human adenovirus, for example, specifically docks subvirus-like particles onto the nuclear pore opening and injects the viral chromosome into the nucleus. Human herpesvirus has a similar virus-specific process for nuclear entry. The existence of a nucleus thus imposes many molecular and evolutionary constraints on eukaryotic DNA viruses. One such constraint may require eukaryotic DNA viruses to coat their DNA in order to protect their genomes during passage through the cytoplasm. Yet as noted above, DNA integration is very common in both bacterial and archaeal DNA viruses but uncommon in eukaryotic DNA viruses, so clearly some molecular constraints must exist within the nucleus as well.

RNA transcription and splicing and the nucleus

The eukaryotic nucleus contains three classes of DNA-dependent RNA polymerases that lack compelling homology to the RNA polymerases used by any prokaryote. Although there exists some consensus sequence similarity within the catalytic core of the two largest subunits of all DNA-dependent RNA polymerases, this homology is mainly structural and cannot be seen at the amino acid sequence level, suggesting that it results from convergent evolution. Thus, these transcriptional enzymes are distinct for prokaryotes and eukaryotes. Furthermore, in eukaryotes the products of the RNA polymerases must frequently undergo posttranscriptional modifications (such as splicing) prior to functioning as mRNA, tRNA, or rRNA in the cytoplasm. This poses another dilemma for the origin of the nucleus. In order to prevent mistranslation of mRNA or prevent unspliced tRNA and rRNA from entering the cytoplasm, the nucleus must separate the transcription and processing of RNA from the cytoplasmic transport of processed RNAs. In fact, it seems that the nucleus would have needed to exist before the evolution of these posttranscriptional modification events, such as splicing of translated mRNA sequences. For example, eukaryotes splice the pre-mRNA of coding sequences via complex protein-based spliceosomes, whereas existing prokaryotes do not splice within coding regions or use spliceosomes. This suggests that RNA processing did not evolve first in the progenitor prokaryote; rather, it evolved later, after acquisition of the nucleus. Thus, it would seem logical that the progenitor eukaryotic cell first needed to invent the nuclear membrane in order to allow the evolution of introns, at least for introns within coding regions. Three types of splicing are known. Group I introns are self-splicing, mobile elements and often code for a DNA transposase protein. Group II introns code for a reverse transcriptase-like protein, and although they can be found in the phages and some tRNA genes of cyanobacteria, they too are absent from most prokaryotes. Group II introns are thought to have originated in the RNA world. Introns are also removed by another process involving a system (the spliceosome, an RNA-protein complex) that uses small RNAs to recognize the splice junctions and splice the RNA (after capping and polyadenylation). All three of these intron systems are mainly absent from prokaryotic cells (but several are present in prokaryotic viruses). Furthermore, genomic analysis now suggests that bacteria have never had introns in any of their coding genes. Curiously, in chloroplasts, which are considered to have originated from symbiotic prokaryotes, cytosolic glyceraldehyde-3-phosphate dehydrogenase (GAPDH) protein has an intron in a location similar to that of a nuclear gene, suggesting intron invasion from the nucleus to the plastid, after plastid colonization. Thus, no bacterium has the molecular characteristics for either RNA polymerization or splicing that would make it look like the possible progenitor to the eukaryotes. However, as discussed below, prokaryotic phages and DNA viruses of unicellular algae do have both of these characteristics and frequently code for and conserve spliced RNA of various types, including mRNAs with splicing in the coding sequence (e.g., thymidylate synthase in T4).

Modification of RNA at 5′ and 3′ ends and the nuclear membrane

Eukaryotes cap the 5′ ends of their mRNAs with 7-methylguanosine and add poly(A) sequences to the 3′ ends. Although some bacteria can also attach short 3′ poly(A) tails to their mRNAs, bacteria use a poly(A) polymerase that is distinct from the eukaryotic poly(A) polymerases, as only the eukaryotic ones are all members of the polymerase beta superfamily. In addition, bacterial polyadenylation of mRNA decreases its chemical stability and does not increase mRNA half-life, as it does in all eukaryotes. The resulting 5′- and 3′-modified eukaryotic mRNAs are then transported through nuclear pore structures, which reside in the nuclear membrane or cage. This membrane is itself distinct from the plasma membrane and is dissolved after S phase (DNA replication) and subsequently reformed at late anaphase/telophase of the cell cycle. No such division-associated membrane dissolution-reformation process is known for prokaryotes. In addition, all of the complicated molecular modifications of mRNA and nuclear RNA are highly conserved in eukaryotes but absent from all prokaryotes. Thus, these traits of mRNA modification appear to have been rapidly acquired de novo during the evolution of the nucleus and cannot now be identified in any existing prokaryotic cell.

In conclusion, we cannot identify the prokaryotic cell that might have symbiotically provided the eukaryotic nucleus. This leaves us with several unsatisfactory options. One option is to conclude that the progenitor single cell life form of the eukaryotic nucleus came from a domain of life distinct from *Bacteria* and *Archaea* and that all members of this domain have become extinct. However, this conclusion seems to go against phylogenetic evidence that suggests that the genes unique to this putative predecessor are as old as the *Bacteria* lineage itself. And it suggests that the only surviving cellular descendants of such distinct cells would be the current eukaryotes. Another, even less appealing, possibility is that the complex molecular distinctions of the eukaryotic nucleus all arose after the symbiotic fusion of the two progenitor prokaryotic cells and that evolution rates subsequently underwent a major acceleration, resulting in a huge increase in genetic novelty. For example, certain bacteria, such as *A. tumefaciens*, are now known to contain more than one chromosome and sometimes also host large linear plasmids. These plasmids appear to be prone to acquisition of new genetic information in the form of prophages and addiction modules. A multichromosome system with increased rates of evolution could have developed from a progenitor such as this. (Incidentally, some spirochetes have second linear chromosomes that show clear relationships to viral genomes.) If eukaryotes did arise out of symbiotic fusion of two progenitor prokaryotic cells, then it would be necessary not simply to increase evolution rates but to massively accelerate evolution in order to allow the development of all the other eukaryotic molecular traits characteristic of the nucleus. If such a massive increase in evolution could be attained, then it might also explain the great genetic change and massive genetic morphing that must have caused the predecessor bacterial mitochondrial genome to transfer many of its genes to the eukaryotic nucleus.

Even when allowing for a massive increase in evolution, one of the complex eukaryotic traits—the highly conserved mitotic spindles and their associated tubulin—still poses a major problem for the known mechanisms that could have created such large-scale genetic novelty. How did the tubulin-based system for chromosomal segregation originate so quickly? The possibility has been suggested that the bacterial FtsZ protein, involved in chromosome segregation, might have evolved to become the microtubular proteins involved in chromosome segregation. Since the bacterial FtsZ protein resembles tubulin structurally, has some low but discernible sequence similarity to tubulin, and can be assembled into tubular sheets, this idea seems viable on the surface. However, closer examination of this hypothesis shows major problems. For this to have happened—for a simple bacterial protein to have evolved into a complex and distinct molecular system for chromosome segregation—the scale of change needed is well beyond that which can be explained by any existing process. In general, the evolution of prokaryotic systems into eukaryotic systems would involve so much complex change that it would defy explanations based on accepted Darwinian processes, such as genetic mutation, duplication, and recombination. In the specific case of the bacterial FtsZ protein, its protein sequence is well conserved in all prokaryotes but is very different from the tubulin sequence in eukaryotes. Thus, it must have assumed a role in eukaryotes that is different from its conserved role in prokaryotes. None of the prokaryotic lineages can be identified as a predecessor of the eukaryotic tubulin.

One would need to propose that just after the prokaryotic nuclear symbiosis, the rate of adaptation and genetic change was transiently much greater than it currently has been measured to be. However, proposing such a transient but enormous increase in the rate of evolution following fusion of the progenitor cells after almost 2 billion years of stable prokaryotic life on earth is problematic. How and why could this happen? The problems posed by the origin of the tubulin system are actually less daunting than the problems posed by the origins of the other eukaryotic characteristics of the nucleus, such as pore complexes and replication, transcription, and splicing systems. Thus, we are left to choose from several very improbable scenarios. Finally, there are other distinctions that also require explanations: eukaryotes frequently have diploid (or sometimes polyploid) chromosomes, eukaryotes generally have sexual meiotic reproduction involving haploid gametes (the soma-germ line dichotomy), and eukaryotes also can have specialized cells. All of these features have no clear prokaryotic counterparts. And so with all of the above examples, one can understand the depth of the dilemma of explaining the nucleus and hence the conclusion of Smith and Szathmary that we lack a sensible scenario to explain the origin of the eukaryotic nucleus.

The Cytoplasm: Dilemmas aside from the Nucleus

In addition to the distinctions of the eukaryotic nucleus and its associated chromosome structure, other important differences outside of the nucleus also differentiate prokaryotes from eukaryotes. Prominent among these is the existence of the ER and the Golgi complex, a complex system of internal membranes involved in protein processing, modification, and transport. Also distinct and absent from prokaryotes are the tubulins, which in addition to chromosome segregation are also involved in spindle and microtubule formation as well as several basic cytoplasmic processes such as motility (cilium and flagellin function). Eukaryotes also have complex cytoskeletons and actinomyosin systems. Another issue is the distinct nature of the eukaryotic translational system. The various differences between the eukaryotic and prokaryotic translational systems indicate that the eukaryotic translational system could not have come from one prokaryotic source but instead appears to be a mosaic of prokaryotic ancestors. There does not exist an obvious precursor prokaryotic cell that could have provided all of the above systems or even one specific lineage that could have provided the origin of the translation system.

From the perspective of a virus, it seems clear there are major distinctions between prokaryotes and eukaryotes as molecular habitats, and hence the viruses that infect prokaryotes and eukaryotes are similarly distinct. Eukaryotic viruses must be adapted to the much more complex nuclear and cytoplasmic structures of the eukaryotic cell. Yet some of these eukaryotic

DNA viruses have clear phylogenetic relationships to the viruses of bacteria. Might these viruses have contributed to the evolution of the eukaryotic cell?

A Viral Origin of the Nucleus: Do Viruses Have Enough Genes?

I shall now discuss the possibility that a complex DNA virus was involved in the symbiotic origin of the eukaryotic nucleus. This possibility, although it has been proposed on several occasions, has been essentially ignored or dismissed in most earlier reviews of the topic of nuclear origin (see the work of Margulis and Sagan in Recommended Reading). Therefore, a more detailed evaluation is provided here. Simply stated, the hypothesis is that the predecessor of the nucleus was a large membrane-bound DNA virus that persistently colonized a prokaryotic host cell. This colonized host lost its cell wall (resembling a phage conversion event), and subsequently the virus acquired many of the prokaryotic genes (mainly metabolic and translational system genes) in the protonuclear chromosome (similar to the acquisition of transposons and accessory genes by parasitic bacterial plasmids). This view corresponds to the viral-origin hypothesis. As a corollary, this hypothesis also argues that there never existed a free-living progenitor of the eukaryotic nucleus; therefore, the reason that the eukaryotic lineage appears old is because the viral lineages that created it are themselves old (setting aside the problem that high virus evolution rates can confound evolution studies). The idea, however, that a large cytosolic, extrachromosomal DNA virus could have provided all the genes needed for eukaryotes is generally met with skepticism. How could the relatively small genome of a DNA virus have been able to provide all the genes needed to create the eukaryotic nucleus? One point to consider along these lines is that following a successful and permanent host colonization, virus transmission would occur through host cell reproduction. As a consequence, the packaging constraints that would have previously been necessary for assembly into an infectious virion would be lost, allowing an increase in gene content of the formerly viral genome. However, although this increase might allow a rapid acquisition of genes in the protonucleus, the colonizing virus would still have needed to provide a substantial number of complex and interacting genes at the start. The gene content of a large DNA virus ranges only from 150 to just over 900 genes. The largest are dsDNA viruses that infect bacteria (670 kbp, *Bacillus megaterium* phage), algae (560 kbp, *Pyramimonas* phycodnavirus) and, most recently, amoebae (670 kbp, mimivirus). It is interesting that these viruses of algae and amoebae show some clear relationships to each other. The gene content of these dsDNA viruses, although large by viral standards, might seem inadequate to have formed the genetic basis of all eukaryotes. It should be noted, however, that these viruses contain more genes than the genomes of the smallest cellular organisms (mycoplasmas). Nonetheless, this viral gene content is still far smaller than the genome size of most free-living bacteria (about 2,000 genes), let alone the more complex genomes of eukaryotes.

A viral LUCA

Recent sequence analysis of whole genomes of numerous prokaryotic and eukaryotic organisms indicates that the number of genes conserved among all life forms is surprisingly small. Those genes are thought to correspond to those genes carried by the last universal common ancestor (LUCA) that are still found in all cellular life (prokaryotes and eukaryotes). Current estimates are that the LUCA gene set consists of only about 324 genes. Ironically, this conserved LUCA gene set does not include the genes for proteins that replicate the DNA genome, which might be considered fundamental and common to all life. Given its small size, the gene content of the LUCA is well within the range of a large dsDNA virus. A viral genome as the nuclear progenitor would be expected to have its own distinct, noncellular, virus-based lineage of evolution. As a virus, it could also provide a much higher rate of evolution than was present in prokaryotic genomes. Thus, a viral nuclear progenitor could explain both the rapid rate of early eukaryotic evolution and the current absence of a progenitor prokaryotic cell. A competing and perhaps related idea was proposed in 1998 by C. Woese. According to this proposal, the LUCA is not a discrete organism but is instead a pool of exchanging genetic elements maintained by high rates of lateral DNA transfer. Although such a LUCA would not be a specific entity, it can be thought of in terms of "high genetic temperatures" early in the evolution of cells, which had not "cooled" or crystallized into organisms with specific genomes. Therefore, the LUCA would have been a rather diverse community of cells exchanging DNA. Although not addressed by Woese, it is obvious that such a LUCA can also be thought of as necessarily including viruses, since they would represent the originators of the transferred DNA and also the main driving force for "lateral" gene transfer or genome colonization, as was considered in chapter 3. However, the concept of lateral gene transfer between cellular organisms raises some problems concerning the ultimate origins of genes, as considered below.

Possible protoviruses prevailing at the origin of the nucleus

I will now address the issue of which viruses might have been the progenitor to the nucleus. If it is assumed that viral strategies are both old and stable during evolution (such as for the tailed phages, which appear to predate the *Archaea-Bacteria* divergence), it might be possible to identify candidate contemporary virus classes from existing prokaryotic or unicellular eukaryotic populations. Cyanobacteria appear to have evolved just prior to the evolution of the first eukaryote. I have presented the arguments (chapter 1) that persistent viruses, rather than acute viruses, are the most likely sources of the new genetic entities that can become stably associated with their hosts. By considering viruses of cyanobacteria and their closest eukaryotic relatives (e.g., unicellular algae), the possible protonuclear viral agents may be identified. The large DNA viruses that infect unicellular algae (phycodnaviruses and chlorella viruses) show clear relationships to both the bacteriophages and the large DNA viruses of mammals. Because this family of viruses has clear links to both prokaryotes and eukaryotes, it could be of central importance in the evolution of eukaryotes, although it is clear that extant viruses may also have developed and diverged after the nucleus was formed.

What characteristics might be needed for the protonuclear virus? To generate the existing nucleus, we might expect a large dsDNA virus with a linear chromosome, possibly having multiple DNA segments or having multiple origins of replication in a single DNA chromosome that also has eukaryote-like telomere ends. The virus would be nonlytic yet code for its own virus-specific DNA replication and transcription proteins, which would be related to those of eukaryotes, not prokaryotes. The virus should be membrane bound (preferably a double membrane), and its chromosomes should be stoichiometrically coated with small basic polymers, histones, or histone-like proteins. This virus should be able to process RNA [5′ capping, 3′ poly(A) addition, splicing, and transport of RNA through membrane-bound, pore-like structures]. The protovirus would probably be a nonintegrating virus, but with transposases or other DNA mobilization enzymes that would allow for acquisition of host genes. It would have mechanisms (preferably tubulin based) to segregate and package viral chromosomes that could evolve into the tubulin system. Finally, the mode of viral persistence and/or reactivation must be compatible with cellular differentiation, mitotic replication, gamete formation, and sex. On the surface, it might seem like we are asking for far too much genetic complexity in our protovirus. However, surprisingly, all of these characteristics can be found in viruses. A summary of prokaryotic viruses with some of the desired characteristics for the protonucleus is presented in Table 4.1. These viral candidates are evaluated below.

Table 4.1 Possible protonuclear prokaryotic viruses

Type of phage	Phage species	Characteristics
Cyanophage	CPS1, CPS2, S-PM2	dsDNA, DNA polymerase, RNA polymerase (for lytic phage)
Archaeal phage	AFV-1, SIRV-1, TTV1, -2, -3, and -4	Linear dsDNA, telomeres, chromatin, internal lipid Persistent and basal to poxviruses
Mycobacterial phage	L2 and persistent phage species	Extrachromosomal genome maintenance, bidirectional origin
Eubacterial phage	P1, N15	Linear dsDNA, telomeres, DNA polymerase, RNA polymerase, chromatin, internal membranes
B. subtilis-specific phage	P1-like (latent, large, spore associated)	Extrachromosomal genome maintenance, addiction modules

In terms of predecessor prokaryotic host cells and possible symbiosis, several possibilities are apparent. It was suggested by J. A. Lake and M. C. Rivera in 1994 that a gram-negative bacterium may have engulfed an archaebacterium and that this archaebacterium then evolved to provide the nuclear DNA replication system. That a bacterium that had lost its cell wall might be the progenitor was first proposed by Cavalier-Smith in 1987. However, in 1991 M. L. Sogin noted that a tree based on rRNA (not proteins) places eukaryotes at the root of *Bacteria*. It is very difficult to be certain of this placement due to the low confidence in the statistical analysis. One suggestion is that *Archaea* are secondary and specialized, not primary and ancient. Analysis of protein-based trees suggests that *Eucarya* and *Bacteria* are sister groups. This seems to indicate an old eukaryotic lineage but leaves unclear the issue of the likely progenitor prokaryotic host for possible viral protonuclear colonization.

Cyanobacteria would seem to be a likely symbiotic source of chloroplasts and mitochondria, which also implies that they may have colonized eukaryotic hosts after the generation of the nucleus. In fact, evolutionary evidence suggests that the colonization of eukaryotes by chloroplasts may have occurred many (over 30) times, explaining the diversity of the C3 and C4 photosynthesis systems found in higher plants. As cyanobacteria are now thought to have evolved relatively near but prior to the origination of eukaryotes and as they also seem to have contributed to the origination of the chloroplastids, it is worth considering what type of contemporary viruses known to infect cyanobacteria are possible candidates for the protonuclear virus.

Cyanophages are clearly related to bacterial phages, and both lytic and lysogenic versions of these viruses are abundant in the oceans. Cyanophages CPS1 and CPS2, as well as S-PM2, show close similarity to capsid assembly proteins of T4 phage. S-PM2 also encodes a T4-like gp49 recombination endonuclease protein. It is worth restating that the morphogenesis of T4 phage and cyanophages is exceedingly similar to the morphogenesis of human herpesvirus 1, which strongly supports a common lineage between these evolutionarily distant viruses. In addition, this group of cyanophages codes for virus-specific DNA and RNA polymerases. However, like the T-even phage (but unlike bacteria), these virally encoded DNA and RNA polymerases clearly resemble those of eukaryotes. More recently, and rather surprisingly, S-PM2 has also been shown to contain the two genes that are central for photosynthesis (D1 and D2). It is thought that these virus-specific genes may allow phage-infected cyanobacteria (*Synechococcus*) to overcome excess light-mediated damage (photoinhibition) to the photosynthetic complex in the host bacteria (discussed further below). Thus, this extant family of cyanophages could provide a good starting point for the origin of the eukaryotic nucleus. Nevertheless, these viruses are mainly lytic agents with no membrane, and thus they seem to lack some of the other essential nuclear components.

Archaea, which include the extreme thermophiles and halophiles such as *Sulfolobus* spp., are also important to consider as sources of possible protonuclear viruses. Based on phylogenetic analysis of 16S rRNA sequences, some biologists have proposed that the *Archaea* may be more related to the eukaryotes than are the *Bacteria*. Perhaps more compelling is the observation that the amino acid sequence of archaeal E1-1α contains an 11-amino-acid insert present in the sequence of the homologous eukaryotic protein. This sequence is absent from the sequence of the homologous bacterial protein. Thus, there is reason to think that *Archaea* might be the most likely source of potential protonuclear viruses. As mentioned in chapter 3, *Archaea* are known to host distinct viruses, relative to the viruses of all other prokaryotes, that have especially distinct morphologies, such as *Acidianus* filamentous virus 1 (AFV-1), which has a double tail, or the droplet-shaped SIRV-1. Overall, these viruses have many of the desired characteristics of ideal protonuclear virus(es). This includes a general and strong tendency to establish nonlytic, chronic, and persistent infections with dsDNA viruses. Furthermore, these infections are highly prevalent and often mixed, containing mainly viruses with linear dsDNA genomes. Thus, the maintenance and coordination of complex sets of persistent linear dsDNA genomes are very common in these hosts. In addition, many of these viruses are enveloped and are continuously shed from infected cells. Thus, not only do they have the desired membrane, but also they have a membrane-associated export system. Their genomes are exceedingly unique and can be highly diverse. Most of the viral proteins, including replication proteins, are unique to these viral lineages. Thus, these families of viruses represent a very large and dynamic source of genetic novelty needed to drive the eukaryotic radiation. In terms of specific viral characteristics, the recently sequenced AFV-1 is especially noteworthy. This linear dsDNA virus uses eukaryote-like TATA promoters to regulate transcription. Of even greater relevance, it has small, directly repeated sequences on the ends of its DNA that are very similar to the telomere sequences at the ends of eukaryotic chromosomes but unlike any telomere-like sequences found in prokaryote genomes. Furthermore, phylogenetic analysis indicates that the AFV-1 gene that codes for DNA polymerase is basal to and likely ancestral to the major groups of eukaryotic large DNA viruses, including the phycodnaviruses (chlorella virus), the poxviruses, and *African swine fever virus* (ASFV) (an insect-transmitted DNA virus). This basal relationship is of special interest (as discussed below) because these viruses appear to have essentially all the desired characteristics for a protonuclear virus. Table 4.2 presents a summary of viruses that might have provided these needed characteristics.

In addition to AFV-1, other archaeal viruses and cells have interesting characteristics worth considering. For example, SIRV-1 also has a linear dsDNA genome but with covalently closed ends, a feature of the large eukaryotic DNA viruses that is not typical of phages. Like other viruses of *Archaea,* this virus is not lytic. Therefore, its genetic capacity to persist pro-

Table 4.2 Distinctions of the eukaryotic nucleus and putative viral origins[a]

Characteristic	Virus(es)
Membrane-bound separation of DNA and transcription from translation	Vaccinia virus cores, TTV1 (thermophilic archaeal host), ASV[b] (eukaryotic host)
DNA stoichiometrically bound as chromatin with small basic proteins	Cytoplasmic DNA viruses, TTV1, phycodnaviruses, ASFV
Linear chromosomes with repeats on ends and multiple origins	Cytoplasmic DNA viruses, AFV-1, N15 phage concatenated (not prokaryotic)
Distinct DNA replication proteins and initiation control	Phycodnavirus progenitor, phage T4, phage lambda
Distinct RNA polymerases I, II, and III	ASFV progenitor to all three forms
mRNA capping, polyadenylated RNA, introns	Phycodnaviruses, ASFV (viral modification of nucleic acids)
Nuclear pores	Vaccinia virus (ATP-dependent extrusion of processed mRNAs from core)
DNA synthesis associated with membrane acquisition, tubulin transport	Vaccinia virus (de novo core assembly, cytoplasmic transport, extrusion)

[a]The nucleus origin is not virus colonization of the host genome but full displacement; there is not bacterial-viral horizontal transfer. Also, there is no direct progenitor (symbiont) to the eukaryotic cell-virus progenitor (LUCA).
[b]ASV, avian sarcoma virus.

vides a good starting point for the possible evolution of the nucleus. Perhaps more interesting are *Thermoproteus tenax* virus 1 (TTV-1), TTV-2, TTV-3, and TTV-4 of *Crenarchaeota*. These viruses have linear dsDNA genomes with stoichiometrically bound and highly basic DNA binding proteins. Thus, they might provide a molecular basis for the evolution of eukaryotic chromatin. In addition, the capsids contain both internal and external lipid envelopes, and both temperate and lytic versions are known. Therefore, the viruses infecting *Archaea* have many, if not most, of the features that would make these agents attractive candidates for the origins of the protonucleus. Recently, a new phylum of *Archaea*, *Nanoarchaeota*, has been proposed. These species have very small cells (e.g., *Nanoarchaeum equitans*, which has a 490-kbp genome) that appear to live as symbionts on the surfaces of larger "mother" archaeal cells (e.g., *Ignicoccus*). This observation suggests the existence of *Archaea* that have genomes as small as those of some of the large DNA viruses but that persist on the surfaces of other cells, rather than infecting them as do viruses. The relationship of these cellular genomes to each other and the potential participation of persistent genetic parasites in this relationship have not yet been evaluated.

Since they lack cell walls, mycoplasmas are often thought of as the most likely source for the host cell that was colonized by the protonucleus, leading to the development of the eukaryotes. Mycoplasma virus L2 is a quasi-spherical, enveloped virion containing a circular dsDNA genome. Members of this virus family can integrate their viral DNA into host cell genomes. However, L2 infection of *Acholeplasma laidlawii* host cells leads to an episomal, noncytocidal productive infection in most cells, with the possible involvement of two origins of DNA replication. Virus early expression is followed by establishment of lysogeny in all (or most) infected cells. This cytosolic characteristic would be good for the putative protonuclear virus. However, the small size of the L2 DNA genome (11,965 bp) and the absence of extensive virally encoded DNA replication and transcription systems seem problematic for suggestions that this family of virus alone gave rise to the nucleus.

Bacterial viruses with broad host range

PDR1 virus (related to φ29; see below) is an intriguing candidate for the putative protonuclear virus. It has a broad host range and is a dsDNA tectivirus with an internal membrane. P1 is a PDR1-related, tailed polyhedral virus that also has a broad host range, although it specifically infects *Mycoplasma pulmonis*. P1 has a linear dsDNA genome with inverted terminal telomere repeats. This virus has many of the needed characteristics for a protovirus, including the virally encoded DNA-dependent DNA polymerase, DNA-dependent RNA polymerase, single-stranded DNA (ssDNA) and dsDNA binding proteins, and an internal membrane. It is especially interesting that the viral DNA and RNA polymerases are more similar to the eukaryotic counterparts than are the related prokaryotic genes. However, this family of virus follows a mostly lytic or acute life strategy, and they seem to lack the constellation of genes needed for stable host colonization. As indicated below, there are other related phages that latently infect spores of *Bacillus subtilis*. Thus, this family of P1-related phages remains a strong candidate for the protonuclear virus.

With respect to extant viruses of prokaryotes, we cannot now be certain as to which is most related to the putative protonuclear virus. Several strong candidates have been identified. Thus, it seems clear that viruses infecting prokaryotes still retain most of the features that would be required for this germinal role. However, another way to evaluate the viral-origin hypothesis is to examine existing cytoplasmic eukaryotic DNA viruses to see if they retain basal characteristics expected of a protonuclear virus.

Vaccinia virus: the best-studied eukaryotic DNA virus

The best-characterized cytoplasmic DNA viruses are *Vaccinia virus* and the other related members of the poxvirus family, as well as some members of the insect iridoviruses. These viruses have a multiple membrane with an internal core structure containing the viral chromatin. This multiple-membrane arrangement is inherent to and conserved among these viruses. The virus loses the outer membranes after entry but retains the viral core structure. These membraneless core structures, which reacquire membranes later in the virus life cycle, clearly resemble mininuclei. Within them is contained the viral DNA-dependent RNA polymerase, which transcribes the viral dsDNA genome, modifies the resulting mRNAs with 5′ caps and 3′ poly(A) tails, and extrudes these mature transcripts into the cytoplasm through as-yet-uncharacterized structures. Interestingly, another primitive member of the large DNA virus family (which includes only those DNA viruses that can infect both insects and mammals) is ASFV. This virus codes for a DNA-dependent RNA polymerase that, according to phylogenetic analysis, is basal to all three classes of eukaryotic DNA-dependent RNA polymerases. It is worth emphasizing that no prokaryotic DNA-dependent RNA polymerase is a member of the clade of eukaryotic RNA polymerases (I, II and III), let alone basal to all three eukaryotic clades. Only that of ASFV seems to hold that basal position. This cytoplasmic DNA virus also codes for an enzyme that caps the 5′ ends of mRNAs. Similar to the case with ASFV RNA polymerase, both the ASFV capping enzyme and the capping enzyme from *Paramecium bursaria Chlorella* virus 1 (PBCV-1) (the DNA virus of *Chlorella*-like unicellular algae, discussed below) have been shown by phylogenetic analysis to be basal to the mRNA capping enzymes of all eukaryotic cells. These two viruses also code for poly(A) polymerases, which attach adenylate residues to the 3′ ends of mRNAs. The extrusion of mature vaccinia virus mRNAs into the cytoplasm occurs via an ATP-dependent process through as-yet-undefined exit structures on the viral core membrane. A schematic outline of the replication cycle of vaccinia virus is shown in Fig. 4.1.

Vaccinia virus mRNAs become associated with tubulin and ER membranes soon after synthesis and export. These associations are involved in the cytoplasmic translation of viral proteins. It is especially intriguing that the vaccinia viral core structures become wrapped in ER-derived membranes and that the synthesis of viral DNA occurs within these mininuclear, membrane-bound structures. DNA synthesis initiates from viral telomere end repeat sequences and can result in concatenates of long, multi-ori DNA structures. Resolution of these concatenated structures involves the telomere sequences and must occur prior to the packaging of single chromosomes into new virions and the subsequent tubulin-mediated transport of membrane-wrapped viral structures to the plasma membrane. At this point, DNA synthesis stops and the virion-surrounding membranes unwrap, probably via the action of viral kinases. A maturation of the virion structure then occurs in which it acquires a second plasma membrane. Finally, an association with actin occurs, and an actin polymerization-dependent motility system facilitates the exit of virions at the plasma membrane. These extracellular virions can then infect nearby cells. In conclusion, it can be seen that for vaccinia virus, DNA synthesis is directly linked to membrane acquisition and subsequent membrane loss (resem-

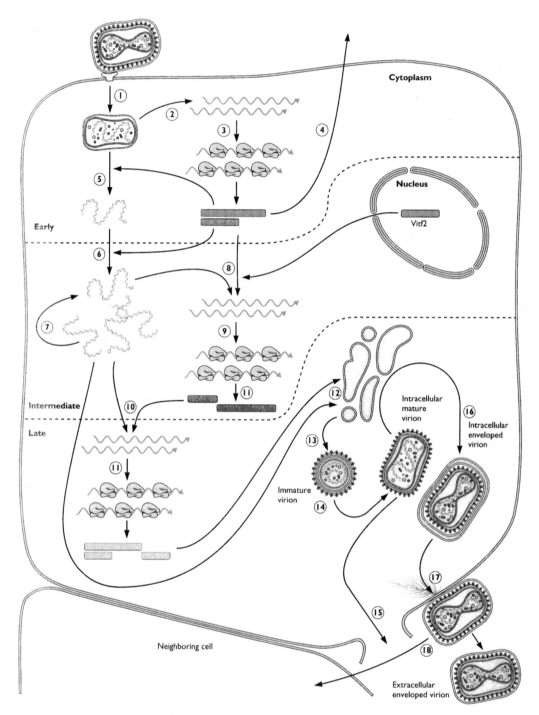

Figure 4.1 Schematic diagram of the life cycle of vaccinia virus replication. Reprinted from S. J. Flint, L. W. Enquist, V. R. Racaniello, and A. M. Skalka, *Principles of Virology: Molecular Biology, Pathogenesis, and Control of Animal Viruses,* 2nd ed., p. 831 (ASM Press, Washington, D.C., 2004), with permission.

bling S phase). Furthermore, the daughter chromosomes (viral cores) are transported via tubulin action.

The resemblance between these poxvirus processes and the activities of a cycling eukaryotic nucleus is striking and clear. This similarity encompasses most of the events and mechanisms that are characteristic of a mitotic nucleus. Viral transcription is fundamentally segregated from translation. The viral DNA is linear and chromatin-like and has telomeres. Viral mRNAs undergo host-like 5′ and 3′ processing and are exported. A dissolvable, multimembraned coat is associated with the synthesis of viral DNA. Viral proteins bind both tubulin and actin, affecting their polymerization and mobilization functions. Thus, it is clear that viral genes are directly involved in and dependent on motility. The tubulin-associated transport of immature viral cores is also associated with the resolution of multiple viral genomes and the acquisition of a second membrane. These events and processes encompass most of the features that distinguish eukaryotes from prokaryotes and add powerful support to the hypothesis that viruses could have provided the origin of the eukaryotic nucleus. However, it might be argued that the selective pressures on eukaryotic DNA viruses would lead to adaptations in which viral molecular strategies resembled those of the host, and hence this similarity between virus and host could be evidence for virus-host coadaptation or convergent evolution. Yet we know of clear examples in which these viral processes differ fundamentally from those of the host, such as the protein capping of 5′ mRNA ends by picornaviruses, the completely distinct DNA polymerase and DNA synthesis processes of adenoviruses relative to that of the host, and the existence of single-stranded viral nucleoproteins (polynucleic acids with covalently attached polypeptides) that are used as templates for transcription and replication. All of these other (non-nucleus-like) distinct virus strategies are equally old according to phylogenetic analyses. Clearly viruses do not need to be host-like to function properly, even to perform host-like functions. Other eukaryotic DNA viruses that use host-like DNA replication processes differ from their hosts, having, for example, highly conserved core viral replication proteins such as the T-antigen of polyomaviruses (or early genes of all papovaviruses) that lack cellular analogues. This leaves us with the question of why the cytoplasmic DNA viruses are so similar to their hosts in all these mechanisms. Two other points should now be made. First, a protonuclear virus offers a solution to the dilemma of the origin of the nucleus. If we eliminate the possibility that viruses were able to provide the origin of the nucleus, we also eliminate the solution to the dilemma of the missing symbiotic cellular progenitor of the nucleus. Second, the viral proteins involved in the host-like processes generally appear to be basal to those of the host, as described below.

That various viral genes (those for DNA polymerase, RNA polymerase, capping enzyme, etc.) are phylogenetically basal to those of all eukaryotes does not convince everyone that viruses were the progenitors of these genes. After all, the ability of a virus to acquire host genes is well known: the old and popular argument is that viruses "steal" host genes (especially accessory genes), and this accounts for viruses with host-like genes. I mentioned this issue in chapter 1, so it suffices to say here that viral genetic creativity is vast and unsurpassed by that of any other life form. And as noted earlier for phage evolution, new viral genes tend to originate from other viral elements, not host genes. There is the technical problem that due to the much higher rate of virus evolution relative to that of host, it can be difficult to be certain of the relationship between virus gene evolution and host gene evolution. Yet there are numerous examples, especially for persistent viruses, in which virus and host gene trees are highly congruent, indicating similar patterns of coevolution. And in spite of prevailing views to the contrary, phylogenetic analysis does indicate that there are few (or no) clear examples of viruses acquiring core genes from host sources (e.g., the T-antigen example mentioned above). Most of the viral core genes are of an ancient origin, and their lineages are generally monophyletic. Phylogenetically, the core viral replication and transcription proteins (when present) are as well conserved among different viral lineages as any viral gene. This conservation is especially true for the virally encoded DNA polymerase, PCNA, RNA polymerase, and mRNA capping enzymes of DNA viruses. It is in fact the conservation of these core genes that is used to construct the phylogenetic relationships of DNA viruses and, in the case of the DNA polymerase gene, links eukaryotic DNA viruses to prokaryotic DNA viruses. Vaccinia virus DNA polymerase, for example, most closely resembles the DNA polymerase of phage T4. Furthermore, some of these viral lineages are clearly very old, for example, herpesviruses, which still show clear relationships to the T-even phage. Yet the *Herpesviridae* and *Poxviridae* lineages are paraphyletic to the host lineages and do not branch from the host lineage. Taken together, these observations strongly argue that these large DNA viruses are not derived from rogue host replication systems and establish that they have the evolutionary and genetic capacity to have been the origin of the eukaryotic nucleus.

Nuclear pores

The pore structures of the eukaryotic nucleus pose another significant dilemma for the possible prokaryotic origin of the nucleus. These large complex structures have no counterparts in the prokaryotic world. However, pore structures would not seem to be nearly as problematic for the viral-origin hypothesis. We know that among extant eukaryotic DNA viruses, some mammalian DNA viruses, such as adenoviruses and herpesviruses, specifically dock onto the nuclear pore structure in order to facilitate nuclear entry of viral chromatin and thereby initiate infection. Clearly these viruses are highly adapted to nuclear pore function and seem to use the pores as internal receptors. Functionally homologous "pore function" has also been noted, in that vaccinia virus cores extrude mRNA from viral core or chromatin structures via an ATP-dependent process. This indicates the existence of some virally based process that transports mRNA from the point of synthesis into the cytoplasm. In terms of prokaryotic viruses, the idea that viruses could have led to the creation of novel pore structures is not without evolutionary precedent. As indicated in chapter 3, bacterial viruses frequently use various types of pores as toxins, which are also components of immunity modules and which compel the host to maintain the persistent virus. In addition, bacteriophages (like lambda) use holins as small membrane proteins that accumulate on the membrane and then, at a specific time, program membrane permeability for the release of lytic virus. More than 100 distinct viral holin genes are known and can be organized into over 30 orthologous groups. Clearly, this system is highly diverse. Another related point is that the baseplate at the tailed end of bacteriophages clearly resembles a pore in function. It is a highly complex multiprotein structure that attaches to the host cell receptor, generates a hole in the host cell membrane, and injects the viral nucleic acid. Furthermore, the proteins making up bacteriophage baseplates and receptors are probably the most diverse of all phage and bacterial proteins. Baseplate and receptor evolution in phages and prokaryotes is a highly dynamic process. It may even be more dynamic than previously thought. For example, recent reports of phages from *Borrelia* have identified a family of phages that has evolved a reverse transcriptase-mediated system that can mutagenize the mRNA of the baseplate receptor gene, followed by integration of the variant gene into the host genome. This process generates sufficient diversity in terms of distinct receptor proteins to allow the rapid adaptation of phages to a new or altered host. This is a most remarkable phage reverse transcriptase-based system for the generation of protein diversity, and it clearly resembles the adaptive immune system of vertebrates (though it must be much older phyogenetically). All this enormous virally based gene diversity is used simply to make pores in bacterial cells in order to allow entry of viral DNA, suggesting that viruses make good candidates for the progenitor of nuclear pores as well.

Virus-mediated covalent marking of nucleic acids

One might pose the question about the origin of the viral enzymes that add modifications to viral RNA [5′ cap, 3′ poly(A)]. How might we justify the view that these processes first occurred as a viral, and not cellular, process? In many cases [such as for the poly(A) enzyme of vaccinia virus], these viral enzymes show no similarity to host enzymes. In those cases, it cannot be argued that they could be of host origin. However, in other cases, such as chlorella virus 1, the viral poly(A) polymerase is similar but basal to that found in eukaryotes. Why would viruses modify their nucleic acids in such ways? As was discussed in chapter 3, almost all viruses are known to covalently mark their genomes, RNA, and proteins with various types of chemical modifications. The most common among these is the methylation of various bases in the viral DNA by virus-specific methylases. However, it is also clear that viruses, even prokaryotic viruses, have employed noncovalent DNA binding or chromatin-like proteins (e.g., TTV of thermophiles) to differentially condense or mark their genomes. This marking allows other viral enzymes, often hydrolytic enzymes, to distinguish viral genomes and transcripts from those of the host, as well as to distinguish one virus from another, thereby facilitating the degradation and recycling of these non-self-molecules. Thus, the idea that a virus might have marked its mRNA via 5′ capping and/or 3′ poly(A) addition fits well with known viral molecular strategies of genetic identification.

A Scenario for the Viral Origin of Spindles and Tubulin

Let us consider a specific and significant molecular dilemma in understanding the origin of eukaryotes: that is, the origin of eukaryotic spindles and tubulin. Here I discuss the dilemma from the perspective of a putative viral origin. Prokaryotes do not have a tubulin system, so it has been hard to see how this complex process evolved from prokaryotes. In fact, the tubulin problem is considered one of the major dilemmas for a prokaryotic origin of eukaryotes. Therefore, I will now describe this situation in considerable detail. However, the details presented below are largely circumstantial and may present a burden for some nonexpert readers. Thus, those readers might choose to skip this section.

Cellular FtsZ

As mentioned previously, prokaryotes do have the FtsZ gene, which codes for a protein involved in prokaryotic chromosome replication and segregation. These processes occur via what is believed to be ring-shaped septum and membrane attachment. Structural analysis of FtsZ indicates that it has a shape very similar to that of tubulin and that it also has some regions of discernible but low sequence homology. So, it seems plausible that prokaryotic FtsZ and tubulin are related. Yet while FtsZ is highly conserved in all prokaryotes, there is less than 20% sequence homology to tubulin, and tubulins are highly conserved in all eukaryotes. As a consequence, we cannot identify a prokaryotic cellular progenitor (either functional or by sequence similarity) for tubulin.

Acute phage, FtsZ, and chromosome segregation

Bacterial viruses, such as φ29 (a T7-like virus of *B. subtilis*), code for FtsZ-like proteins as well. These phage FtsZ-like proteins can have functions distinct from those seen for bacterial FtsZ in bacterial host cells. Furthermore, these phage proteins also display some biochemical and structural characteristics that make the phage proteins appear more similar to tubulin than to bacterial FtsZ. φ29 is a linear dsDNA phage with short terminal repeats and covalently attached terminal proteins. This phage also codes for very abundant small ssDNA binding and dsDNA binding proteins, which are essential for DNA replication, as well as coding for a DNA polymerase (a type B polymerase) and a late-expression DNA-dependent RNA polymerase. As with φ29, the P1 gene encodes an early protein that is similar to bacterial FtsZ. It is thought that this P1 protein may bring the viral DNA polymerase to membranes via the telomeric ends of phage DNA. This core viral protein is needed for the initiation of DNA replication and for the attachment of the replication complex to the membrane, which is necessary for segregation. Furthermore, this P1 protein is able to form much more "tubulin-like" polymerized tubular and tertiary sheet structures than can the host bacterial FtsZ protein. However, the φ29 and P1 proteins are still rather different from tubulin and thus may not be the direct progenitor of tubulin. Nonetheless, that prokaryotic viruses can code for such proteins clearly raises the possibility that tubulins may also have a viral (phage) origin. φ29 is a relatively small DNA phage (15 kbp) and is mainly lytic, so it does not appear to be a good candidate by itself to have evolved the eukaryotic tubulin structures. In addition, this phage clearly has a non-host-like DNA replication system, which resembles that of adenovirus in replicating DNA by a 5′ protein-primed mechanism using a DNA polymerase related to that of adenovirus. Thus, φ29 would not seem to be the likely direct progenitor of the tubulin system of the nucleus.

FtsZ and phage immunity

Although φ29 might not be the direct progenitor of the tubulin system of the nucleus, φ29 and P1 might well represent a remnant of how phages can link DNA replication to tubulin. There is compelling evidence for the existence of a large number of unassigned members of φ29-related *Podoviridae* infecting a wide range of bacteria. In addition, there is strong evidence that FtsZ-related proteins in such phages are directly important for virus persistence. For example, the host bacterial FtsZ protein is very frequently a target of various prophage immunity genes. Numerous *Bacteria* have sequences related to the Kim region of lambdoid prophages, which corresponds to DicF RNA. This RNA is antisense to the FtsZ mRNA; thus, it inhibits cell division and appears to be part of an addiction module (see chapter 3 for the role of phage addiction modules in host evolution). Most often, temperate phages use antisense RNAs as an antitoxin to a second stable viral death gene. However, some of these DicF-like RNAs do not affect bacterial host cell division. Thus, the intended target for these RNAs is not the replication of the bacterial host itself. It seems more likely that these RNAs instead target the replication of other persistent or acute viruses. Thus, it seems more likely that they are involved in immunity and under the selective pressure associated with immunity. The similarity of flanking regions of DicF RNA to the immunity region of P4 further suggests a role in immunity.

φ29, FtsZ, latency, and sporulation

B. subtilis can produce sporulating bacterial cells, which are of special interest in evolutionary biology. Sporulation resembles both the sexual gametogenesis of eukaryotes (and hence relates to the soma-germ line dichotomy) and the differentiation of committed cells, which is otherwise absent from most prokaryotes yet common in eukaryotes. In the case of φ29 infecting *B. subtilis*, vegetative φ29 replication is inhibited during cellular sporulation, and the virus is incorporated into sporulating cells in a latent state. φ29 is reexpressed and replicates following germination of the latently infected spore. This is an example of latent infection by a DNA virus with many of the characteristics needed for it to have been the protonucleus, including a virally encoded FtsZ protein. Furthermore, this latent infection is associated with characteristics that resemble both sexual reproduction and differentiation as seen in eukaryotes. Although φ29 itself may lack the genetic carrying capacity to have been the sole or direct progenitor of the tubulin system of the nucleus, various other temperate and pseudotemperate (nonintegrated) phages of *Bacillus* (SP-beta and SP15, which establish extended latent infections) are known to have much larger genomes—up to 385 kbp—and are frequently superimmune to other phages. As discussed in chapter 3, interactions between acute φ29 and these latent phages are very likely to provide mechanisms able to suppress φ29 replication, thereby providing the missing mechanism of persistence as well as other features, including membranes and additional genes. Although these nonlytic *B. subtilis* phages are not well studied, some pseudotemperate phages of *B. subtilis* (which continue to make virus without pathology) are known to express viral genes that increase bacterial sporulation frequency and, intriguingly, also express insecticidal crystal proteins. Clearly these viruses manipulate and compel the basic host cell differentiation programming, as well as provide a survival advantage to latently infected bacterium-plant symbionts. Because these phages need to maintain a latent infection for survival, their fitness is linked temporally to spore survival and germination.

Defectives, viral defense, and tubulin origins

Prophages, such as those that correspond to DicF-like RNAs, are frequently defective. Hence, they often are considered to no longer function as viruses, and viral issues are no longer considered to be involved. However, in chapter 3 I noted important examples in which seemingly inactive or defective prophages can strongly enhance virus persistence and also affect the outcome of host evolution and survival of other acute or persistent virus infections. The example of the P4 defective virus and the nondefective virus P2 is a case in point. Given the high prevalence of prophage-like genetic elements that correspond to DicF-like RNAs, it seems likely that the phage FtsZ-like protein (and the defective prophage that encodes it) may be under selective pressure with competing or latent phages. The phage FtsZ-like gene (and its antisense) may compose elements of an addiction module whose purpose would be to ensure the continued prophage colonization of the host. Thus, it can be proposed that some distant φ29-like virus was able to create a novel version of its FtsZ-like gene. This tubulin-like gene could have resulted in a more efficient system for the extrachromosomal persistence and segregation of linear viral chromosomes. Eventually, this host colonization became permanent, resulting in a superimposed virally mediated system for chromosome replication and segregation. Eventually, eukaryotic tubulin evolved from this viral system.

Viruses of Microalgae

As noted previously, the viruses that are now commonly found in unicellular eukaryotes such as microalgae are of special interest in the evolution of eukaryotes because they may shed light on the relationships of large DNA viruses to the eukaryotic host and its evolution. Microalgae constitute the earliest representative of *Eucarya* for which there is clear sedimentary fossil evidence. Microalgae exist as either free-living forms or endosymbionts (zoospores) of other species, such as *Paramecium* spp. Microalgal species are abundant: it is estimated that as many as 100,000 species of marine algae exist, which would account for up to 40% of photosynthesis on our planet, contributing about 10^{12} tons of cell wall per year or 10^{11} tons of cellulose per year to the biosphere. Microalgae therefore represent a substantial part of the earth's biomass. Bacterium-like DNA viruses are known to exist for many species of unicellular algae. *Micromonas pusilla* is the best-studied free-living microalga and has a simple sexual cycle. Algae of the genus *Chlorella* are the most widely distributed and frequently encountered throughout the water habitat of Earth. *Chlorella* species undergo mitotic division, and most are free living. Ensymbiotic versions called zoochlorellae are also well known. *Chlorella* species have cell walls made of lipopolysaccharide, which chemically resemble the cell walls of gram-negative bacteria. *Chlorella* species have mitochondria, Golgi complexes, and ERs and are photosynthetic, containing chloroplasts. Viruses are known for more than 44 taxa of eukaryotic algae (sometimes referred to generically as chlorella viruses). Similar to phages but unlike DNA viruses of animals, the chlorella virus virion remains external after injecting viral nucleic acid into the host cell. The virion is not taken into the cytoplasm, in contrast to the case for essentially every other eukaryotic virus. The *Chlorella* 16S RNA of both plastids, chloroplasts and mitochondria, is more similar to that of the plastids of cyanobacteria and purple bacteria than to that of the plastids of other organisms, strongly suggesting that these plastids were both derived from symbiotic free-living photosynthetic bacteria. However, plastid RNA genes appear to be composed of mosaics of particular bacterial lineages.

Phycodnaviruses are phage-like

PBCV-1 is the prototype for the phycodnavirus family, which includes chlorella viruses. The *Chlorella*-like algae that host PBCV-1 include both free-living unicellular algae and zoospores. As mentioned above, zoospores are algae that live symbiotically within *Paramecium* and other eukaryotic hosts, providing them with photosynthesis. Viruses of *M. pusilla*, which is a free-living microalga, are also known and well studied. The viral genomes are linear dsDNA (330,742 bp) with closed hairpin ends (similar to vaccinia virus). PBCV-1 contains 376 predicted coding regions, 40% of which clearly resemble those corresponding to other known prokaryotic and eukaryotic proteins. Structurally, phycodnavirus virions resemble the virions of animal iridoviruses and show some sequence homology to iridovirus capsids. The great majority of viruses of microalgae are related large dsDNA viruses. However, a few RNA viruses have also been observed, such as a rod-shaped RNA virus (tobacco mosaic virus-like) reported for *Chara corallina* microalgae. Algal DNA viruses are similar to bacterial phages in many ways, although they tend to be generally larger and more complex than most bacteriophages. Unlike many eukaryotic viruses, PBCV-1 has a high particle-to-PFU ratio (25 to 50% of particles are infectious), indicating that these viruses undergo efficient virion assembly. In this characteristic, they are more reminiscent of bacteriophages, which also show high particle-to-PFU ratios. Also like phages, phycodnaviruses generally have high levels of methylated DNA bases. An additional phage-like characteristic is the ability of phycodnaviruses to digest an opening in the host cell wall and inject the viral genomes. No other eukaryotic virus appears to operate in this phage-like way. Also phage-like is that phycodnaviruses code for numerous restriction/modification enzymes. In fact, these restriction enzymes are the only example to date of eukaryotic restriction/modification systems. Other phage-like features of phycodnaviruses include the presence of transposons, mobile introns, and phage-like DNA repair systems. Yet in spite of all these similarities to phages, in many other respects phycodnaviruses are much more like eukaryotic viruses and eukaryotic hosts than they are like prokaryotes (discussed below).

Natural history of microalgae, their phycodnaviruses, and symbiosis

In spite of much virological study, the natural history of phycodnaviruses and their microalgal hosts is not well understood. Particles that closely resemble phycodnaviruses are exceedingly abundant and can be found in surface waters of the oceans and freshwater at levels that range from 1×10^{11} to 5×10^{11}/liter. The studies of phycodnavirus populations have frequently been done with the aim of biological control of oceanic algal populations and their blooms. Such blooms can devastate the oceanic ecology by killing other species due to oxygen depletion. Clearly these viruses represent a major and natural constituent of the aqueous habitat. With respect to the symbiotic algae, it is interesting that the *Paramecium* hosts of algal zoospores may prevent phycodnavirus access to and infection of the symbiotic algae: these symbiotic algae are not susceptible to PBCV-1 infection when the algae are within their protozoan host. However, zoospore algae can frequently grow as free-living cells, which PBCV-1 does infect and kill. The natural biology of this relationship is not well understood in that it is not clear how algae colonize their *Paramecium* hosts in nature. However, the evolutionary implications of a host species escaping acute virus parasitization by becoming engulfed by another and very different cell are quite intriguing. If the engulfing cell is sufficiently different from the symbiont (such as lacking the same viral receptors), then the engulfed cell would be surrounded by an alien cell type and be shielded from any acute viruses. This relationship may well define a virus-based natural selection pressure that can drive a virus-susceptible host into an initially parasitic relationship within another cellular species simply to escape from prevalent acute viruses. This engulfment by another cell, if stable, could provide a selective pressure to initiate the evolution into a symbiotic relationship between the two cell types, without the need for one cell to provide a clear advantage to the other cell. Such a "virus escape" idea might also apply to the origination of symbiotic eukaryotic plastids (such as chloroplasts and mitochondria), which appear to have originated from free-living bacterial organisms. These free-living plastid ancestors might also have been driven into the early "aplastid" eukaryotic cell to escape lytic cyanophages, which are prevalent in the oceans. In support of this idea, plastid sequence data clearly show that plastids have nucleotide word frequencies that do not avoid restriction/modification or palindromic sequences. Yet the genomes of all known free-living prokaryotes avoid such palindromic nucleotides. As restriction/modification is a major bacterial system for immunity to or addiction by phages, all prokaryotes are under pressure by lytic viruses to maintain restriction/modification systems. Thus, the lack of restriction word avoidance in plastids suggests that the selective pressure to avoid cyanophage viruses was absent after cell engulfment. However, as discussed in chapter 5 in the section on Fungi, DNA viruses that colonize mitochondria are known and are prevalent in some situations, but such viruses are not controlled by restriction/modification systems.

Life in the sun and life after death

Another very intriguing biological feature of phycodnavirus and marine phages (cyanophages) that has broad implications for evolutionary biology is the ability of these viruses to respond to UV light damage. At the ocean's surface, where the majority of the microbiological flora resides, UV inactivation from sunlight is the most significant cause of bacterial, algal, and phage death and turnover. In addition, photosynthetic organisms, such as cyanobacteria and *Chlorella* green algae, undergo photoinactivation of photosynthesis. This occurs when excess light damages the D1 and D2 proteins of the photosynthetic reaction centers, resulting in decreased photosynthesis. The intense light levels that can exist at the ocean's surface have strongly affected the genetic makeup of oceanic viruses. Accordingly, most marine viruses and phages appear to have a half-life of less than 1 day, mainly due to intense light levels. To counteract this, cyanobacterial phages (such as S-PM2) have very efficient light-dependent repair machineries for UV damage. As mentioned above, they also encode phage versions of the D1 and D2 photosynthetic proteins that would presumably restore the photosynthetic capacity of phage-infected cyanobacteria in excess light. Because these viral genes have several mobile introns within them, they can be clearly distinguished from the corresponding host genes. Within the phycodnaviruses, there also exist various genes that aid infected algae in dealing with excess light energy. Several specific adaptations for repairing the damage to proteins and DNA caused by UV light are known. In addition, phycodnavirus replication itself can be affected by light, such as with *M. pusilla* viruses, which have light-dependent reproduction and fail to induce severe disease in the dark.

Surprisingly, phycodnaviruses have been shown to be capable of replication even in UV-killed host cells. For example, PBCV-1 can infect and replicate at reduced but significant levels in lethally damaged cells. This restored replication is due to the expression of various virus-specific repair enzymes that can resurrect the capacity of the cell to synthesize macromolecules. A similar capacity has long been known for bacterial phages. Besides restoring the damaged cell, these viruses are inherently much more resistant to UV killing than the host cells due to the much smaller genome target size. However, even UV-mediated inactivation of virus does not necessarily prevent subsequent virus replication. This is because a UV-killed virus can still replicate through a process known as multiplicity reactivation. Multiplicity reactivation occurs if there is a sufficiently high ratio (or multiplicity) of virus to the host cell, such that one host cell is infected with numerous virions. In this way, even if each of these virions has sustained a lethal UV hit in part of its genome, it may still be capable of expressing some subset of genes. If these expressed genes either result in genome repair or allow functional complementation to the expressed genes of other damaged viruses, then virus replication will be restored by the combined action of the damaged viruses. The selective advantage of such a "resurrection" capacity seems clear and large. It provides a selective circumstance in which the coordination of otherwise defective genetic elements is strongly favored by being inherently dependent on the complementation of these otherwise UV-killed genomes. Therefore, the defective mixture must be able to cooperate as a set in order to reconstitute virus replication. Such a "set" cooperation has some resemblance to the hypercycle quasispecies cooperation mentioned in chapter 2. This process is also clearly similar to a group selection process that has been considered and dismissed as implausible by most evolutionary biologists. But in the context of UV-killed viruses, group selection must operate on a population of otherwise dead virus genomes, not on individual virus genomes. In addition to acute viruses, such group selection of otherwise defective virus could also apply to viruses that colonize the host genome. Like for the persistence of defective prophages discussed in chapter 3, a host colonized by a mixture of defective prophages could be stable but could also express complementing phage genes with the combined capacity to produce virus. The apparently stringent conservation of virus-specific repair genes in DNA viruses and phages and some defective prophages supports the idea that repair capacity is indeed highly selected in natural virus populations. Thus, neither the death of the host cell nor the death of the individual virus is sufficient to exterminate the survival potential of such a virus system. The implications of this are mind-boggling: a mixture of dead viruses may still persist in its potential for life. Only viruses are known to have such a complementing, resurrecting capacity, as no other biological entity can resurrect itself after certain death by mixing defective genomes. Most DNA virus families not only have highly conserved DNA repair genes but also conserve their ability to recombine genomes and can do so at very high efficiencies. In fact, it is this very ability of defective genomes to recombine into an infectious virus that has been put to practical use: this is the method used to experimentally generate recombinant viruses. It is also precisely due to this complementation and recombination capacity that some virologists worry that otherwise unavailable viruses, such as smallpox, might be reassembled from subgenomic parts for the purpose of bioterrorism.

Similarity of Phycodnavirus Repair Enzymes to Eukaryotic Enzymes

I have noted above many clear similarities of the phycodnaviruses to the viruses of prokaryotes. For example, PBCV-1 encodes a UV-induced, DNA repair enzyme that is clearly similar to the product of the T4 gene *denV*; there are no known cellular enzyme homologues (prokaryotic or eukaryotic) that resemble this protein or its mechanism of DNA repair. However, there are also an equal number of compelling similarities between phycodnavirus genes and those of eukaryotes, including most of the core virus genes. In the context of repair, PBCV-1 superoxide dismutase (SOD) is thought to protect against sunlight-induced reactive oxygen and most probably extends the life of infected cells in damaging sunlight. This viral enzyme is of the aerobic form, which according to phylogenetic analysis is basal to the SODs of eukaryotic cells but similar to those found in the large DNA baculoviruses of insects. Furthermore, this enzyme shows no similarity to those of prokaryotic cells. However, similarity between PBCV-1 SOD and the SOD found in lysogenic bacteriophage Fels-1 is apparent. A dendrogram showing a phylogenetic analysis of PBCV-1 SOD is shown in Fig. 4.2.

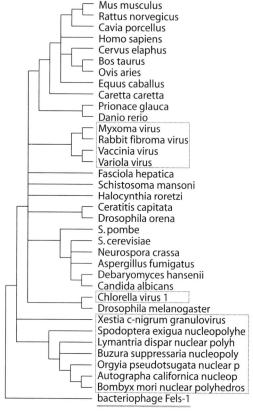

Mus musculus
Rattus norvegicus
Cavia porcellus
Homo sapiens
Cervus elaphus
Bos taurus
Ovis aries
Equus caballus
Caretta caretta
Prionace glauca
Danio rerio
Myxoma virus
Rabbit fibroma virus
Vaccinia virus
Variola virus
Fasciola hepatica
Schistosoma mansoni
Halocynthia roretzi
Ceratitis capitata
Drosophila orena
S. pombe
S. cerevisiae
Neurospora crassa
Aspergillus fumigatus
Debaryomyces hansenii
Candida albicans
Chlorella virus 1
Drosophila melanogaster
Xestia c-nigrum granulovirus
Spodoptera exigua nucleopolyhe
Lymantria dispar nuclear polyh
Buzura suppressaria nucleopoly
Orgyia pseudotsugata nuclear p
Autographa californica nucleop
Bombyx mori nuclear polyhedros
bacteriophage Fels-1

Collapsed SOD tree rooted with bacteriophage

Figure 4.2 Phylogenetic analysis of the SOD gene from PBCV-1. Shown is a collapsed neighbor-joining tree. Viruses in the lowermost dotted box are nuclear polyhedrosis viruses.

DNA polymerase

Numerous other PBCV-1 genes (such as for DNA polymerase) also show a related pattern of similarity to eukaryotic genes and to viruses of eukaryotes and prokaryotes, but not to prokaryotic cellular genes. PBCV-1 does not encode its own DNA-dependent RNA polymerase, as does ASFV (discussed further below). However, it does encode a DNA polymerase. The PBCV-1 DNA polymerase is a highly conserved core enzyme that has been used to identify other members of both the phycodnavirus and phaeovirus families (see below). Natural PBCV-1 isolates conserve this core gene sequence but often differ by having acquired additional but unknown accessory genes. An extensive phylogenetic analysis of PBCV-1 and related DNA polymerase has been reported by Victor DeFilippis and me and is shown in Figs. 4.3 and 4.4. Figure 4.3 shows all the genes in GenBank whose products had extensive amino acid sequence similarity to PBCV-1 DNA polymerase. The alignment identifies conserved blocks of sequence and presents the sequences from the smallest to the largest. The set included the genes for all the replication proteins of eukaryotes, as well as for DNA polymerases of the major

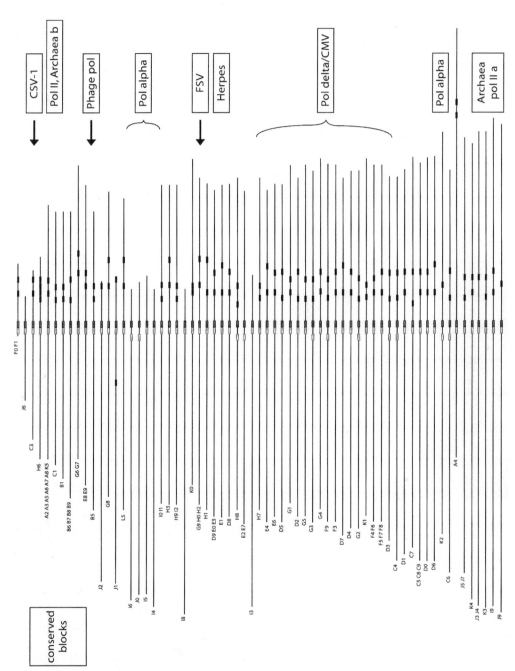

Figure 4.3 Alignment of DNA polymerase genes with strong similarity to the PBCV-1 DNA polymerase gene. Highly conserved blocks of sequence are indicated by rectangles. Sequences are arranged from smallest to largest. Reprinted from L. P. Villarreal and V. R. DeFilippis, *J. Virol.* **74**:7079–7084, 2000.

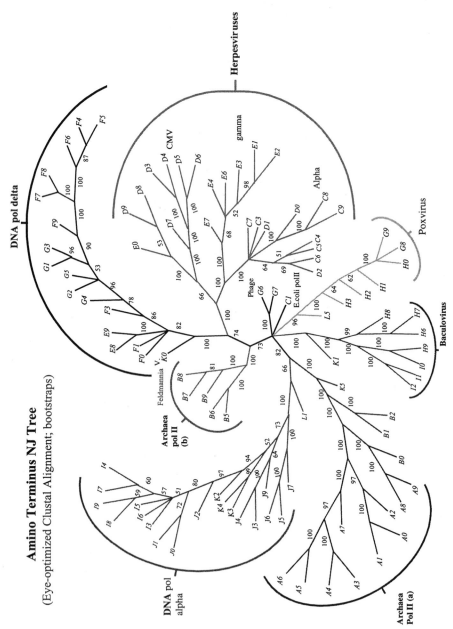

Figure 4.4 Phylogenetic dendrogram of DNA polymerase genes similar to the PBCV-1 DNA polymerase gene. Shown is a neighbor-joining (NJ) tree. Reprinted from L. P. Villarreal and V. R. DeFilippis, *J. Virol.* **74:**7079–7084, 2000.

classes of large DNA viruses of eukaryotes. The PBCV-1 DNA polymerase gene is seen at the top and encodes one of the simplest DNA polymerases in the entire set. Figure 4.4 presents the results of a phylogenetic analysis that suggests an evolutionary tree for the set of DNA polymerases. This dendrogram indicates that the PBCV-1 DNA polymerase (and related phaeovirus DNA polymerases) are basal to the DNA polymerase beta (extension polymerase) of all eukaryotes. Yet it is most similar to the DNA polymerase of the human herpesvirus family. In terms of prokaryotes, this polymerase is most similar to those of the T-even phage but is distantly related to DNA polymerases of *Archaea*. PBCV-1 DNA polymerase is not related to the replicative-extension polymerase in *Bacteria*. Thus, this viral DNA polymerase occupies a basal position in the eukaryotic phylogenetic tree and resembles the progenitor of all eukaryotic extension polymerases.

RNA modification

The early PBCV-1 transcripts are polyadenylated, but late RNAs are not. In this the virus mRNA is both eukaryote-like and prokaryote-like. The viral poly(A) polymerase does not resemble those of prokaryotes but does resemble eukaryotic poly(A) polymerases. In addition, PBCV-1 mRNAs are 5′ capped. The viral capping enzyme is related to those of yeasts and is also basal to all eukaryotic RNA capping enzymes.

Biosynthetic enzymes

Viruses are not generally considered to contribute much to host metabolic enzymes or metabolic activity, as they are not considered to require any virus-specific metabolic activity. Yet, unlike those of most other viral families, a large number of PBCV-1 genes (total of 12) encode enzymes that synthesize or metabolize sugars, lipids, and polysaccharides. This includes a hyaluronan synthase protein, which accumulates on the outside of infected cells. Hyaluronan is of special interest because it was previously thought only to be found in (and characteristic to) vertebrate species, along with some capsules of pathogenic bacteria. Intriguingly, when the human genome was sequenced, this gene was identified as one of the few clear examples of what appeared to be horizontal gene transfer, presumably from bacteria to vertebrates, since it was absent from lower eukaryotes. However, phylogenetic analysis shows this PBCV-1 gene to be basal to and the likely ancestor of all three versions of the eukaryotic gene. This result best supports a viral, not bacterial, origin for this mainly eukaryotic gene. PBCV-1 also encodes a chitosanase gene that may be packaged into the virion. This enzyme makes a linear homopolymer that is a normal component of fungal cell walls, insect exoskeletons, and crustacean shells but is rarely found in algae. Of considerable interest, PBCV-1 makes its own glycosylating enzyme, found in the Golgi complex and ER, which is not present in its algal host cell and is also not a common constituent of algae. Also, a putative cellulose synthase gene is present as a PBCV-1 open reading frame. Thus, PBCV-1 seems to encode a surprisingly large number of biosynthetic enzymes that represent synthetic pathways usually associated with eukaryotic cells. In many respects, PBCV-1 appears to span the boundary of prokaryotes and eukaryotes. Even PBCV-1 promoters seem to span the prokaryotic-eukaryotic boundary, in that these viral promoters are unique and able work well in both higher plant and bacterial cells.

Introns

Finally, there is the issue of introns. Nineteen of forty-two viruses that infect *Chlorella* contain short, nuclear, spliceosomally processed introns in a viral DNA repair gene (these are U2-type GT-AG introns). Interestingly, the intron sequences are more conserved than exon sequences, seemingly at odds with the protein domain-exon shuffling hypothesis. The highly conserved DNA polymerase gene also has a related intron, but it is found in all strains and its greatest sequence conservation is within the exons. Clearly, viral introns do not evolve faster than exons in chlorella viruses.

In conclusion, PBCV-1 appears to span the discontinuity between prokaryotes and eukaryotes. It has prokaryotic characteristics but also has numerous genes and processes that are characteristic of and basic to eukaryotic organisms but which are absent from prokaryotes. Furthermore, the PBCV-1 versions of eukaryote-specific genes and elements appear to be more basal than those found in eukaryotic cells.

Viruses of Filamentous Brown Algae: Multicellularity and Sexual Reproduction

Eight species of filamentous brown algae are currently known, each of which harbors its own species-specific DNA virus known as a phaeovirus (*phaeo*, Greek for brown). These viruses are clearly related to the phycodnaviruses but differ in several important molecular characteristics. *Ectocarpus* species viruses and *Feldmannia* species viruses (FsV) (virus of *Feldmannia simplex*) are the most well-studied members of these viral families. The biology of these viruses and their hosts differs substantially from that of PBCV-1 and its unicellular host. Unlike the strictly lytic relationship that phycodnaviruses have with their *Chlorella*-like hosts, phaeoviruses are persistent genomic parasites and are passed in a Mendelian fashion to infected host offspring. In addition, the host brown algae have a much more complex life and sexual cycle, which involves diploid states (not simply haploid, as for microalgae). Brown algae also differentiate sex structures and produce mobile gametes. The phaeoviruses have circular DNA, not a linear DNA with snapback repeat ends as does PBCV-1. The phaeovirus *Ectocarpus siliculosus virus 1* (EsV-1) occurs worldwide and infects its host in all areas.

Furthermore, the complex sexual cycle of the host is linked to virus replication. Host algae grow in a vegetative haploid state, which can grow male and female gametes that can fuse to form diploids. These diploid forms grow the filamentous forms and can produce diploid spores. These diploid spores can undergo meiosis to subsequently make haploid meiospores. It is during the production of diploid spores that *Ectocarpus* species viruses and *Feldmannia* species viruses are produced from germ line-infected hosts in quantities of 1×10^6 to 5×10^6 PFU per cell. These resulting viruses degrade host nuclei and assemble progeny virus extranuclearly. During diploid spore formation, no host cell wall synthesis occurs, so virus is released by the stimulus that releases spores or gametes. Phaeoviruses only infect the wall-less, free-swimming spores or gametes of algal hosts. In natural populations, infections can be highly prevalent. The gametangia (or sporangia) are mobile gametes (or spores) that are frequently virus infected. In some host species, all individuals are infected. The extremely high levels of virus production during sporulation can disrupt the host sexual cycle, essentially rendering the host asexual. In this feature, it seems that virus reproduction overrides host sexual reproduction. The virus does not grow during vegetative growth of the host. EsV-1 infects *Feldmannia* zoospores, but it does not multiply and cause malformations. Thus, there is clear species specificity and possible viral interference to this virus-host relationship. EsV-1 does not affect the host's rate of photosynthesis or growth. The EsV-1 DNA becomes integrated into the host chromosome as a normal and essential part of the virus reproductive life cycle, a feature that is rarely observed for eukaryotic DNA viruses.

The phaeovirus genome and latency genes

EsV-1 has a 335,593-bp linear dsDNA genome with terminal inverted repeat sequences. This genome codes for 231 predicted proteins, only 28 of which are similar to entries in GenBank. EsV-1 differs from PBCV-1 in that it contains no tRNA genes, no poly(A) polymerase, and no capping enzyme; it has no introns or DNA-dependent RNA polymerase gene. However, it does code for a bacterium-like sigma factor for RNA polymerase. It also differs from PBCV-1 and bacteriophages in that one-third of the DNA is noncoding. This noncoding DNA corresponds to both repeated and nonrepeated sequences. The repeated sequences are similar to poxvirus 4 ankyrin repeats and may contain SET-like genes (also known as histone chaperone TAF-Iβ) involved in protein-protein interacting domains (possibly used for chromatin remodeling). These types of genes are absent from PBCV-1 and are suspected to be involved in coordinating the latent-to-lytic transition of EsV-1. EsV-1 also encodes several signal transduction proteins, including six hybrid histidine kinases, which are rare in bacteria but commonly found in two-component transduction systems of eukaryotes. These genes are also suspected to regulate latency in EsV-1, since PBCV-1 lacks them. Interestingly, EsV-1 codes for an H1-like histone and an RCF small-subunit protein. The EsV-1 DNA polymerase is much more like that of *Feldmannia* viruses than either the host or PBCV-1, so this core enzyme appears to define a common virus lineage. EsV-1 also has a PCNA gene (which is PBCV-1-like). Interestingly, EsV-1 has a bacterium-like transposon with an open reading frame that codes for a factor related to plant defense proteins and pathogenesis, PR-5. This transposon has a phage-like transposase (integrase) as well as a *Lactococcus* phage-like anti-repressor of the lysogenic cycle, and thus it establishes a clear relationship of EsV-1 to phages. It is assumed that some of these EsV-1 proteins must function to link virus reactivation and replication to host sexual reproduction. This intimate link of EsV-1 reactivation to host sexual reproduction is especially intriguing because a similar link was observed with phages that latently infect *B. subtilis* spores (discussed above). Given the old evolutionary lineage of EsV-1, it seems possible that this virus system was involved in the origination of this host process (sexual reproduction) as well.

The Red Algae

Algae, which are mainly photosynthetic, are the major life form in the oceans. By fixing carbon dioxide into organic molecules, they provide the major carbon source for the food web and thus are critical for the energy flow of most oceanic life. In particular, algae are used as food by marine micrograzers such as protozoa, ciliates, nematodes, and microinvertebrates (usually microlarval forms). The preceding sections of this chapter have focused mostly on eukaryotic green microalgae and brown filamentous algae and their viruses, which constitute the majority of algal species. However, there also exists another distinct phylum of eukaryotic algae, the red algae. Red algae (Rhodophyta) are eukaryotic and have the nuclei, mitochondria, chloroplasts, ER, and Golgi complex that are characteristic of all eukaryotes, although flagella are notably absent. Their nuclei can be rather small; in fact, for some red algae the nuclei are smaller than the single plastids. However, in many respects red algae seem either to be a sister group to all other eukaryotes or possibly to be the oldest eukaryote. This is mainly due to the nature of the chloroplasts (as well as being supported by rRNA analysis). Distinctive features of red algal chloroplasts are their disorganized, unstacked thylakoid photosynthetic membranes and the occurrence of phycobilin pigment granules, which give them their distinctive red color. In these characteristics, red algae more closely resemble cyanobacteria than do green algae. As consequences of these pigments and chloroplast organization, red algae can tolerate a wider range of light levels than any other group of photosynthetic plants. Furthermore, they can also thrive in relatively deep water (up to 268 m) under low light, as well as in shallow tropical waters under intense light. In addition, evolutionary links between green and red algae established by 5S rRNA analysis are tenuous, consistent with a sister group relationship in which the red algal sister group would lack any other out-groupings. However, red algae have an uneven fossil record, which limits geological consideration of this issue. Species of red algae are not nearly as numerous as those of other algae, and it is estimated that they account for only 1 to 2% of all algal species. Rhodophyta are the least studied and understood of all algal groups.

Red algae and viruses

There is considerable interest in red algae as a group since they are responsible for the toxic red algal blooms (red tides) that can be so destructive to shellfish, fish, and marine mammals. A well-studied species responsible for toxic blooms is a member of the Raphidophyceae, *Heterosigma akashiwo* (Hara and Chihara). Curiously, these red tide bloom populations have a highly clonal character and are known to frequently terminate rapidly. There is now strong evidence that such toxic blooms can be terminated by the production of lytic virus specific to this species of red algae. Early observations using electron microscopes showed that the terminations of red algal blooms are often associated with the induced production of virus-like particles. Subsequent studies have isolated a lytic, large (202-nm) icosahedral DNA virus, *Heterosigma akashiwo* virus (HaV), that is able to lyse specific strains of Raphidophyceae. The virus appears to replicate in the protoplasm of infected cells, like the poxviruses. Molecular details about these viruses are still lacking. Since the initial identification, numerous other strains of HaV have also been isolated, establishing a broad viral diversity in natural settings. Currently, it is felt that production of lytic virus is usually associated with the termination of red algal blooms. In other words, it appears that these infections have large effects on natural host population dynamics. Although these lytic infections are species specific, many natural strains of *Heterosigma* species are resistant to infection. These resistant clones have been seen to develop at the termination of blooms. The mechanism of this resistance has not been determined, but this behavior is very reminiscent of the establishment of lysogeny by bacterial phages and subsequent lytic phage immunity. However, species-specific, nonlytic persistent infections by HaV have not been investigated. Furthermore, it is clear that HaV is not the only type of virus that can infect red algae. A rod-shaped virus that can form hexagonally packed inclusions in the ER has been observed in *Audouinella saviana*. This virus seems likely to be an RNA virus. Since all isolates of this algal species (but not related species) seem to harbor this virus, it may be a highly ubiquitous but species-specific persistent virus. In addition, various red algae are known to harbor ssDNA plasmids that bear clear similarity to geminiviruses, in that they have covalently attached initiator proteins. Such rolling-circle viruses are well known in various other host kingdoms (bacteria, plants, and animals) to either be dependent on (e.g., satellite viruses) or interfere with the replication of acute and sometimes larger DNA viruses. Although persistent viral agents of red algae can clearly be expected to also interact with and/or compete with acute viral agents of the same host, these issues have not yet been examined in Raphidophyceae.

Transferred or infectious nuclei of red algae

The nuclei of red algae are typical of eukaryotes in most respects, except for their uniformly small haploid size (1 to 3 μm, compared to 3 to 10 μm for higher eukaryotes). However, there is one additional striking characteristic that applies to all red algae that should be considered: nuclear migration. All algae that are not strictly haploid seem to cycle between haploid and diploid states in association with sexual reproduction. This alternative ploidy feature was discussed in some detail above with respect to the sexual reproduction of filamentous brown algae and the production of genomic phaeovirus, but it also applies to many fungi (see chapter 6). In red algae, the formation of the diploid cell during sexual reproduction occurs by the migration of the nucleus from a donor haploid cell to a recipient haploid cell via a primary pit connection, which provides a cytoplasmic bridge between adjacent cells. The transferred nucleus then replicates in the recipient cell. However, besides sexual reproduction there are many other examples of nuclear migration in red algae, involving vegetative cells from the same organism that are not derived from a common cell division and that can result in heterokaryons. The transfer of replicated nuclei can be on a very large scale, resulting in cells that contain hundreds or thousands of nuclei, sometimes arranged into hexagonal arrays under the surface of the plasma membrane of the recipient cell. This results in striking geometric patterns of 4',6'-diamidino-2-phenylindole (DAPI) fluorescence. In addition, nuclear transfer between different species to form heterokaryons has also been well established. Furthermore, and unique to red algae, the transfer of nuclei between different species of red algae can be parasitic. In some cases, the parasitic nuclei fuse with the nuclei of the host. In other cases, however, the parasitic nuclei undergo rapid replication and are transferred throughout the host in a most infectious process that can also spread to new hosts. All of these nuclear transfers have several features in common, including the migration of newly replicated nuclei to the plasma membrane, the formation of a pit connection from the "parasitic" cell to the host, and the migration of the parasitic nuclei into the new host (be it from the same individual, another individual of the same species, or another individual of a different species of organism). The behavior of these parasitic nuclei is clearly virus-like. In fact, most of these nuclear migration processes are highly similar to those described above for the movement and transmission of poxviruses and some viruses of hyperthermophiles. This distinctive virus-like nuclear characteristic is found in essentially all red algae and may well relate to the biological origin of the nucleus from a virus. Another kingdom of organism that is also commonly able to transfer nuclei between cellular hosts is the Fungi (discussed in chapter 5).

Overall, the viruses that infect algae appear to have most of the characteristics that would be needed to span the prokaryotic and eukaryotic kingdoms. They are both lytic and latent, and the latent life cycle is tightly linked to host sexual reproduction. Analysis of the viral DNA polymerases suggests that two families, the phycodnaviruses and the phaeoviruses, are clearly related to each other but that their corresponding lytic and latent lifestyles have endowed them with distinct gene sets. Curiously, herpesviruses are among the closest algal virus relatives. The algal viruses also have a clear relationship to herpesviruses, poxviruses, baculoviruses, and ASFV. Yet all of these viruses, as noted previously, appear to have evolved from the AFV-1-like viruses of thermophiles. These algal viruses have numerous genes that are bacterium-like or eukaryote-like. This mixture of prokaryotic and eukaryotic genes is not a general rule for other DNA viruses, even those found in the oceans. For example, white spot syndrome virus is a large DNA virus that infects shrimp (see chapter 5) and has essentially no genes that are similar to those of bacteria and only a few (10%) genes that are similar to those of eukaryotes, even though its DNA polymerase does show similarity to phycodnavirus and herpesvirus DNA polymerases. Taken together, the characteristics of these red algal viruses are strong support for the viral-origin hypothesis for the eukaryotic nucleus. The capacity of large DNA viruses for large-scale generation of genetic novelty is well established, and the possibility that this virus-based genetic creativity can colonize the host, resulting in the origination of the eukaryotic nucleus, can now be well supported.

Recommended Reading

Evolutionary Dilemma: Ancestor and Nucleus

Cavalier-Smith, T. 1991. The evolution of prokaryotic and eukaryotic cells, p. 217–272. *In* E. E. Bittar (ed.), *Fundamentals of Medical Cell Biology*, vol. 1. *Evolutionary Biology*. Jai Press Inc., Greenwich, Conn.

Cavalier-Smith, T. 1975. The origin of nuclei and of eukaryotic cells. *Nature* **256**: 463–468.

Cavalier-Smith, T. 2002. The phagotrophic origin of eukaryotes and phylogenetic classification of Protozoa. *Int. J. Syst. Evol. Microbiol.* **52**:297–354.

Kyrpides, N., R. Overbeek, and C. Ouzounis. 1999. Universal protein families and the functional content of the last universal common ancestor. *J. Mol. Evol.* **49**: 413–423.

Lake, J. A., and M. C. Rivera. 1994. Was the nucleus the first endosymbiont? *Proc. Natl. Acad. Sci. USA* **91**:2880–2881.

Maynard Smith, J., and E. Szathmary. 1995. *The Major Transitions in Evolution.* W. H. Freeman Spektrum, New York, N.Y.

Poole, A., and D. Penny. 2001. Does endo-symbiosis explain the origin of the nucleus? *Nat. Cell Biol.* **3**:E173–E174.

Sogin, M. L. 1991. Early evolution and the origin of eukaryotes. *Curr. Opin. Genet. Dev.* **1**:457–463.

Woese, C. 1998. The universal ancestor. *Proc. Natl. Acad. Sci. USA* **95**:6854–6859.

Symbiotic Theory

Margulis, L., and D. Sagan. 1997. *Slanted Truths: Essays on Gaia, Symbiosis, and Evolution.* Copernicus, New York, N.Y.

DNA Polymerase Issues

Bernad, A., A. Zaballos, M. Salas, and L. Blanco. 1987. Structural and functional relationships between prokaryotic and eukaryotic DNA polymerases. *EMBO J.* **6:**4219–4225.

Braithwaite, D. K., and J. Ito. 1993. Compilation, alignment, and phylogenetic relationships of DNA polymerases. *Nucleic Acids Res.* **21:**787–802.

Edgell, D. R., H. P. Klenk, and W. F. Doolittle. 1997. Gene duplications in evolution of archaeal family B DNA polymerases. *J. Bacteriol.* **179:**2632–2640.

Forterre, P. 1999. Displacement of cellular proteins by functional analogues from plasmids or viruses could explain puzzling phylogenies of many DNA informational proteins. *Mol. Microbiol.* **33:**457–465.

Forterre, P., and H. Philippe. 1999. Where is the root of the universal tree of life? *Bioessays* **21:**871–879.

Spicer, E. K., J. Rush, C. Fung, L. J. Reha-Krantz, J. D. Karam, and W. H. Konigsberg. 1988. Primary structure of T4 DNA polymerase. Evolutionary relatedness to eucaryotic and other procaryotic DNA polymerases. *J. Biol. Chem.* **263:**7478–7486.

Wang, T. S. 1991. Eukaryotic DNA polymerases. *Annu. Rev. Biochem.* **60:**513–552.

Wang, T. S., S. W. Wong, and D. Korn. 1989. Human DNA polymerase alpha: predicted functional domains and relationships with viral DNA polymerases. *FASEB J.* **3:**14–21.

The Viral-Origin Hypothesis

Bell, P. J. 2001. Viral eukaryogenesis: was the ancestor of the nucleus a complex DNA virus? *J. Mol. Evol.* **53:**251–256.

Filee, J., P. Forterre, and J. Laurent. 2003. The role played by viruses in the evolution of their hosts: a view based on informational protein phylogenies. *Res. Microbiol.* **154:**237–243.

Filee, J., P. Forterre, T. Sen-Lin, and J. Laurent. 2002. Evolution of DNA polymerase families: evidences for multiple gene exchange between cellular and viral proteins. *J. Mol. Evol.* **54:**763–773.

Forterre, P. 2002. The origin of DNA genomes and DNA replication proteins. *Curr. Opin. Microbiol.* **5:**525–532.

Takemura, M. 2001. Poxviruses and the origin of the eukaryotic nucleus. *J. Mol. Evol.* **52:**419–425.

Villarreal, L. P. 1999. DNA virus contribution to host evolution, p. 391–420. *In* E. Domingo, R. G. Webster, and J. J. Holland (ed.), *Origin and Evolution of Viruses.* Academic Press, San Diego, Calif.

Villarreal, L. P., and V. R. DeFilippis. 2000. A hypothesis for DNA viruses as the origin of eukaryotic replication proteins. *J. Virol.* **74:**7079–7084.

Viral Defectives and Group Selection

Szathmary, E. 1992. Viral sex, levels of selection, and the origin of life. *J. Theor. Biol.* **159:**99–109.

Szathmary, E., and L. Demeter. 1987. Group selection of early replicators and the origin of life. *J. Theor. Biol.* **128:**463–486.

Vaccinia Virus as a Mininucleus

Mallardo, M., E. Leithe, S. Schleich, N. Roos, L. Doglio, and J. Krijnse Locker. 2002. Relationship between vaccinia virus intracellular cores, early mRNAs, and DNA replication sites. *J. Virol.* **76:**5167–5183.

Mallardo, M., S. Schleich, and J. Krijnse Locker. 2001. Microtubule-dependent organization of vaccinia virus core-derived early mRNAs into distinct cytoplasmic structures. *Mol. Biol. Cell* **12:**3875–3891.

Moss, B., and B. M. Ward. 2001. High-speed mass transit for poxviruses on microtubules. *Nat. Cell Biol.* **3:**E245–E246.

Tolonen, N., L. Doglio, S. Schleich, and J. Krijnse Locker. 2001. Vaccinia virus DNA replication occurs in endoplasmic reticulum-enclosed cytoplasmic mini-nuclei. *Mol. Biol. Cell* **12:**2031–2046.

φ29 and Tubulin

Bravo, A., and M. Salas. 1998. Polymerization of bacteriophage variant phi29 replication protein p1 into protofilament sheets. *EMBO J.* **17:**6096–6105.

Serna-Rico, A., M. Salas, and W. J. Meijer. 2002. The *Bacillus subtilis* phage φ29 protein p16.7, involved in φ29 DNA replication, is a membrane-localized single-stranded DNA-binding protein. *J. Biol. Chem.* **277:**6733–6742.

The Phycodnaviruses

Van Etten, J. L., M. V. Graves, D. G. Muller, W. Boland, and N. Delaroque. 2002. Phycodnaviridae—large DNA algal viruses. *Arch. Virol.* **147:**1479–1516.

Van Etten, J. L., and R. H. Meints. 1999. Giant viruses infecting algae. *Annu. Rev. Microbiol.* **53:**447–494.

The Phaeoviruses

Delaroque, N., I. Maier, R. Knippers, and D. G. Muller. 1999. Persistent virus integration into the genome of its algal host, *Ectocarpus siliculosus* (Phaeophyceae). *J. Gen. Virol.* **80**(Pt. 6):1367–1370.

Delaroque, N., D. G. Muller, G. Bothe, T. Pohl, R. Knippers, and W. Boland. 2001. The complete DNA sequence of the *Ectocarpus siliculosus* virus EsV-1 genome. *Virology* **287:**112–132.

Cyanophage

Mann, N. H., A. Cook, A. Millard, S. Bailey, and M. Clokie. 2003. Marine ecosystems: bacterial photosynthesis genes in a virus. *Nature* **424:**741.

Red Algae

Cole, K. M., and R. G. Sheath. 1990. *Biology of the Red Algae.* Cambridge University Press, Cambridge, England.

Douglas, S., S. Zauner, M. Fraunholz, M. Beaton, S. Penny, L.-T. Deng, X. Wu, M. Reith, T. Cavalier-Smith, and U. G. Maier. 2001. The highly reduced genome of an enslaved algal nucleus. *Nature* **410:**1091–1096.

Microscopic Aquatic Organisms and Their Viruses

Introduction

The viruses of microscopic aquatic eukaryotes and their relationship to the evolution of their hosts is a topic that has historically received little attention. Given the importance of such aquatic organisms to the origin of higher life forms, this chapter considers their virology in some detail. Later in this chapter, I also consider the virology of lower and higher Fungi since, as indicated in chapter 1, fungal evolution is of central importance for the evolution of animals and higher plants. The aquatic microorganisms can be defined operationally as being of the size range from 50 to 500 μm and thus include both large unicellular eukaryotes, such as protozoa, and microinvertebrates, or the microscopic larval forms of invertebrates. Invertebrates are covered in subsequent chapters. These organisms include many that feed by grazing on algae. However, with only some exceptions, the viruses of these organisms and their hosts present little apparent medical or agricultural risk; thus, studies of them have generally not been well supported financially. Because of this, it is possible that our current understanding of these organisms and their viruses is limited or distorted by the relatively few examples that have been investigated in greater detail and that there may yet exist other virus-host relationships that await discovery. Some of these organisms, such as *Giardia* (diplomonads) and *Trichomonas* (Parabasalia), are clearly rather primitive versions of eukaryotic cells in that they lack mitochondria or, in the case of the dinoflagellates, lack histones. Generally, these organisms are not thought to be of the same lineage that led to multicellular eukaryotes and appear to have diverged early from that lineage. Other microscopic aquatic organisms, although more like higher eukaryotes, still represent lineages that diverged early in evolution of multicellular eukaryotes, and most of their genes appear to have evolved in parallel to those of higher eukaryotes. These organisms include the protists, ciliates (e.g., *Tetrahymena*), kinetoplastids (e.g., trypanosomes), euglenids (e.g., *Eu-*

glena), Apicomplexa (e.g., *Plasmodium*), and plant "fungus" (heterokonts, genetically distinct and not true Fungi). The microscopic eukaryotic lineages that are more related to multicellular eukaryotes include the algae (discussed in chapter 6), *Dictyostelium* (social Amoebozoa), and the true Fungi (e.g., *Saccharomyces cerevisiae* and *Neurospora*), the last two of which are most related to animal lineages.

Protozoa and the Persistence of dsRNA Viruses

Protozoan organisms are most often found in aquatic environments. Therefore, they are clearly exposed to the large quantities of viral agents known to exist in all aquatic habitats. As mentioned, these aquatic viruses have morphologies that mainly correspond to those of phage (i.e., icosahedral capsids with tails, containing DNA) and phycodnaviruses (large double-stranded DNA [dsDNA] icosahedrons), although a significant but small subset includes a mixture of various other viral morphologies (e.g., small icosahedrons and rods). In spite of this common immersion in pools of mainly diverse types of DNA-containing viruses, the orders of Protista are not typically associated with infections by large DNA viruses but are instead more often associated with RNA virus infections. There may be a caveat to this situation, since our observations may be biased towards acute viruses. For example, the recent discovery of a very large DNA virus of amoebae (a mimivirus) may signal the existence of a larger number of such inapparent viruses. In some cases, however, sufficient investigations have been completed to indicate that these are general relationships between viruses and hosts. One clear and overall virus-host pattern is that a wide range of these microscopic eukaryotes are frequently infected with related families of dsRNA viruses that have small icosahedral morphology, but these infections appear to be mainly persistent or latent and are nonpathogenic. This is in stark contrast to viruses of microalgae (chapter 4) and insects, among which many examples of strictly lytic dsDNA viruses are known. As I have already discussed the microalgae and filamentous algae, in this chapter I collectively consider the nonalgal aquatic microscopic species that make up a rather diverse set of eukaryotic organisms, which includes the protists, ciliated protozoa, dinoflagellates, and the lower and higher Fungi.

Persistence, sex, and reproductive isolation

As in previous chapters, this chapter also examines the best-studied examples of virus and host to consider virus-host interactions. The main types of viruses to be considered are dsRNA viruses (*Partitiviridae* and *Totiviridae*), single-stranded RNA (ssRNA) viruses of Fungi, and linear dsDNA viruses of Fungi. All of these viral agents are generally persistent in their protist hosts but are not prevalent in either bacteria or algae. As mentioned in chapter 1, persistence tends to be highly host species specific due apparently to the need to closely coordinate the virus with host regulatory systems. Persistence tends to superimpose onto the host mechanisms for virus maintenance and competition with and exclusion of other viral agents, and frequently these systems of maintenance are addiction modules. Also as discussed earlier, virus persistence is often linked to and transmitted during host reproduction. Thus, the reproduction of latent viruses can be directly associated with the sexual reproduction of the host, as has already been noted in chapter 4 for the phaeoviruses of filamentous brown algae and the spore-infecting phage of *Bacillus subtilis*. These general characteristics of persistent infections are all also clearly evident in various protists and their viruses. In many cases, it appears that persistent infections of protists by viruses are highly prevalent in nature. In such circumstances, we can expect that virus-virus competition may also be prevalent and may likewise provide a selective pressure for the persistent virus to exclude host colonization by a competing virus. Since such exclusion or competition of viral systems often involves addiction modules, including toxin and immunity genes that are harmful or lethal to uninfected hosts, we can expect these colonizing viruses to also provide a selective pressure that can eventually reproductively isolate infected from uninfected host organisms. Along these lines, it is clear that some of these persistent agents can result in harmful maternal effects during host sexual reproduction. This situation may be rather generalized and appears to occur across a broad array of viruses and hosts. Sometimes, this maternal harm involves virus reactivation, mitochondrial infection, or egg-associated virus production. For example, sexual induction of lysogenic viruses in bacteria occurs when the male cell harbors a prophage expressing immune functions but the recipient bacterium is not latently infected or immune, resulting in lytic virus induction. Other examples include the phaeoviruses of filamentous algae, killer viruses of Fungi, gypsy viruses of *Drosophila*, and ascoviruses of locusts. All of these are examples in which harm to the offspring results following sex between infected and uninfected sexual partners, such that uninfected females or eggs reactivate virus and harm the host. Only infected females (or recipients) produce viable offspring from mating with infected males. A specific example of this is the killer viruses of various yeast species (these are generally species and strain specific, as discussed below). Examples of sex-linked virus effects from other protist and fungal species are numerous, and some are presented below. This sex-related virus harm is a characteristic that seems to especially apply to hosts that have been colonized (infected) by various retroviruses, such as the gypsy virus in *Drosophila* species, and has been called a "maternal effect." However, colonization of a host genome by an endogenous retrovirus (ERV) can result in permanent alterations to the germ line of a particular host. Examples of genome-wide stable retrovirus colonization, such as that of the Fungi kingdom, are also presented. It will be seen that these ERV colonization events frequently represent points of bifurcation and divergence in host lineages, some of which maintain these ERV agents, while others do not.

Other cryptic phenotypes of persistence

In other persistent infections, the mechanism of virus persistence and maintenance and its consequence to the infected host are not clear because the agents appear to be very cryptic. These cryptic viruses may appear to be defective (not encoding gene products for virion production) or be without obvious addiction module characteristics. Yet even in these circumstances, it is likely that the presence of such cryptic agents in the host modifies the host biological outcome of infection. Satellite viruses (such as those found to be associated with yeast killer viruses) are examples of cryptic viruses that can affect the replication potential of related killer viruses and appear to allow a more stable persistence. In other cases, more direct effects on the host can be seen, such as the longevity phenotype that can result from *Podospora* mitochondrial infection with the pAL2 virus (referred to as linear plasmids) and seemingly extends the life span and transmission potential of the infected host. Thus, like for bacteria colonized by temperate phages, a main consequence of cryptic virus infection can be to affect the host potential for other virus interactions. As virus-virus interactions are inherently conditional (thus potentially cryptic) and seldom evaluated in laboratory experiments, the literature has only isolated examples of such relationships. Yet in the entire kingdoms of Fungi and Protista, we see a surprisingly high prevalence of persistent cryptic viruses. Most fungi from natural isolates harbor persistent viruses. It seems likely that this prevalence of cryptic viruses relates in some way to the mainly fused hyphal life strategy of Fungi, which provides much opportunity for the transmission of cryptic or silent viruses. It is striking that there are few, if any, clear examples of lytic viruses infecting protists, unlike the situation observed for microalgae or as is described for insects in chapter 7. A number of cryptic plant viruses that contain two dsRNA segments have also been observed. These cryptoviruses lack movement functions and are vertically transmitted. Also, they appear to have few phenotypic effects on host cells, although they can affect the outcome of infections with other viruses.

Mineralized Algae: a Viral Paucity

The biology of green microalgae and the filamentous brown algae is discussed in chapter 4 and will not be repeated here. However, the term *algae* covers a diverse set of organisms that have distinctly different characteristics. In fact, historically the term *algae* included the prokaryotic cyanobacteria. The current usage, however, is restricted to eukaryotic organisms. The mineralized algae are algae, such as diatoms, that form a shell or exoskeleton which is composed of minerals taken in from the ocean. The most common minerals used for these shells are Ca^{2+} and Si^{2+}, which are actively pumped into cells as soluble ions from seawater and deposited at membrane interfaces onto the cell exterior to make a solid mineral exterior, usually in plate patterns. The capacity of algae to make mineralized exteriors arose at the Precambrian-Cambrian boundary and rapidly evolved through the 40 million to 50 million years after the Cambrian period. By that time, there had already evolved a large number of species of these organisms, whose fossils are visible in shale and diatomaceous earth. These algae are especially efficient at fixing CO_2 and are thought to have been responsible for lowering the CO_2 content of the early atmosphere. They continue to contribute a large quantity of fixed CO_2 and mineral sediment (especially calcium) to the biosphere and remain very important on a global scale. Diatoms are responsible for the main sediment of the ocean and have contributed massive mineral production to these sediments. Thus, these organisms are both very abundant and of major significance to the early evolution of eukaryotic life.

Origins of diatom photosynthesis

Diatoms compose the brown algae, which are distinct from both the red algae and green algae and have distinct (brown-pigmented) light harvesting systems and proteins. Diatoms have large central nuclei, and their Golgi complexes are involved in shell synthesis. Recent phylogenetic studies of diatom (*Cyanophora*) chloroplast genomes (which encode about 210 proteins) indicate that they appear to be sister groups to the red algae, their closest relative: 45 genes were identified as being conserved among cyanobacteria, red algae, and diatoms. These data have further suggested that the progenitor organisms of both of these lineages were likely cyanobacteria. Diatoms are distinct from other algae in that the genes for light harvesting are nuclear, not chloroplast borne. This observation supports the idea that the early evolution of photosynthesis was nuclear, followed by the migration of photosynthetic genes into the genomes of the chloroplasts (and other plastids). These diatom photosynthetic proteins are translated in the cytoplasm and then imported across a distinctive membrane that surrounds the diatom chloroplast. This scenario is consistent with the proposal in chapter 4 that the eukaryotic nucleus appears to have evolved from the stable colonization of an extragenomic DNA virus of a cyanobacterial host, followed by the migration of bacterial metabolic and other genes into the protonuclear viral genome. The light harvesting genes exist in loci that clearly appear to have evolved by gene duplication. Although the occurrence of duplicated and transposed sequence in the diatom genome is not well known, recent success at introducing foreign plasmid DNA into diatom genomes has shown that the integrated DNAs are frequently observed in tandem duplications. This establishes that a repeat-induced point mutation (RIP)-like "antiviral" system, which is known for some Fungi (discussed below), does not operate in diatoms to prevent the acquisition of duplicated DNA in their genomes.

Microscopic studies fail to find diatom viruses

Due to the highly recognizable and species-specific shape of the diatom mineral shells, which is well preserved in sedimentary material, shale deposits of these organisms have been intensively studied by transmission and scanning electron microscopy. These studies have demonstrated that some species (such as testate amoebae) have been morphologically stable for millions of generations. In spite of the prevalence of phycodnaviruses in nonmineralized microalgae, however, no phycodnavirus, or any other virus for that matter, is known to infect any member of the mineralized algal species. Specific examples of diatoms for which there are no reports of viruses are *Navicula pelliculosa* and Foraminiferida. This striking situation could simply be due to the lack of a systematic search for viruses in these organisms. These algae can have hard, glass-like shells composed of silica deposited in very highly structured shapes. Thus, entry and release of virus could be highly restricted events. These silica-based shells have been under intense scrutiny as possible sources of biologically nanofabricated, highly structured silicates, and they have often been extensively examined by electron microscopy. It therefore seems likely that virus-like particles (VLPs) or subparticles would have been observed by now, such as was the case with the VLPs observed during EM examination of protozoa (discussed below). Yet such reports are noticeably absent from the literature on mineralized algae. Although it remains possible that the lack of virus observation is simply due to a lack of looking, it seems more likely that lytic virus replication or high-level latent virus reactivation is, at the very least, rare or possibly absent from these species. This raises the possibility that the hard mineral shell may be a very efficient barrier to the entry and exit of viruses and might relate to the long-term stability or non-dynamic morphology of these organisms. Whether diatoms have virus-like, inapparent genomic parasites or their defective genetic transposon relatives has not yet been determined and needs to await the sequencing of one of these genomes as well as the sequencing of possible extragenomic elements.

Protista: a Diverse Kingdom

Protists are a very diverse collection of microscopic eukaryotes. The most primitive members of these organisms are diplomonads (*Giardia*) and Parabasalia (*Trichomonas*), which lack mitochondria, Golgi complexes, and endoplasmic reticula. A large proportion of protist organisms, including those that are more primitive, are binucleate. Overall, these organisms clearly appear to support infections with viruses, mainly nonsegmented dsRNA viruses that also tend to establish persistent infections, often with continuous virus shedding. The accumulation of VLPs in the cytoplasms of protists has often been reported, and transmission to uninfected hosts has also been demonstrated in some cases. The two best-studied members are *Giardia lamblia virus* (GLV) of *Giardia lamblia* and *Trichomonas vaginalis* virus (TVV) of *Trichomonas vaginalis*. These viruses are related to each other.

However, the RNA polymerase gene of TVV is more related to those of viruses that infect *S. cerevisiae* and *Leishmania* spp. These viruses belong to the family *Totiviridae,* which includes a number of dsRNA viruses that infect protozoa and are parasitic to animals and that infect the Fungi and are parasitic to plants. Susceptible protozoa include *Leishmania, Eimeria, Giardia,* and *Trichomonas* spp. Susceptible Fungi include *Saccharomyces, Ustilago, Aspergillus,* and *Thielaviopsis* spp. These totiviruses show little pathology in their hosts, exist in low copy numbers, and are vertically transmitted. Viruses enter cells by endocytosis and are clearly similar to other eukaryotic viruses in this regard. Progeny virus is continuously secreted via peripheral vacuoles. Several totiviruses can act as helpers to satellite dsRNA viruses, some of which code for and can secrete killer toxins that kill sensitive, uninfected hosts. A virus similar to GLV can be found in the benthic larvae of crayfish, which is used for aquaculture. In crayfish, however, the virus is lytic, not latent.

Protist mitochondria and medical studies

The more developed protists (not diplomonads) undergo sexual reproduction and tend to be diploid. They are motile via flagella, engulf food, and are abundant in aquatic habitats. These protists also have mitochondria, but interestingly, these mitochondria are often very unusual, almost bizarre relative to those of higher eukaryotes, with characteristics such as replicating their linear DNA genomes via rolling circular or protein-primed processes. Such replication processes are used by many ssDNA viruses, but are not used by any free-living organisms, to replicate their genomes. Although they are not themselves photosynthetic, some protists harbor symbiotic zoospores of algae that are photosynthetic, as mentioned in chapter 4. Examples of this are *Paramecium bursaria* (a ciliate) and *Hydra viridis* (a coelenterate), both of which harbor symbiotic microalgae that provide photosynthesis and are themselves often able to support infection by chlorella virus. This is interesting in that these symbiotic algae, but not the protist hosts, can be infected with phycodnaviruses, as discussed in chapter 4. As mentioned, protists are often parasitic to other organisms (such as humans and other animals), and hence they have been well studied from the perspective of human diseases, especially intestinal and mucosal diseases. Because of this medical importance and as a result of extensive experimental evaluation, these agents have been well observed, making it more likely that we have already discovered the most prevalent virus-host interactions.

Prevalent persistent infections of protists

As mentioned above, latent or inapparent virus infection seems to be a normal or common situation in most protozoan species. Early EM-based studies in the 1970s found that VLPs were made by most of the protozoan species examined. Possibly the best examined of these species is *Entamoeba histolytica*. Several physical versions of VLPs were reported, including small (40-nm) ovoid VLPs. All protozoan strains that were examined seemed to harbor such particles, which were made in large numbers especially in early sporozoites but not in other cell types. These agents thus appear to be latent viruses and have also been called hereditary viruses, since attempts to clone or cure species of VLPs have generally failed. Subsequent experimentation has established that some VLPs are clearly authentic viruses, as they could be used to infect and lyse other permissive cells. Other VLP forms are also seen, including some that are filamentous and form beaded-string structures in the nucleus. Some of these agents are also associated with toxin production, but they have not been characterized.

LRV-1 and origins

One of the other best-characterized protozoan viruses infects the protozoan parasite *Leishmania*. Isolates of *Leishmania* are frequently persistently infected with a 32-nm virus particle (*Leishmania RNA virus 1* [LRV-1]) having a single-segment dsRNA (5.2 kb). This RNA contains two overlapping open reading frames (ORFs): one for an RNA-dependent RNA polymerase and the other for a capsid protein. Old and New World-specific versions of *Leishmania RNA virus* (13 strains; 12 LRV-1 and 1 LRV-2) are known, which are highly conserved and phylogenetically congruent with their corresponding hosts, suggesting long-term stability. A survey of *Leishmania* isolates showed that 12 of the 71 isolates examined harbored a virus related to LRV-1. Most of these virus-harboring isolates were from the Amazon Basin. The RNA polymerase of LRV-1 is most similar to that of *Saccharomyces cerevisiae virus L-A (L1)* (ScV-L-A), a yeast killer virus, and thus LRV-1 seems to represent a predecessor to the yeast virus. Consistent with this basal placement of LRV-1 relative to the yeast viruses, the four dsRNA virus types that persistently infect fungi and protozoa appear to have a common lineage. These virus types include *Ustilago maydis virus H1*, TVV, and LRV. Many strains of *T. vaginalis* are infected with TVV, and the presence of the virus is directly associated with host phenotypic variation involving surface antigen switching. Thus, persistent TVV infection can result in a specific host phenotype. Infection with LRV predates the divergence of host *Leishmania* species, which further supports the old and stable nature of this virus-host relationship. However, because most leishmaniae are infected, it has been difficult to study the consequences of infection since there is no uninfected host for comparison. There are a few distinct *Leishmania* species that lack any virus infection, but these species seem unable to support heterologous virus infections. These virus-free species apparently arose from infected predecessors due to lost persistent infections of the ancestor organisms. In addition to *Leishmania*, other parasitic protozoa are also known to harbor viruses. The dsRNA viruses found in *G. lamblia* (GLV) are clearly related to the dsRNA viruses of yeasts and other Fungi. Like many of the viruses described above, GLV is continuously shed from infected cells. *Cryptosporidium parvum* protozoa (parasitic to the gastrointestinal tracts of mammals) are frequently infected by a two-segment dsRNA virus. Previously, these bipartite partitiviruses were mainly found in Fungi and plants. In *C. parvum,* the virus is mainly found in the cytoplasm of sporozoites; it is not found in other species of *Cryptosporidium*. Strikingly, there are only a few examples of phycodna-like viruses of protozoa in the literature, which is in contrast to the situation for algae and protozoa.

Ciliated protozoa

Protists can be considered to include several major groups of organisms. The ciliated protozoa (phylum Ciliophora) include species of *Tetrahymena, Paramecium, Euplotes,* and *Oxytricha,* which all have dual nuclei in vegetative cells. One of these nuclei is a small diploid and transcriptionally silent micronucleus that maintains the germ line, and the other nucleus is a large vegetative transcriptionally active macronucleus that has elevated copy numbers of active genes but has lost numerous intergene sequences. These genome-wide rearrangements of macronuclear DNA are a striking difference between ciliates and most other organisms. In addition, with the separation of nuclear fates, we see the first evolutionary separation of the germ line and from the soma. Many ciliated protozoa are members of the hypotrichs, which undergo multiple rounds of DNA synthesis (endoreduplication) during the generation of the macronucleus. During sexual reproduction, the micronucleus is able to undergo meiosis, losing three of four of the resulting haploid nuclei. Also, during sexual conjugation, one of these resulting haploid micronuclei is transmitted via a cytoplasmic bridge into another haploid cell, where two haploid nuclei fuse into one diploid nucleus. This process of nuclear transport and transmission from one cell to another via a cytoplasmic bridge is most intriguing and is also very reminiscent of the infection-like transmission of nuclei between cells that is widely seen in red algae. In a sense, the dimorphic nuclei of ciliates also resemble the two nuclei of heterokaryons (as in Fungi), which can also result from nuclear transmission. It is interesting that in some cases, the nuclei of the recipient cells undergo enlargement and degrade their DNA, reminiscent of macronuclear DNA changes.

The Macronucleus of Hypotrichs

The macronucleus of the numerous hypotrich species undergoes continuous rounds of DNA replication (endoreduplication) without intervening mitoses, resulting in the overreplication of DNA. This initially generates polytene chromosomes, followed by vesicle formation within compartments of the DNA and by excision and loss of much of the intragenic DNA. The presence of these initially polytene chromosomes distinguishes hypotrichs from other ciliates. The macronucleus can be contrasted with the micronucleus in that it becomes overpopulated with gene-sized DNA fragments, which are highly expressed, and it reaches a terminal state in that it becomes unable to continue the lineage of the organism. In the macronucleus, the intragenic excised sequences (IESs) are small repeat sequences that constitute a significant portion of the genome. These IESs are precisely excised and degraded. Furthermore, the resulting fragments are 0.4- to 20-kbp DNA and are present in high numbers (15,000 to 40,000 per macronucleus). These linear sequences acquire telomere sequences (short tandem repeated sequences) via de novo DNA synthesis by an error-prone reverse transcriptase (RT). The RT activity is present at high levels during the somatic phase, but it quickly diminishes with transition to the sexual phase (during conjugation and autogamy). This macronucleus thus undergoes a terminal transition to a state of amplification and high gene expression, acquires a new molecular genetic identity, and also loses DNA regions and withdraws from participation in the germ line DNA.

The excision of IESs from the macronucleus of ciliates (*Paramecium*) is under epigenetic control, and thousands of IESs undergo excision from germ line DNA during development. During the formation of the somatic macronucleus, genomic DNA undergoes about 6,000 deletion events, which involve sequences containing long terminal repeats (LTRs). The Tlr1 element is one of the better studied of these eliminated LTR sequences and consists of a 13-kbp sequence with an 825-bp LTR. These are clearly transposon-like structures that resemble viruses in their excision. In some striking cases (such as the actin gene of *Oxytricha nova*), the initial subgene segments are scrambled and out of coding order but are reassembled along with differential and sequential IES deletions to form a functional contiguous and correctly ordered gene in the macronucleus. These IESs are short, noncoding elements with direct repeats and thus closely resemble transposons. In fact, longer repeat sequence elements (Tec1, TBE1, and Tlr1) can be found in the genomes of various ciliates (*Euplotes*, *Oxytricha*, and *Tetrahymena*). These longer elements not only have the IESs as inverted terminal repeats at their ends but also code for transposases. Thus, they clearly resemble functional genetic parasites. It seems likely that these longer, less defective versions of the IESs may have been the original colonizers of ciliate genomes, leading to the evolution of the much smaller IES direct repeats. Interestingly, a single copy of the longer IES in the macronucleus prevents excision in the newly developing macronucleus during sexual germination. This is a maternal chromosome effect that probably works by pairing of homologous nucleic acids.

Random DNA amplification and telomere addition

Tetrahymena probably represents the best-studied example of a ciliate that undergoes macronuclear formation. The two nuclei (macro- and micro-) of *Tetrahymena* differ significantly. In stark contrast to the micronuclear tight control of DNA replication, the macronucleus is able to replicate (endo-reduplication) and amplify not only specific regions of ribosomal DNA (rDNA) but also most any DNA sequence. The macronucleus will even allow lambda DNA to replicate when this DNA is microinjected directly into the macronucleus. Furthermore, the injected lambda DNA also acquires telomeres, which are added via an RT-like terminal transferase activity. These telomeres appear to inhibit end-to-end fusion of linear DNAs, such that the resulting lambda DNAs replicate as linear DNAs to high copy number, resulting in up to 20,000 copies per nucleus. The single micronuclear copy of rDNA undergoes DNA replication in the macronucleus, resulting in 9,000 copies of the 21-kbp repeat derived from chromosomal copying during conjugation. This rDNA replicon can also be maintained as a linear plasmid. The distinct difference in control of DNA replication between the micronucleus and the macronucleus is striking. The high-level overreplication of macronuclear DNA can attain levels similar to those attained by some DNA viruses.

RNA splicing

Loss of RNA sequences from self-splicing also occurs in the macronucleus. The group IB introns were, in fact, first identified in *Tetrahymena* and were found in nuclear rRNA genes. *Tetrahymena*'s group IB introns (splicing with no protein factors) have conserved the same catalytic fold as the ribozyme of T4 (thymidylate synthase introns). Thus, there seems to be a clear relationship between bacterial phages and *Tetrahymena* in intron processing. Such a high degree of similarity has led some to suggest that it is the result of a common ancestry. The hallmark of these self-splicing introns is a 16-nucleotide consensus sequence. This element also resembles the consensus sequences found in viroid RNAs, such as the potato spindle tuber viroid. It is also interesting that the group I introns found in *Chlorella* viruses, mentioned in chapter 4, are very similar to those in found algae, yeasts, and *Paramecium* cells, suggesting that these *Chlorella* viruses could represent the ancestor of these cellular introns or that a related virus mediated the spread of these introns into all these protist lineages.

Origins of dimorphic nuclei

The dimorphic character and variable DNA content of the two nuclei of ciliated protozoa raise some interesting issues with respect to their origins and possible relationships to DNA viruses. The diploid micronucleus transmits the germ line but maintains strict control of DNA replication in that it only allows each replicon to replicate once per cell cycle. In this regard, it is similar to the nuclear regulation of DNA synthesis present in most eukaryotic nuclei. However, the micronucleus is transcriptionally inactive, which is not typical of germ line nuclei in higher eukaryotes. In contrast, the macronucleus has relaxed cell cycle control of DNA replication but is highly active in transcription. What might account for the origins of this dual nucleus strategy? It is worth recalling that I have already argued that the nucleus itself may have evolved from a persistent infection with a large DNA virus (chapter 4). A micronucleus that is silent or repressed would clearly resemble a persistent or latent genome of a DNA virus. This repressed transcriptional state closely resembles the chromatin-repressed DNA that would be expected for a latent virus. Thus, the link of this silent nucleus to germ line transmission would also fit the general tendency for latent viruses to associate with sexual reproduction. Furthermore, the transcriptional activation and overreplication of the macronucleus during sexual reproduction also resemble virus-like behavior. In this regard, the macronucleus closely resembles the lytic reactivation of a latent DNA virus, in that it is characterized by DNA amplification outside of cell cycle control and subsequent degradation of the nonamplified DNA sequences, followed by high-level, global transcriptional activation of the replicated DNA sequences, with correspondingly high-level protein expression. These events all clearly resemble the replication and late gene activation of a DNA virus. Furthermore, the fate of such a macronucleus is essentially terminal (i.e., lytic) in that the macronucleus is a dead end, destined to be degraded and not able to contribute to the germ line. Explaining the origins of all these characteristics, especially amplification of some DNA but loss of other DNA, presents a major challenge for theories of evolution based on gradual accumulation of favorable point mutations and recombination events. However, if a nucleus can arise from a viral progenitor, then these characteristics themselves could have all come about from elements of the known and prevalent virus life strategies. In this scenario, the various micronuclear characteristics resemble those of a latent DNA virus, whereas the various macronuclear characteristics closely resemble those expressed during the lytic reactivation of a persistent virus for high-level virus production. It could thus be proposed that both the micronucleus and the macronucleus closely resemble known and prevalent life strategies of DNA viruses.

Why are there so few DNA viruses of ciliates?

A ciliate macronucleus, with unregulated DNA replication, and highly active and terminally committed gene expression, would seem to provide an excellent habitat for the general amplification of viral DNAs and virus replication. And, in fact, we know that lambda DNA injected into micronuclei is amplified very well in such nuclei. However, in spite of this seemingly inherent capacity to replicate foreign DNA and maintain high levels of gene expression, there are very few examples of nuclear DNA viruses that infect hypotrich species. This may be related to the frequently parasitic lifestyle of protozoa, in that their host cells may shield them from exposure to many DNA viruses. Yet symbiotic algae are also shielded from phycodnavirus by their hosts but can clearly support numerous DNA viruses. It is seems more likely that the dimorphic life strategy and the wide distribution of IESs in the chromosomes, along with a system for IES excision (protection) and DNA degradation of non-telomere-containing DNA, might pose major barriers for any DNA viruses that colonize ciliates. Any DNA virus that finds itself in a micronucleus undergoing endoreduplication and subsequent DNA degradation of the macronucleus is likely not to survive the dual nucleus process. This would especially pose a viral barrier if, due to the sexual cycle, the virus must first infect the micronucleus. This is because a DNA virus would need to infect the micronucleus prior to host genome amplification and silently persist in this nucleus until macronuclear formation. The dimorphic nuclear life strategy may, in fact, be a powerful mechanism to strip out rogue DNA replicons that have infected the nuclei by requiring that parasitic DNA not amplify in the micronucleus. In addition, it appears that permissible amplification in the macronucleus is also subjected to transposon-mediated excision and degradation. In considering how such a dimorphic life strategy might have originated, I proposed above the possibility that a lytic persistent virus system could have originated both the micronuclear and macronuclear structures. However, it might also be suspected that an additional colonization event by a second parasitic element must have been involved, which could contribute the transposon needed to elimination intergenic DNAs. Such a process could have resulted from selection for the need to compete with and eliminate other genetic parasites. The most likely candidate for the second parasitic element would be a virus related to the Tec1 or Tlr1 transposon. The resulting colonized genome would then have required the presence of the transposon LTRs to protect the replicons from DNA degradation and loss. Perhaps this system now prevents the colonization of ciliate organisms by any DNA viruses. However, as mentioned above, the large majority of the extant viruses of these ciliate species are latent dsRNA viruses whose primary habitats are the cytoplasms and the translational systems of the host cells, not their nuclear systems. It seems clear, therefore, that although ciliate species may have developed systems that exclude most DNA viruses, they remained highly susceptible to dsRNA viruses.

Dinoflagellates

Dinoflagellates represent another distinct and old form of microscopic, oceanic eukaryotic life. However, dinoflagellates are sufficiently different from all other eukaryotes to be considered a sister group. These organisms are responsible for toxic red tides, which result from the synthesis of domoic acid, an analogue of glutamic acid that irreversibly binds glutamic acid receptors on neurons. Some of the dinoflagellates have features of both plant and animal cells in that they are both photosynthetic and mobile, being able to move towards light. In addition, some members of this order, such as sea fire (*Gonyaulax catenata*), are also able to emit light. The ability to photosynthesize and move towards light is reminiscent of photosynthetic cyanobacteria. The photosynthetic capacity of dinoflagellates seems to have developed early in the evolution of these organisms, which may account for their very unusual chloroplasts. The individual genes of these odd chloroplasts are carried by minicircular plasmids, a situation unique to dinoflagellates. Dinoflagellates are frequently symbiotic with other organisms, such as the coelenterates that can build coral reefs only with the cooperation of the symbiotic algae. Dinoflagellates are clearly a distinct order of life and are distinguished from all other eukaryotes in that they have no histones or nucleosomes on their chromosomes. Instead, they have variable numbers of chromosomes that contain four basic chromosomal proteins, which are present at 1/10 the mass of DNA and not at the 1:1 mass ratio seen in all other eukaryotes and protozoa. The chromatin DNA is organized into right-handed double helical bundles. Also distinct from other eukaryotes, the nuclear membrane does not dissolve during mitosis. Dinoflagellates are binucleate and undergo sexual reproduction. However, the dinoflagellate nuclei are not dimorphic, and they do not appear to undergo the DNA changes (amplification and deletion) noted above for the micro- and macronuclei of ciliates. Dinoflagellates have enormous genomes, which are 1 to 10 times the size of the human genome. Seventy-five percent of the dinoflagellate genome has low copy complexity, 18% has intermediate repeat complexity, and the rest has very simple sequence complexity. Thus, it appears that the dinoflagellates have been highly colonized by repeated elements. Also distinct, the dinoflagellate rRNA is plastid encoded (SSU).

DNA viruses

Unlike for the hypotrichs, and in apparent support of the notion that dimorphic nuclei may protect against DNA viruses, DNA viruses of dinoflagellates are known. In the case of the dinoflagellates that are symbiotic to coral-building organisms, there is evidence that they can be killed by virus infection. These coral-building dinoflagellates appear to be lysed by an icosahedral, dsDNA-containing virus. Interestingly, this virus is often latent and is induced following exposure to increased temperatures. Some observations suggest that many of the coral-building dinoflagellates may be latently infected. It appears that the latent viruses can become lytic, leading to death of the dinoflagellate and bleaching of the coral.

Other DNA viruses of dinoflagellates are also known. A virus infecting a shellfish-killing dinoflagellate (*Heterocapsa circularisquama* virus) was recently isolated from Japanese coastal waters following initial observations of VLPs by EM. This virus is also an icosahedral dsDNA virus about 200 nm in diameter that lacks a tail. This size is similar to that of a poxvirus. The virus was found in large numbers in "viroplasmic structures" in the cytoplasms of dinoflagellate host cells. The virus was lytic in 18 strains of *H. circular-*

isquama but not in 24 other phytoplankton species. Thus, like those of microalgae, lytic DNA viruses of dinoflagellates are established and prevalent. Curiously, no other types of virus (e.g., filamentous, dsRNA- or ssRNA-containing, etc.) have been reported for these organisms. Thus, the dinoflagellates are in stark contrast to both the protists and ciliates in that they are prone to infections (often latent) with DNA viruses.

Fungi

The Fungi are of special interest from the perspective of evolutionary biology since they appear to have evolved well after many of the protists discussed above. Fungi are thus not representatives of the earliest eukaryotes. However, it is now accepted that the animal lineage has a monophyletic origin that shares ancestry with the Fungi. Thus, it seems likely that the molecular characteristics of Fungi and their relationships to viruses could identify molecular characteristics that also led to the evolution of animals. One in five of all known fungal species is an obligate symbiont colonized with green algae or cyanobacteria, such as the lichen-like *Ascomycota* (which includes 98% of lichenized species). This symbiotic relationship meets carbohydrate requirements of the fungi when they are colonized by a CO_2-fixing symbiont. Current analysis suggests that the lichenized fungal species evolved early, followed by multiple losses of symbiosis. Furthermore, it appears that all the higher forms of Fungi evolved from such symbiotic species, which later became autonomous. For example, *Penicillium* and *Aspergillus* appear to have been derived from such lost symbionts.

Overall viral patterns

Overall, there are some clear patterns for Fungi and their viruses. And these patterns are distinct from those of the other microscopic aquatic organisms. Infection of natural fungal populations with viruses is in many cases exceedingly common. dsRNA viruses, ssRNA viruses, dsDNA viruses, retroviruses, and even prions have all been frequently found in Fungi. In the case of retroviruses, Fungi represent the simplest organisms which are known to broadly support this virus family, although interestingly, these retroviruses all lack *env* sequences. Most (possibly all) natural isolates of fungal species (aside from yeasts) harbor persistent and inapparent viral infections. The great majority of these persistent infections are due to dsRNA viruses. In stark contrast to eukaryotic algae, dinoflagellates, and prokaryotes, few fungal species have been observed to be infected by nuclear or large dsDNA viruses, except for some very interesting agents that infect fungal mitochondria. In this regard, fungal plastids are especially unusual compared to all other eukaryotes in that their mitochondria are frequently infected by either dsRNA or DNA viruses. Although for the most part systemic infections of Fungi with dsRNA viruses are not pathogenic, they are frequently associated with toxin genes and killer phenotypes, which are pathological to nearby uninfected hosts. This situation clearly resembles the addiction module persistence strategy described for phages and bacteria in chapter 3. As with most persistent infections of lower organisms, transmission (both infection and production) of fungal persistent dsRNA viruses is vertical or frequently associated with the sexual reproduction of the host and in some cases directly associated with the mating type of the host. In addition to the highly common infections with dsRNA viruses, some specific lineages of Fungi can be infected with linear dsDNA viruses (often called plastids).

Lower Fungi

Lower Fungi represent a rather diverse and polyphyletic assemblage. Many are zoosporic (especially aquatic Fungi), such as *Oomycetes, Plasmodiophorales,* and *Thraustochytriales,* which grow from small sexually reproduced mononuclear spores. Lower Fungi (such as Phytophthora) are defined by their relatively simple mycelia, in which the separation between cells tends to be via simpler structures. They also have fruiting bodies that similarly tend to be simpler than those of higher Fungi. Many simple Fungi, such as *Rhizidiomyces* spp., have motile mononuclear zoospores. Zoospores appear to represent the earliest evolutionary versions of Fungi and are morphologically close to algae. Higher Fungi have septate-reticulate mycelia and large complex fruiting bodies, and they are composed of multihyphal structures that tend to be capable of prolonged survival. Higher Fungi (such as the ascomycetes) are much more diverse than lower Fungi. In addition, many lower Fungi are asexual, whereas the higher forms tend to be sexual. Fungi can be both diploid and polyploid, but there is a tendency, like with algae, for Fungi to be predominantly in haploid states and to cycle between diploid and haploid states via sexual reproduction. Fungi have notably small nuclei (1 to 3 μm) relative to the 3- to 10-μm nuclei for most eukaryotes. And Fungi lack the chromosomal plate characteristic of mitosis. Another unusual feature of most fungal nuclei is that their mitoses are closed; that is, the nuclear membrane does not dissolve during mitosis (similar to the case for dinoflagellates). The DNA content of the fungal genome is rather small relative to that of other eukaryotes, and the occurrence of repeated sequences in some cases (such as *Neurospora*) is very limited.

Modular, hyphal, and long-lived organisms

With the evolution of Fungi, we have the first clear example of the creation of nonmotile, modular hyphal organisms, as well as the development of individual organisms that can have very extended life spans. Although not all Fungi are hyphal, this is by far the most common fungal morphology. It is estimated that only 1% of species of Fungi are yeastlike (and these tend to live on plant surfaces). This is in contrast to unitary organisms (like animals and most bacteria), which have set morphologies for the juvenile and adult forms. Modular organisms grow by branching (tree- or rootlike) processes and thus have pliable morphologies that can adapt to the local circumstances, such as growing towards food, invading new habitats, and growing away from toxins. Such modular organisms are essentially clonal, and some, such as deuteromycetes (*Aspergillus* species and *Candida albicans*), are also asexual (or parasexual). Some Fungi can generate very large, often clonal organisms via mycelia that grow by invading adjacent habitats. However, this growth characteristic, in which new cells are physically in continuity with the parent, creates a difficulty for the definition of fitness, since growth and survival can occur with little reproduction of independent progeny. Even in cases where the continuity of parent and offspring is broken, the progeny are often clonal and most often haploid. What, then, defines reproduction and fitness in this modular circumstance? It would seem that the continued existence of a metabolizing organism with the potential for growth and reproduction would need to be considered. This growth characteristic is especially evident in some specific cases, such as shown in 1992 by M. L. Smith et al. with the *Armillaria bulbosa* fungal mycelia (mats) that have been found in Canadian forests. One such mat was estimated by aerial observation of ring growth patterns to cover about 15 ha (0.15 km^2) and to be about 1,500 years old. Other very large and old fungal mats are also known. The Fungi that grow within the stones of the Antarctic continent were proposed by R. C. Prince in 1992 to be the longest-lived organisms on earth. Such longevity has been referred to as the Methuselah factor. This long-life strategy has been referred to as a K-selected life strategy, which would indicate that such species are under competition for prolonged periods for limited resources. It is interesting that long-lived fungal species tend to be diploid, which is otherwise uncommon in most filamentous Fungi.

A network of transmissible nuclei

Fungal hyphae form interconnected networks, and such connections can result in many nuclei that reside in one shared, communicating cytoplasm, a characteristic shared by some species of red algae. These hyphae grow by nuclear division and hyphal extension at the tips. When growing tips encounter other parts of the same or a different organism, they can either be repelled or be attracted. Those that are attracted can undergo fusion (anastomosis and septal degeneration). This especially occurs between the same or similar species, although this seems to be an uncommon outcome in some natural settings. The interconnected characteristic of filamentous Fungi is expected to provide a very attractive and possibly unique habitat to viral agents, as it could allow rapid access to the entire cellular network without the need to make extracellular virus. Some higher Fungi (such as ascomycetes and basidiomycetes) can undergo self-fusion of either the same or different parts of the organisms. The hyphal characteristic of Fungi is associated with another somewhat unique biological characteristic, that of heterokaryon formation. In addition, some Fungi form stable dikaryons, and mating types exist in many species. Self-fusion is a unique and widespread biological characteristic. Sometimes self-fusion can result in protoplasmic degradation via induced phenolic compound oxidation. Such fusion events can be followed by nuclear replacement events (involving very rapidly motile small nuclei that move through hyphae), in which nuclei in the recipient hyphal compartment are often destroyed and replaced by daughter mitotic nuclei from the donor. This situation is most reminiscent of the nuclei of transmissible parasitic red algae, as described in chapter 4. Mitochondria can also undergo elongation and move through hyphae, but not via the same process that moves nuclei; also, they are not generally transferred into fused hyphae. It has been suggested that such invasive "male-like" behavior of the transferred nuclei allows the nuclei to leave behind mitochondria, or rogue mitochondria, and other cytoplasmic parasites that have colonized the host network of cells. It has also been proposed, with some experimental support, that the vegetative incompatibility system is another system of self-identification and may have developed to limit the spread of cytoplasmic genetic parasites. Although it is known that mitochondria can sometimes be horizontally transmitted, vegetative incompatibility clearly limits such transfers and also affects the transmission efficiency of virus-like parasites.

The invasive characteristic of fungal nuclei has also been called "parasexual" because it allows a form of sexual colonization or exchange (but only sometimes with heterokaryon recombination) without the typical sexual process. This parasexual characteristic can even apply to nonsexual fungal species, which may help explain how such otherwise clonal and haploid species can maintain genetic diversity. In field isolates, fungal individuality is rather rare due to frequent formation of heterokaryons following hyphal fusion, although this is not typical of *Podospora* and *Neurospora*.

Sex types, incompatibility, and nuclear interactions

Neurospora crassa has 10 loci that make heterokaryons incompatible; thus, self-recognition systems are clearly under strong selection in this species. The mating type is one of these incompatibility regions. In addition, mating type switching can also occur, although this switching is a highly atypical situation for filamentous Fungi. *Neurospora*, for example, does not switch mating types, as do some of the well-studied yeast species (e.g., *S. cerevisiae*). When switchable mating types do exist, it is generally the case that there are one stable type and one type which is silent but switchable. The mating type locus can often be repressed via DNA methylation and heterochromatin formation. The incompatibility mating function of *Neurospora* mtA1 is related to that of *S. cerevisiae* α1; thus, some common processes seem to apply to diverse Fungi. These genes are needed to be able to respond to pheromones. Mating type genes can often encode the pheromone precursors that undergo protein processing, very similar to the processing of insulin-like hormones. These mating type genes can also resemble addiction modules, in that wrong (nonself or noncomplementing) combinations can induce a damaging response. In dikaryons and diploids, the combination of two nuclei switches the developmental pattern of both nuclei, suggesting that *trans* gene interaction occurs and affects opposite *cis*-acting elements between the two chromosomes. Within *Neurospora,* there are four heterothallic species that do not yield fertile offspring in interspecific crosses. However, not all Fungi undergo nuclear fusion. The best-studied (but possibly overemphasized) mating type system is the MAT system of *S. cerevisiae,* which unlike filamentous Fungi does not differentiate between two sexes. Two mating type versions are known, **a** and α; α represses **a**. These are both expressed in haploids, but when combined in a heterokaryon diploid, they act together to make a diploid. The *S. cerevisiae* mating type elements are small genomic regions (600 to 700 bp) that code for *trans*-acting DNA binding proteins containing HMG box motifs. The diploids express a complementing gene pattern that induces pheromone synthesis and mitosis, generating haploids by mitosis of 2N cells. This situation has a clear resemblance to a reactivation program between otherwise defective elements. During type switching, the silent mating type copy is transposed into an active site via a transpositional activation that clearly resembles that of Mu phage, the P2-P4 satellite phage system, or *Borrelia* phage. This transposition allows early (unmethylated) DNA replication to occur and activates transcription. Targeted transposition involves flanking repeated elements. Yet some Fungi prevent repeat element expression or transcription. Therefore, it is clear that the RIP system of *Neurospora* (discussed below) is not present in these yeast species, as it would not tolerate such a duplicated sequence. However, it is interesting that the use of DNA methylation to repress mating type bears a clear resemblance to methylation induced postmitotically (MIP) suppression systems (described below). Mating type switching thus resembles a transpositional reactivation (invasion) from a state of silence (persistence).

Asexual Fungi: phenotypic diversity and repeat elements

Some yeasts are asexual. Due to its medical importance, the best studied of these asexual yeasts is *C. albicans,* which appears to always remain as a diploid. In this state, it resembles the transient dikaryon of the sexual phase induced prior to spore production in other Fungi, such as *Nadsonia* species. Although it is asexual, *C. albicans* has a highly switchable colony morphology, which is also closely associated with human pathogenicity. This switching can be induced by UV light irradiation, but it does not appear to be a mutational event. Instead, it appears to involve a transpositional and recombinational process involving silent regions of heterochomatin adjacent to telomeres, but it is not directly related to the MAT system of *S. cerevisiae* described above. It is also possible that the mechanism involves changes to chromatin structure and committed expression of particular gene sets. One idea is that a system of homology-dependent gene silencing could be involved in the switching system to induce new gene sets. Regardless of the specific mechanism, which remains unknown, this is clearly a complex and highly adaptive system that affects cell morphology, protein synthesis, antigenicity, and drug sensitivity and involves persistent and otherwise silent genetic elements that can be activated. Although the mechanism of function is currently unknown, it is interesting that the *C. albicans* genome is known to harbor several classes of moderately repeated sequences (such as 27A, Ca3, Ca7, CARE-1, CARE-2, and RPS1), because such repeats are clearly not found in all Fungi.

RIPs, repeats, and fungal genomes

Genomes of filamentous Fungi are rather small, about 30 Mbp, which is 10-fold greater than the size of the *Escherichia coli* genome but 1/100 the size of the human genome. It is most interesting that some Fungi (such as *N. crassa*) are essentially devoid of repeat sequence DNA but that others, such as the lower Fungi (e.g., *Phytophthora* spp.), can have up to 18 to 53% of their genomes as repeat elements. This of special interest if we consider that such repeats could be the remnants of early colonization by genetic parasites. *Neurospora* and other Fungi, such as *Ascobolus* spp., have actually developed molecular systems that effectively prevent the accumulation of repeat sequences. The best-studied such system, which is in *N. crassa,* is the RIP (repeat induced premeiotically, now called repeat induced point mutation) system, which efficiently induces point mutations in repeat sequences. This process is induced in one of the dikaryon nuclei during the sexual cycle when two haploid nuclei of opposite mating types come together to share a common cytoplasm. In one sexual cross, between 8 and 80 point mutations (CG-to-AT transitions) have been measured to result from RIP of one copy of a gene following sexual (but not vegetative) reproduction. *Ascobolus* (also a haploid) induces high-level DNA methylation (MIP) in repeated sequences. (DNA methylation is generally at low levels in most Fungi.) The origin and mechanism of this process are not well understood: because the process occurs on a microscopic cellular level, it is difficult to investigate the biochemistry. However, as it is induced only premeiotically following the fusion of the two gamete nuclei, it resembles a nucleic acid surveillance system derived from some genomic element that precludes other, related elements. This would be along the lines of the DNA methylation/restriction system of phages and bacteria, which can also repress extra or related copies of the sequence. This RIP-like system is not seen in other Fungi, such as *Aspergillus* or *Podospora,* so it is not a universal characteristic. In addition, other mechanisms of homology-dependent gene silencing have also been described, such as quelling, which is known to posttranscriptionally silence repeated sequences in *Neurospora*. Thus, suppression of repeat elements is clearly a redundant and well-maintained phenomenon in these species. However, this observation makes clear that the accumulation of genetic parasites, such as retroposons, is not an essential feature for the adaptation and survival of specific fungal genomes. Thus, systems that prevent retroposon accumulation have been successfully developed by some organisms. *N. crassa* in particular appears to have a wide array of genome defense systems. *Neurospora* has no intact mobile genetic elements, and only 10% of its genome is repeat DNA, much of which is rDNA in the nucleolus. Curiously, the genome of *N. crassa* is about 40 Mbp (about 10,000 ORFs), only 25% smaller than that of *Drosophila melanogaster* and much larger than that of *S. cerevisiae* (12 Mbp), both of which lack the RIP system.

Repeated rDNA and Tad

Although *Neurospora* has few repeated sequences, the clear exception is the rDNA repeats found in the nucleolus. Repeated rDNA of *Neurospora* is present in 185 copies in chromosomal DNA. This is quite unlike what was described above with the unique genomic rDNA in *Tetrahymena* that amplifies in the macronucleus. As with most higher organisms, *Neurospora* rDNA is organized into the nucleolus as a distinctive structure (which was absent from red algal species). It has been proposed that the organization of rDNA copies into a nucleolus may have evolved to protect these repeat elements from a RIP-like system that would otherwise destroy the repeated sequences, although this does not explain the absence of extranuclear rDNA amplification as in ciliates. Retroposons are generally absent from the genomes of Fungi. However, it is known that *N. crassa* can have a 7-kbp Tad, a retroposon repeat element that encodes an RT. This element is unique in lacking LTR. Also, when present, Tad is found as a low-copy-number or unique element, and most field isolates of *N. crassa* lack the element altogether. Tad was found in an isolate from the Ivory Coast, which is why it is not common in field isolates. In other words, Tad appears to be a recent colonizer of Ivory Coast *Neurospora*. However, most *Neurospora* genomes have numerous copies of highly mutated (RIP) and dysfunctional Tad sequences. The presence of these degenerate sequences indicates that prior (or possibly ongoing) Tad colonization has occurred but has been rapidly and effectively inactivated in the *Neurospora* genome. This presence of degenerate Tad further suggests that there exists a natural but unknown source of Tad, which has continued to colonize the *Neurospora* genome. In stark contrast to *Neurospora*, *Cryptococcus neoformans* has a mating type locus that is composed of 60 kbp of DNA, which has both interspersed unique and repetitive elements and encodes the pheromone precursor peptide. It seems clear that the presence of or prevention of repeat elements varies considerably among specific fungal lineages. Such variation suggests that at the origin or radiation of fungal lineages, a peculiar and lineage-specific pattern of genome colonization by these elements occurred. Along these lines, it is interesting that most filamentous Fungi have low copy numbers of the L1 long interspersed repetitive element (LINE)-like element (which lacks terminal repeats with identical 3' ends). This element is maintained in plants, invertebrates, and mammals, all of which appear to have descended from a fungal predecessor.

The Fungal Viruses

Penicillium stoloniferum was used to identify the first source of interferon at Eli Lilly Laboratories in the early 1950s. An agent named statolon was identified from cultures of *P. stoloniferum* as possessing interferon-inducing activity after it was observed that *P. stoloniferum* inhibited vesicular stomatitis virus plaque production in mouse L cells. Subsequent investigation established that this agent resulted from high-level production of dsRNA by a persistent and inapparent dsRNA mycovirus that had infected the *P. stoloniferum* isolates. Strains of *Penicillium funiculosum* were also observed to harbor mycoviruses. Subsequently, the list of dsRNA viruses found in mold and other higher Fungi grew, and it now seems that essentially all such species are inapparently infected with related viruses. That is, ubiquitous and persistent virus infection is the rule in higher Fungi. Many of these mycoviruses have been well studied and are discussed below. However, the situation may be different for lower Fungi relative to higher Fungi. In keeping with the aim of examining virus evolution from the perspective of host evolution, I first focus on the lower Fungi and the various dsDNA viral agents that have been described for these species.

Linear dsDNA viruses of lower Fungi

For marine Fungi, such as *Spartina alterniflora*, there are no reports of viruses in the current scientific literature. If such viruses are present, it seems most likely that they are inapparent or require induction. However, without direct evidence, this remains speculation. Other aquatic Fungi are clearly known to harbor viruses. For example, *Rhizidiomyces* (a lower aquatic fungus in the *Hypochytriales*) zoospores were demonstrated in heat shock experiments to carry viruses. These fungal zoospores are parasitic to members of the order *Oomycetes* (e.g., *Achlya* and *Saprolegnia*). When standard lab strains of *Rhizidiomyces* zoospores were heat shocked (during experiments conducted in the early 1980s), it was observed that the resulting zoospores would no longer infect their hosts. Zoospores failed to mature, and instead their nuclei produced large numbers of VLPs. In addition, VLPs were commonly observed following heat shock in many other lower Fungi. This VLP assembly occurs in the nucleus and resembles herpesvirus assembly. However, the VLP capsids are smaller than herpesvirus capsids, being similar in size to adenovirus capsids. Like adenoviruses, the icosahedral VLPs in the zoospores contain linear dsDNA. Other observed VLPs were cytoplasmic. Interestingly, most natural isolates of *Rhizidiomyces* have similar VLPs. In all, eight distinct *Rhizidiomyces* isolates were evaluated, and all were positive for VLPs. When zoospore production was induced by salt treatment or heat shock, all isolates made large numbers of intranuclear VLPs, resulting in the destruction of the zoospores. These isolates could not be cured of VLP production by subcloning. The capacity for virus induction was very stable for long-term passage of cells (up to 10 years). Other lower Fungi (*Rhizidiomyces, Thraustochytrium, Schizochytrium,* and *Paramoebidium* species) may all contain similar inapparent DNA viruses. Clearly the size of these icosahedrons and the DNA content resemble those of adenoviruses, not the much larger algal phycodnaviruses (625 versus 6,340S). No uninfected isolates of lower Fungi have been seen. This viral ecology is seldom mentioned in textbooks about viruses or chapters on fungal viruses, despite the fact that it seems distinct from the viral ecologies of both algae and higher Fungi in that most higher Fungi are persistently infected with dsRNA viruses. The viral ecology of lower Fungi resembles the lysogenic state in bacteria or, more specifically, the relationship between phycodnaviruses (or phaeoviruses) and their filamentous brown algal hosts, including the link of latent virus induction to sexual gamete production. Although ubiquitous, this lower-Fungus-specific family of viruses has not been well characterized. It might be predicted that such persistent agents would need to have genes that compel the host to maintain the infection (such as addiction strategies and modules). In addition, their ubiquitous nature would also imply that competition between host-specific viral agents for host colonization should be fierce. However, essentially nothing is known about the genes of these viruses or how they attain stable persistence.

DNA viruses in the mitochondria of higher Fungi

The virus situation in higher Fungi is clearly distinct from that described above. Higher Fungi are known to frequently support linear DNA plasmids (defective viruses), and all these agents are organelle associated, as they infect mitochondria, not nuclei. Infection of plastids by genetic parasites is essentially unknown for most higher eukaryotes. For example, no animal virus is known for mitochondria. Related linear DNA plasmids, however, are found in many filamentous Fungi, yeasts, and some higher plants. All have terminal inverted repeats with $5'$ covalently attached terminal proteins, although a few have covalently closed snapback ends. These plasmids (e.g., pMC3-2) generally contain two ORFs, which usually correspond to a DNA polymerase and an RNA polymerase, that are clearly similar to those found in adenoviruses but are more distantly related to the linear phages PDR1 and ϕ29; adenoviruses and these two phages also have $5'$ terminal protein DNA priming. A distant phylogenetic relationship to linear plasmids found in soil bacteria (*Streptomyces*) can also be seen. This type of protein-primed polymerase activity is clearly of viral origin and is not found in any host cell. Furthermore, the plasmid DNA replicates via the same process as that of adenoviruses. It is therefore very curious that these mainly mitochondrial plasmids are clearly similar to the mainly nuclear adenovirus-like DNA viruses seen in lower aquatic Fungi. It seems more than coincidental that capsid-encoding nuclear DNA viruses are ubiquitous in lower Fungi but that defective, noncapsid versions of a similar virus family are ubiquitous in the mitochondria, but not the nuclei, of higher Fungi.

Killer phenotypes, addiction, and DNA parasites

All of these linear plasmids can be grouped according to the conservation of their DNA polymerase sequence, such as the prototype pMC3-2. Two major polymerase groups are apparent, and they correspond to those found in yeast species and filamentous Fungi. In *Saccharomyces kluyveri*, the pSKL plasmid has a terminal protein that resembles those in adenoviruses and φ29 of *B. subtilis*. Some yeasts harbor two plasmids. *Kluyveromyces lactis* and other yeast species have multipartite linear DNA plasmids (pGKL1 and pGKL2) that have genes in addition to those for the DNA and RNA polymerases. The additional genes confer a killer phenotype via toxin and immunity genes. Within the pGKL1-pGKL2 plasmid set, pGKL2 codes for the RNA polymerases, while pGKL1 contains the additional genes that are essential for maintenance and immunity. Thus, these two elements are complementary. In this case, plasmid persistence is attained via an addiction module that pGKL1 contains, which includes a three-subunit killer toxin gene and an immunity gene against the toxin. pGKL1 also codes for the DNA polymerase. pSKL is related to pGKL1 in organization but not by hybridization, and it does not confer a killer phenotype. The RNA polymerase encoded by this plasmid is most similar to the 140-kDa subunit of yeast polymerase II. Circular forms of related DNA plasmids (LaBelle and Harbin-1) are also known, but these are uncommon and confined to a few genera of yeasts. Such killer plasmids or viruses have not been reported for any filamentous Fungi. Furthermore, in yeast species, all of the linear plasmids appear to be cytoplasmic, whereas in filamentous Fungi, the linear plasmids are all mitochondrial. Thus, these two groups of plasmids appear to make up distinct types, not just in sequence but also in their relationship to their hosts.

Plasmid phenotypes and relationships to other elements

These linear plasmids have additional similarities to other genetic agents. For example, the plasmid terminal nucleotides that correspond to the origin of replication and the attachment point for the terminal protein clearly resemble the copia transposon sequences, which are so common in insect genomes. In actinomycete, related linear plasmids are also known. In these fungus-like soil bacteria, the plasmids generally code for advantageous abilities, such as nutritional versatility or the degradation of toxins (such as phenols); thus, they clearly have phenotypes. Similar but cryptic elements (with few genes) are called hairpin elements and can be found in plant-pathogenic (but not nonpathogenic) *Rhizoctonia solani*. These plasmids identify a complementing but persistent system of multiple cryptic elements. However, this observation makes another important point. The stable maintenance of a two-genome system (a simple version of a dikaryon) can be attained by a combination of an element with an addiction module and a second suppressing defective that allows persistence.

DNA parasites that affect longevity

Although the majority of the linear plasmids of filamentous Fungi are not associated with killer phenotypes, many do have other phenotypes in their hosts, and some can affect host longevity. In *Podospora anserina,* the pAL2-1 plasmid can integrate into mitochondrial DNA (mtDNA) and, as a consequence, can stabilize mtDNA from age-dependent degradation and induce longevity in the host, allowing colonies to grow to 10-fold-larger radii. Given the evolutionary importance of longevity and hyphal growth to the survival of higher Fungi in large fungal mats, this is a very interesting characteristic for a genetic parasite to bestow on its host. Various forms of these pAL plasmids are seen in natural isolates; of the 78 that have been characterized, all appear to have terminal proteins bound to their 5′ ends. Although these plasmids all have a common central DNA polymerase region, their terminal inverted sequences vary in a host-specific way. It appears that mtDNA with integrated pAL plasmids becomes stable to degradation. But the integrated copies do not suppress the wild-type mtDNA genomes, in sharp contrast to the situation seen in *Neurospora* (described below). Fourteen of the seventy-eight *P. anserina* isolates (taken from natural populations) were shown to carry a pAL2-related longevity-inducing linear plasmid. Curiously, most of the plasmids in these natural isolates were not integrated into mtDNA, and only one strain showed the longevity phenotype. The plasmids that were not inserted into mtDNA were maintained as autonomous linear mitochondrial replicators. The natural isolates also commonly carried multiple mixed plasmids; one isolate was determined to have 12 different plasmids. In addition, giant (50-kbp) tandem plasmids were also reported. However, there was no evidence of plasmid-plasmid interactions. Loss of some plasmids was often seen during sexual transfer, consistent with the previously described predictions to explain the male-like behavior of nuclei to escape genetic parasites. These linear plasmids are inherited maternally, as they are mitochondrial plasmids, and they are readily transmitted among field isolates. The transmission of these linear plasmids is restricted by the vegetative incompatibility of the host, which was shown to decrease (but not eliminate) transmission rates 10-fold.

Fungal mitochondria

The association of various genetic parasites with the mitochondria of filamentous Fungi is a rather unique feature of Fungi. It is worth considering what is known about fungal mitochondria that might illuminate this unique situation. The well-studied ascomycete mtDNAs range from 19 to 115 kbp, which is much smaller than the mtDNAs of higher eukaryotes. Fungal mtDNA contains two rDNA subunits, tRNAs, and introns, which are sometimes infectious (e.g., *N. crassa* introns). Unlike some fungal nuclear genomes, fungal mtDNA can tolerate the presence of repeated DNA sequences. For example, *N. crassa* mtDNA contains a series of GC-rich palindromes. Also, unlike other mtDNAs, fungal mtDNA recombines readily, as do the parasitic DNA plasmids of these mitochondria. Clearly, mtDNA is not subjected to the RIP system or another repeat-limiting system of nuclear DNA. In *P. anserina*, rogue mtDNA replicons can occur by intron invasion into the *cox* gene, leading to senescence. Yet integration of the pAL2 DNA plasmid can lead to stabilization of mtDNA and extension of the host's life span. In addition to these linear DNA plasmids, fungal mitochondria also seem to be highly prone to colonization by other types of genetic parasites. For example, dsRNA-based VLPs are often found in mitochondria. In *Neurospora*, Mauriceville and Varkud retroposon-like elements invade and disrupt mtDNA. As to why these mitochondria are so prone to genetic parasites, there is currently no clear answer. It seems likely that the capacity of fungal mitochondria to undergo recombination might provide a good molecular habitat for the linear and circular DNA plasmids and the retroposons. However, this feature would not be expected to support colonization by dsRNA viruses. Perhaps the answer lies outside of the mitochondria. It might be that the ubiquitous presence in Fungi of homology-dependent gene silencing systems, which operate both transcriptionally and posttranscriptionally (such as quelling and RNA interference), would provide a molecular habitat that would prohibit many nuclear and cytosolic genetic parasites, thereby increasing the likelihood of mitchondrial genetic parasites.

Neurospora and senescence

N. crassa is also commonly colonized by mitochondrial genetic parasites. Linear dsDNAs, such as the Kalilo agent, are the best studied of these elements in *Neurospora*. These mitochondrial agents strongly affect host longevity. However, in the case of *Neurospora*, longevity is always decreased (i.e., senescence is induced) and not extended, as in *Podospora* with the pAL2 plasmid. *Neurospora* mitochondrial genetic parasites include not only these linear DNA plasmids but also retroposon elements, such as the Mauriceville and Varkud retroplasmids and elements related to the pFOX plasmid of *Fusarium oxysporum*. These mitochondrial retroposon elements use an unusual and possibly primitive form of RT activity which is not primer dependent and may represent a transition between RNA polymerases and DNA polymerases. In this case, the primer 3' end is similar to a tRNA. Fungal mitochondria also have group II introns, which are self-splicing, code for RT, and are invasive. All of these agents (linear plasmids, retroposon elements, and invasive introns) have a tendency to induce erratic growth and senescence in infected *Neurospora*. Similar to linear DNA plasmids described above, the linear Kalilo DNA plasmids are 9-kbp linear DNAs that code for adenovirus-like DNA and RNA polymerases. The terminal inverted repeat ends are the most diverse parts of these sequences and show a phylogeny that is congruent with the host nuclear DNA. Transmission of these elements among species is common, but mating type vegetative incompatibility reduces transmission rates 10-fold, suggesting that vegetative incompatibility may function to limit such parasite transmission. The Kalilo plasmids and related plasmids are widely distributed across the *Neurospora* genus.

Yeast retroposons

Yeast species also have other types of retroposons, belonging to the Ty3/gypsy and Ty1/copia families. Unlike *Neurospora* Tad, a LINE-like element, these yeast elements are more typical retroposons and have associated LTRs. These retroposons contain *gag* and RT, protease, RNase H, and integrase-like genes. In keeping with the general absence of fungal viruses that have natural extracellular routes of transmission (e.g., DNA plasmids and mycoviruses, described below), the yeast retroposons also do not contain the *env* gene, which encodes the main external structural proteins of the retrovirion. These yeast retroelements appear to represent the first instance of genome colonization by defective ERVs, although the numbers and diversity of these elements are very much smaller than those which occur in plants and animals. For example, the LINE-like elements (Tad, CRE1, and SLACS of trypanosomes) are present in relatively low (unique) genomic copy numbers in some Fungi.

Mycoviruses

As previously mentioned, mycoviruses of Fungi are generally latent and highly prevalent in numerous fungal species. The majority of these viruses have dsRNA genomes and, like most fungal viruses, lack an extracellular transmission phase.

Totiviridae, as mentioned for Protozoa, have a single dsRNA genome. The L1 virus of yeast, which establishes a permanent persistent infection, is the prototype totivirus. Totivirus agents can be found in nine genera of Fungi and four genera of Protozoa (e.g., *Giardia* virus and LRV-1). Totiviruses can also encode killer phenotypes. *Partitiviridae* have multipartite dsRNA genomes and can be found in an additional nine genera of Fungi. Only *Candida*, which lacks an apparent sexual cycle, and *Podospora* have no known dsRNA viruses. In all cases, only persistent infections are observed.

Killer viruses

In various yeast hosts, the totiviruses (like the DNA plasmids described previously) can code for killer phenotypes via a diverse set of toxins, which usually kill by membrane disruption (although some, e.g., K28, can also stop DNA synthesis). The LA killer dsRNA viruses are the best-studied yeast killer agents. These killer agents code for immunity proteins in addition to the matched toxins; hence, they carry addiction module gene sets. A major coat protein is also generally encoded. However, as described below, many of these agents are cryptic or satellites to other agents and have reduced coding capacity. Killer dsRNA viruses were first discovered in *S. cerevisiae*: in 1963, E. A. Bevan and M. Makower described the killer characteristics and the corresponding VLPs that were produced (see Schmitt and Eisfeld [1999] in Recommended Reading). Killer viruses were later shown to be ubiquitous in natural and lab isolates of *S. cerevisiae*. Curiously, some field studies show a relatively low prevalence (1 to 5%) of these agents in natural yeast species.

Killers and beer

Yeasts have been extensively studied due to the economic importance of their fermentation properties as used in the brewing and baking industries. The very large industrially grown cultures of yeasts are susceptible to infection by the killer agents present in wild yeast isolates. These infections pose a serious problem and can cause major disruptions to production. It is interesting that the best defense the brewing industry has developed against exogenous yeast genetic parasites has been to colonize the industrial yeast cultures with protective versions of killer viruses. This is operationally very similar to the dairy industry's use of *Lactobacillus*-mediated fermentation of milk products, which are also threatened by wild phage and are best protected by the immunity of various persistent lactophages.

The LA killer virus contains a 4.7-kbp L1 dsRNA, which can undergo encapsidation into icosahedral particles that are present at several thousand particles per cell. This RNA encodes a major coat protein and an RNA-dependent RNA polymerase. Like for dsRNA phages and reoviruses, transcription of the LA dsRNA template occurs within viral core particles. The LA RNA polymerase shows clear similarity to dsRNA phages and reovirus RNA polymerases, but not to the polymerases of ssRNA viruses. Core particle transcription results in the production of uncapped viral mRNAs, which might be a target of the S1 host defense system. However, it seems likely that virion-associated transcription could shield the viral transcription system from the host silencing systems that target transcription. In some cases (K1 killer strains), a second satellite dsRNA virus exists (e.g., M1 satellite to L1), which codes for preprotoxins and toxin-specific immunity proteins. The toxins encoded by the M1 satellite (1.9-kbp dsRNA) are dimeric, exocellular toxins. These toxins are alone (in the absence of the corresponding immunity protein) kill sensitive yeast strains. Additional satellites (M2, M3, M28, etc.) that encode additional killer toxins are also known. Thus, there is clearly much natural diversity in these killer and satellite systems. In the laboratory, yeast strains are usually used as haploid killer strains and can frequently have multiple killers in different clones. The dsRNA viruses' genes can also be involved in linear DNA plasmid incompatibility, indicating some clear interaction between dsRNA agents and dsDNA agents. The details of this interaction have not been explored, however. No recombination is seen in these dsRNA viruses. In nature, they appear to be maintained by vertical transmission, but it is likely that hyphal fusion is a major process of transmission for hyphal viruses. Defective interfering versions of these satellite viruses naturally occur, and these defectives generally lack the preprotoxin coding sequence. The effects of defectives on host colonization by the infectious versions of the viruses are not always clear, but in the case of M satellites, their absence results in massive increase in LA dsRNA and particles. However, the infected cells remain healthy, and thus these defectives clearly can interfere with the similar viruses. In this regard, the defectives appear to allow a more stable form of low-level persistent host infection. Curiously, replication of these defective RNAs has a dependence on host genes (requiring the nonessential SK1 superkiller genes) that is different from that of replication of the helper viruses, indicating that defectives have a distinct albeit poorly understood relationship to their hosts. Ironically, some of these host SK1 genes are members of a genetic system that is thought to search out and destroy uncapped mRNAs.

The killer toxin, when produced by these satellite viruses, undergoes processing and secretion. The toxin acts by binding to the glucan cell wall receptor and kills sensitive strains via plasma membrane interaction and altered permeability. Immunity is not well understood but appears to be mediated by masking of this receptor. Similar toxin-immunity systems are seen in *Yarrowia, Aspergillus,* and *Penicillium* filamentous species. Thus, unlike the linear dsDNA viruses, these dsRNA viruses show killer phenotypes in both yeast Fungi and filamentous Fungi. Virus transmission is likely to occur via hyphal fusion in filamentous species, but for yeast species transmission is not well studied in natural settings. During mating, however, the yeast cell wall is dissolved, and it is likely that spheroplasts can then be infected. Consistent with this hypothesis, laboratory infection has been achieved during the mating of haploid strains in the presence of killer virus, resulting in acquisition of the killer phenotype by a fraction of the offspring. A curious distinction among these dsRNA viruses is that they package positive-strand ssRNA, which suggests that these fungal viruses are a distinct order of virus.

Although dsRNA viruses are common in many Fungi, not all of these agents contain capsid genes or make particles (such as chestnut blight virus, described below). These agents appear to be more similar to defective versions than to the killer viruses described above. And it is not clear if they can exploit other less defective dsRNA viruses for packaging or if they provide interfering capacity to the host cell. However, since viral packaging and replication sequences are generally one and the same, the use of other helper viruses for packaging would require these defective satellites to maintain sequence similarity to potential helpers. This is not always seen. In all cases, however, infection appears to be persistent and generally inapparent.

ssRNA circles

ssRNA viruses of Fungi are also known, although they are not nearly as common as dsRNA viruses. Of these viruses, the 20S replicon has been best studied. This replicon is very different from all other RNA viruses in that it is a circular RNA. Thus, it requires internal initiation of translation, since it would not produce capped mRNA. Interestingly, this might also protect it from an SK1 defense system. These 20S RNA virus particles are induced under conditions that also induce sporulation (e.g., high temperature and acetate medium) and can under such conditions be amplified up to 10,000-fold. In this characteristic, they are similar the nuclear DNA viruses of the lower aquatic Fungi that are induced during sporulation, as described above.

Fungal dsRNA and partitiviruses

In filamentous fungi, dsRNA virus infections are also prevalent; they were first described in 1948 during commercial production of mushrooms. These viruses show clear similarity to viruses found in other protists, such as GLV and TVV (a virus associated with host phenotypic variation). In filamentous Fungi like *Penicillium* and *Aspergillus,* one can also find partitiviruses, which have a bipartite genome with two segments that are very related to plant viruses. Multipartite viruses of filamentous Fungi are also known. In the case of *Ophiostoma novo-ulmi,* virus infection involves greater than seven dsRNA species, only one of which codes for an RNA-dependent RNA polymerase, which resembles the mitochondrial RNA polymerase. It is not known how the transmission of such multipartite genomes is coordinated. Unlike the situation with influenza and other multipartite viruses, the absence of an extracellular phase would not allow copackaging of these genetic components. In *Cryphonectria parasitica,* a dsRNA agent transmitted via hyphae and meiotic spores is found in the mitochondria. This virus codes for an RNA polymerase that is similar to those found in RNA phages.

Multipartite CHV-1 and fungal hypovirulence

Some of the best-studied and most interesting of the fungal dsRNA viruses are the viruses of the chestnut blight fungus. Like most fungal viruses, these viruses lack an extracellular phase. In addition, the dsRNA does not contain a capsid gene, but it still becomes membrane enclosed. These bound structures are sites of RNA polymerase activity. These viruses came to attention of biologists because of their ability to induce hypovirulence in some infected fungi. The chestnut blight fungus, *C. parasitica,* is native to the oriental species of the chestnut tree and was first identified as a major problem in North America in 1904 after accidental introduction. The fungus also spread to Europe, where it had a similar devastating effect on the European species of the chestnut tree. However, in the 1950s a markedly reduced virulence of the fungus was noted in some trees, and the infecting fungal strain was called "hypovirulent." Hypovirulence was associated with decreased asexual spore formation and other phenotypes. It was shown to be due to a cytoplasmically inherited factor later identified as Cryphonectria hypovirus 1/EP713 (CHV-1/EP713). The tripartite dsRNA genome of this virus (composed of M, L, and S segments) has unusual termini, 3′ poly(A) (for the L segment) and 5′ poly(U) (for the M segment), and codes for two gene products, one an RNA-dependent RNA polymerase and the other a helicase, both of which undergo proteolytic self-cleaving. The virus shows relationships to plant potyviruses. Defectives of these viruses are very common. Like the other fungal viruses, CHV-1/EP713 is transmitted via hyphal fusion. Infected fungi show various phenotypic alterations, such as changes in growth rate, female infertility, reduced asexual sporulation, and changes in gene expression, including that of genes involved in signal transduction. Field studies show considerable variation in these dsRNAs, including substantial diversity in the RNA polymerase sequences as well as a diversity in the effects on host gene expression. In Europe, about half of the *C. parasitica* field isolates are infected, and virus transmission between types and isolates occurs in most pairings, indicating that vegetative incompatibility did not prevent virus transmission. Because of the clearly beneficial effect on chestnut trees, there has been considerable effort to use these viruses for biological control of the fungal disease. These efforts have been successful in Europe. However, in spite of considerable effort and clear European success, much less success has been attained using these viruses to control fungal disease in American chestnut trees, except for in some areas in Michigan.

The most accepted theory to explain this situation is that the viruses already colonizing the American *C. parasitica* population are more prevalent and diverse than those found in the European fungi. These diversely infected American populations severely limit the spread of the hypovirulent European viral strains, such as CHV-1/Euro7. Some direct measurements of existing viral diversity and its relationship to virus spread support this theory. This suggests that virus-virus competition (spread in the face of competitor persistence) is a major issue with respect to the fitness of the hypovirulent virus. It would also appear to identify a major element of viral fitness as it applies to a realistic ecological setting. That is, overcoming the seemingly limited and "selfish" genetic effects of defective hypoviruses that have colonized hosts can have a crucial role in the ecological outcome. It

Table 5.1 Aquatic microorganisms and their viruses[a]

Microorganisms	Feature(s)	Viral agents	Virus relationship
Protista			
Lower			
Diplomonads (*Giardia*)	Binucleate; no mitochondria, Golgi complexes, or ER	dsRNA cytoplasmic VLPs (GLV—totivirus) TVV	Asymptomatic, phenotypic switching
Parabasalia (*Trichomonas*)			
Higher			
Entamoeba histolytica	Mitochondria, ER (odd linear DNA forms, protein-primed 5′ DNA)	VLPs of several sizes very common (e.g., sporozoites)	Asymptomatic, lots of virions Two ORFs; polymerase, capsid
Leishmania	Most are parasitic	LRV-1 (dsRNA)	Fungus-like (ScV-L-A)
Ciliated protozoa: *Tetrahymena,* *Paramecium, Euplotes*	Dual nuclei, dimorphic, silent micronucleus, active macronucleus	dsRNA VLPs (latent) No DNA virus but amplify phage DNA	Some species (hydras) harbor symbiotic algae with *Chlorella* virus
Mineralized algae	Diatoms: glassy shells; brown photopigment	No virus reports No VLPs	Many evolutionarily stable forms
Dinoflagellates (toxic tides—domoic acid)	Binucleate: not dimorphic, no histones, no nucleosomes, photosynthetic, motile, big genomes	Nuclear VLPs (DNA) Pox-like; cytovirion RNA virus unknown	Latent temp-induced lytic (coral bleaching)
Fungi			
Lower (*Rhizidiomyces*)	Marine fungi, motile zoospores	VLPs from heat shock	Noncurable
Higher	Small nuclei (genome), closed	Ad-like, nuclear	Zoospore associated
Yeast forms	mitosis, hyphal communicating	dsRNA (totivirus),	Cytoplasmic (LA killers)
Hyphal forms	cytoplasm, all infected	ssRNA, dsRNA (Statolon)	Mitochondrial
Podospora, Neurospora		Linear DNA (Ad-like)	Mitochondrial (aging)
		pAL2-1, pGLK1, Kalilo	RIP-restricted LTRs

[a]ER, endoplasmic reticulum; Ad, adenovirus.

seems likely that mycovirus gene functions have evolved that provide this essential ability to compete with other persistent viruses, thereby allowing successful persistence or transmission. Such functions would appear to be nonessential (accessory) when evaluated in hosts free of viral colonization. Yet in the natural ecosystem, viral colonization of Fungi is essentially inevitable, making the ability of one virus to counter prior colonization by another virus an essential phenotype for the natural habitat. In this case, like with the T4 *rII* gene, it is not simply the host that determines the fitness of a virus, but other competing viruses as well.

The various aquatic microorganisms, their general characteristics, the types of viral agents they support, and the relationship of virus and host presented in this chapter are summarized in Table 5.1.

In conclusion, it is worth considering why all of these distinct characteristics of fungal virus-host interactions (i.e., linear plasmids, killer phenotypes, mitochondrial infections, distorted senescence, and ubiquitous dsRNA colonization) are mostly peculiar to the fungal orders of organisms and yet are generally absent from the plants and animals that are descendants of Fungi, as well as the algae or prokaryote predecessors. Perhaps these genetic parasites are an integral part of the successful genetic milieu of the Fungi themselves.

Recommended Reading

Protozoan Viruses

Widmer, G., and S. Dooley. 1995. Phylogenetic analysis of *Leishmania* RNA virus and *Leishmania* suggests ancient virus-parasite association. *Nucleic Acids Res.* **23:**2300–2304.

Viruses of Fungi

Dawe, A. L., and D. L. Nuss. 2001. Hypoviruses and chestnut blight: exploiting viruses to understand and modulate fungal pathogenesis. *Annu. Rev. Genet.* **35:** 1–29.

Dawe, V. H., and C. W. Kuhn. 1983. Virus-like particles in the aquatic fungus, Rhizidiomyces. *Virology* **130:**10–20.

Ghabrial, S. A. 1998. Origin, adaptation and evolutionary pathways of fungal viruses. *Virus Genes* **16:**119–131.

Ginsberg, H. S. 1984. *The Adenoviruses.* Plenum Press, New York, N.Y.

Interferon and Fungi

Ellis, L. F., and W. J. Kleinschmidt. 1967. Virus-like particles of a fraction of statolon, a mould product. *Nature* **215:**649–650.

Fungal and *Neurospora* Biology

Davis, R. H. 2000. *Neurospora: Contributions of a Model Organism.* Oxford University Press, New York, N.Y.

Gow, N. A. R. 1994. *The Growing Fungus.* Chapman & Hall, New York, N.Y.

Fungal Incompatibility and Fungal Longevity

Glass, N. L., D. J. Jacobson, and P. K. Shiu. 2000. The genetics of hyphal fusion and vegetative incompatibility in filamentous ascomycete fungi. *Annu. Rev. Genet.* **34:**165–186.

Hirsch, P., F. E. W. Eckhardt, and R. J. Palmer, Jr. 1995. Fungi active in weathering of rock and stone monuments. *Can. J. Bot.* **73:**S1384–S1390.

Prince, R. C. 1992. The Methuselah factor: age in cryptoendolithic communities. *Trends Ecol. Evol.* **7:**211.

Smith, M. L., J. N. Bruhn, and J. B. Anderson. 1992. The fungus armillaria bulbosa is among the largest and oldest living organisms. *Nature* **356:**428–431.

Smith, M. L., L. C. Duchesne, J. N. Bruhn, and J. B. Anderson. 1990. Mitochondrial genetics in a natural population of the plant pathogen Armillaria. *Genetics* **126:**575–582.

DNA and Adenovirus-Like Linear Plasmids

Meinhardt, F., R. Schaffrath, and M. Larsen. 1997. Microbial linear plasmids. *Appl. Microbiol. Biotechnol.* **47:**329–336.

Rohe, M., K. Schrage, and F. Meinhardt. 1991. The linear plasmid pMC3-2 from *Morchella conica* is structurally related to adenoviruses. *Curr. Genet.* **20:**527–533.

Linear Plasmids and Longevity

Hermanns, J., and H. D. Osiewacz. 1996. Induction of longevity by cytoplasmic transfer of a linear plasmid in *Podospora anserina. Curr. Genet.* **29:**250–256.

van der Gaag, M., A. J. Debets, H. D. Osiewacz, and R. F. Hoekstra. 1998. The dynamics of pAL2-1 homologous linear plasmids in *Podospora anserina. Mol. Gen. Genet.* **258:**521–529.

Linear Plasmids and Relationship of Killer to Virus

Hishinuma, F., and K. Hirai. 1991. Genome organization of the linear plasmid, pSKL, isolated from *Saccharomyces kluyveri. Mol. Gen. Genet.* **226:**97–106.

Klassen, R., L. Tontsidou, M. Larsen, and F. Meinhardt. 2001. Genome organization of the linear cytoplasmic element pPE1B from *Pichia etchellsii. Yeast* **18:**953–961.

Schmit, M. J., and K. Eisfeld. 1999. Killer viruses in S. cerevisiae and their general importance in understanding eucaryotic cell biology. *Recent Res. Dev. Virol.* **1:**525–545.

Wickner, R. B. 1996. Double-stranded RNA viruses of *Saccharomyces cerevisiae. Microbiol. Rev.* **60:**250–265.

Fungal Gene Silencing

Cogoni, C. 2001. Homology-dependent gene silencing mechanisms in fungi. *Annu. Rev. Microbiol.* **55:**381–406.

Pickford, A. S., and C. Cogoni. 2003. RNA-mediated gene silencing. *Cell. Mol. Life Sci.* **60:**871–882.

A Virus Odyssey from Worms to Fish:
Viruses of Early Animals and
Aquatic Animals

Introduction

In the preceding chapter, I discussed the viruses of Fungi and their particular relationships to both lower and higher Fungi. This chapter explores the virology and host evolution of animals (and later higher plants) which have evolved from Fungi or depended on Fungi for their evolution. As was previously presented, with few exceptions Fungi show the general ability to support both DNA and double-stranded RNA (dsRNA) viruses. However, these fungal viruses, although ubiquitous, are generally lacking in extracellular forms. Since Fungi are accepted as being representative of the predecessors of both higher plants and animals, they represent a most significant point of bifurcation in the origin of these two higher kingdoms of organisms. It can thus be proposed with much confidence that the ancestors of both higher plants and animals were exposed to representatives of the various fungal viruses. The similarities between plants and fungi, especially the filamentous fungi, are clear. Both are characterized by nonmotile branching morphologies that facilitate acquisition of nutrients from the local environment. The common symbiosis between fungi and photosynthetic algae makes it easy to envision how higher plants might have evolved from these simpler fungal ancestors. With respect to dsRNA viruses of fungi, we can also observe clear relationships between the viruses that infect Fungi and those that infect higher plants, consistent with a linkage of viruses to host evolution. Also, many of the molecular processes that the host uses to control virus replication (e.g., nonadaptive gene silencing and RNA interference) can be found in both fungal and plant lineages. However, there are some viral discontinuities between these host lineages. For example (and most curious), the very prevalent large dsDNA algal viruses, the phycodnaviruses, are not found to infect any fungi or higher plant species. Nor are the adenovirus-like parasites of fungal mitochondria found in plants, yet clear relatives of these viruses are common in Fungi and animals. These and other issues concerning the general relationship of viruses to higher plants are discussed in the following chapter. In this chapter, I focus on the emergence of early animals and their viruses.

Dictyostelium and a Paucity of Viruses

From a phylogenetic perspective, *Dictyostelium* spp. are most closely related to the Fungi and Protozoa. However, *Dictyostelium* also displays many of the characteristics of early animals, and so it appears to represent an evolutionary step towards the origin of animals. *Dictyostelium* is considered a social amoeba, a member of the phylum Amoebozoa, which is basal to both the higher Fungi and animals but is not an apparent direct ancestor of animals. One distinction between Amoebozoa and Fungi is that, unlike the filamentous Fungi, which are predominantly haploid and transient diploids, *Dictyostelium* species are stable diploids and are only mitotic haploids during spore formation. In this feature, *Dictyostelium* more closely resembles the higher animals. *Dictyostelium* has been an intensively studied model system due to the social interactions observed for assemblies of cells. Thus, it is most surprising to realize that there are no reports of viruses in the literature for any *Dictyostelium* species. Given that there has been substantial electron microscopic study of *Dictyostelium*, focusing on the formation of prespore vesicles and spore coat protein assembly, it seems likely that the assembly of virus-like particles would have been observed by now. It thus seems that viruses are rare or entirely absent from the studied *Dictyostelium* species. This suggests that *Dictyostelium* may have developed highly efficient systems that prevent colonization by viral agents.

Origins of antiviral systems of *Dictyostelium*

Historically, the presence of antiviral systems in *Dictyostelium* was not well studied. It was known that *Dictyostelium* species have SK18 (superkiller)-related genes, which are used to control dsRNA virus infections in yeast species. Thus, some antiviral genes appear to be present in *Dictyostelium* genomes. However, more recently it has become clear that, like *Caenorhabditis elegans* (discussed below), *Dictyostelium* has the genes needed for an RNA interference (RNAi) response. The RNAi system responds to dsRNA by interfering with and degrading complementary mRNAs. This RNAi response is both amplifiable, through continued synthesis of small RNAs, and transmissible, in that it can spread to nearby cells and transmit the RNA-specific interference. Three genes related to the *rrf* genes of *C. elegans* have been reported to occur in *Dictyostelium*. These genes code for proteins highly homologous to the RNA-dependent RNA polymerase (RdRP) of *C. elegans*, which is involved in amplifying the dsRNA signal. It therefore seems very likely that *Dictyostelium* also amplifies RNAi via cellular RdRPs. The likely fungal ancestors of *Dictyostelium* do not appear to have had this antiviral system. This RNAi system likely accounts for the apparent absence of RNA viruses from *Dictyostelium* species.

As a digression, it would be most interesting to consider the likely origin of the cellular RNAi system. The general virus-like characteristics of RNAi are striking. Besides the amplifiable and transmissible nature of RNAi, the participation of an RdRP is especially virus-like, since this enzyme is the basal replicative enzyme of all RNA viral genomes but is not a basic component of host cells. This functional similarity seems to strongly imply that the RNAi-associated RdRPs may have evolved from a viral ancestor. Yet no apparent sequence homology can now be seen between these cellular RdRPs and known viral RdRPs. However, the complete absence of RdRPs from prokaryotic genomes supports a viral origin as the most likely explanation. In this regard, the ubiquitous occurrence of persistent infections by diverse dsRNA viruses in filamentous Fungi (described in the preceding chapter) and the fact that these fungal cells are the evolutionary progenitors of *Dictyostelium* make it very plausible that cellular RdRPs resulted from the colonization of host cells by some ancient, persistent RNA viral agent that was very successful at excluding competing dsRNA viruses and other viral parasites via the RNAi system. In this scenario, it is worth recalling that stable virus colonization of a host generally requires the acquisition of some viral phenotype that allows or compels persistence and withstands the onslaught of genetic competitors. The RNAi system is known to silence genomic transcripts and degrade those RNAs that show some sequence similarity to the silenced gene. For example, RNAi is known to silence retroposon elements in the *C. elegans* genome, but it does not prevent their reactivation by other signals. Thus, RNAi seems to have the needed characteristics of a system for persistence and preclusion of competitors. However, in addition to the RNA virus that may have contributed the RdRP, another retrovirus must have also been involved by providing the reverse transcriptase (RT) needed to copy this RdRP gene into DNA and integrate it into the host *Dictyostelium* genome.

The *Dictyostelium* genome and genetic parasite colonization

In terms of genetic parasitic elements, *Dictyostelium* clearly has some unique elements that are related both to retroposons and to type II DNA-mediated elements. In contrast to the *Neurospora crassa* genome, about 10% of the *Dictyostelium* genome is made up of such repeating elements. The most abundant and best studied of these elements is the DIRS-1 element, present at about 200 copies per genome. Thus, unlike higher Fungi, such as *Neurospora*, *Dictyostelium* clearly tolerates the acquisition of repeated elements within its genome. However, these *Dictyostelium* intermediate repeat sequences (DIRSs) are most unusual relative to the vertebrate long terminal repeat (LTR) retroposons or endogenous retroviruses: the DIRS LTRs are inverted or "split," and they integrate without creating a duplicated sequence. DIRSs do have several open reading frames (ORFs), one of which encodes an RT. This RT shows a clear sequence relationship to those of the Ty3/gypsy elements, which are vertebrate endogenous retroviruses (Fig. 6.1). However, unlike these retroviruses (but like many fungal viruses), no coat or *env* genes have

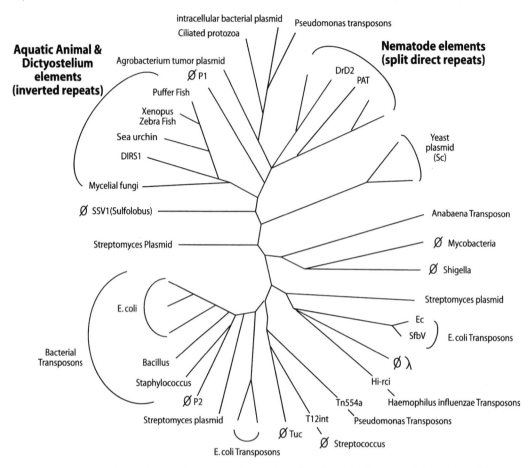

Figure 6.1 Phylogenetic relationship of the lambda-like recombinases found in nematode and aquatic animal genomes. Based on data from T. J. Goodwin and R. T. Poulter, *Mol. Biol. Evol.* **18:**2067–2082, 2001.

been seen in association with DIRSs in the *Dictyostelium* genome, suggesting that these are either defective viruses or viruses that lack an extracellular phase in their life cycle. Other DIRS ORFs are present but of unknown function. It is most interesting that outside of *Dictyostelium*, DIRS-related retroposons (such as Tdd) are found in only in nematodes, sea urchins, fish, and amphibians (discussed below), consistent with a clear lineage relationship among these organisms. Another very distinguishing feature of DIRSs is that they encode a protein clearly related to a lambda-like recombinase. The participation of such a recombinase in DIRS-1 integration would explain the absence of repeats at points of integration. The *Dictyostelium* genome is known to also harbor several other DNA elements, such as DDT. Thus, several transposon families are represented in the genome. It is interesting that there is strong evidence that interactions among these various families of retroposons occur. All of these elements show preference for insertion into loci where other elements of the same family also reside. However, these other interrupted elements lack sequence homology, so homologous recombination does not appear to be involved in this preferential insertion. The reason for this insertional preference has not been studied, but it could result from competition between colonizing elements as they seek to interrupt related competitors, similar to what is seen for some lysogenic phages in bacterial host chromosomes.

Mitochondria free of parasites

The mitochondria of *Dictyostelium* have not been reported to host any genetic parasites, either dsRNA, dsDNA, or retroposon elements. Thus, in this regard they differ significantly from those of the Fungi (and many plants). The *Dictyostelium* mitochondria (56 kbp) are larger than those of Fungi and are relatively devoid of intragenic or repeated sequences. In contrast, the mitochondrial genomes of parasitic nematodes can be very small (e.g., 13,747 bp of circular DNA), and in some cases, these genomes are multipartite.

Dictyostelium virology

Overall, the virology of *Dictyostelium* species is in sharp contrast to that of Fungi or algae, in that *Dictyostelium* species appear not to harbor any known acute or persistent autonomous viruses in their nuclei, cytoplasms, or mitochondria. However, all *Dictyostelium* genomes are clearly colonized by atypical retrovirus-like agents that are not found in Fungi or Protozoa, and these elements show clear relationships to bacterial virus recombinases as well as to vertebrate RT-encoding retroviruses. The biological consequences of these colonizing elements are not yet clear. Their mechanism of persistence, their effect on host fitness or longevity, and their effects on other competing genetic parasites have not yet been evaluated. However, their invariant presence in all *Dictyostelium* lineages may suggest a direct involvement of these parasites in the origin of this lineage. Also, with *Dictyostelium* we have the first candidate for the origin and evolution of the RNAi system of RNA silencing. This defense system offers a possible explanation for the absence of many viral agents from *Dictyostelium*. However, because the RNAi system itself is dependent on an RdRP, it seems most likely that the RNAi system may owe its own origination to a distant persistent genomic parasite derived from a dsRNA virus.

Nematodes and their Viruses

Nematodes are clearly more developed organisms than *Dictyostelium* and represent an important and clear transition towards the evolution of higher animals. The phylum Nematoda is directly under the classification Metazoa and thus represents a basal component of the animal lineage. Nematodes produce most of the cell types typical of animals, including all three germinal layers: endoderm, mesoderm, and ectoderm. The unsegmented nematode body plan is tubular, with bilateral symmetry, typical of higher animals. The tissue types include a neural system, a muscular system, and a digestive tract. Due to these similarities, nematodes have been especially well examined as models for the study of tissue development and differentiation. Consequently, the programming fate for every cell from embryo to adult has been mapped. Overall, these small worms can have two distinct lifestyles: autonomous and parasitic. *C. elegans* is by far the best studied of the autonomous nematodes, and its genome was the first animal genome to be sequenced. Nematodes have a small number of chromosomes (four or five), for a total DNA content of 97 Mbp. Most interestingly, some of the parasitic nematodes have an X/Y chromosomal system for the determination of sex, which represents an early example of this sex strategy, which is common to so many higher animals. Nematodes undergo cell aging and programmed cell death like higher animals and do not appear to show the highly extended life span of some higher Fungi. Parasitic nematodes can also undergo a process of chromosomal diminution involving the loss of DNA sequences. This process is limited to somatic, not germ line, nuclei; thus, it is reminiscent of the macronuclear (but not micronuclear) DNA rearrangements of ciliates.

Parasitic (endoparasitic) nematodes

Endoparasitic nematodes have several distinctions from nonparasitic nematodes and present a global agricultural problem, as they are known to be important vectors for the transmission of many virus infections to crop plants. They function as vectors by allowing viruses to adsorb to specific nematode mouthparts and mechanically transmitting virus to plants during feeding. These plant viruses (such as nematode-transmitted tobravirus) can physically persist but not replicate in the nematode mouth tissue. Some Fungi (e.g., *Polymyxa betae*) can also function as vectors for plant viruses (e.g., the positive-strand single-stranded RNA [ssRNA] benyviruses *Beet necrotic yellow vein virus,* and *Beet soil-borne mosaic virus*) by specific persistence (but not acute replication) in resting motile fungal zoospores. It has been clearly established that various plant viruses have proteins that specifically interact with receptors and other molecules present on their associated fungal or endoparasitic nematode vectors. Thus, the plant-virus-nematode interaction is highly specific to both the virus and the nematode and involves specific virus structural proteins. Nematodes, like Fungi, can clearly be involved as important vectors for viruses. This contrasts with the viral situation for aphids, which are also very common vectors for plant viruses (chapter 7): unlike nematodes, aphid cells replicate their associated RNA viruses. In spite of the frequent association of nematodes with viruses (as vectors for plant virus transmission) and in contrast to the case with Fungi, no persistent or acute viruses of any kind have been described for nematodes. Nor are any parasites of nematode mitochondria known. In these two characteristics, nematodes seem most similar to *Dictyostelium* and distinct from fungal and higher animal species.

Nematode nonadaptive immune systems

It seems likely that nematodes have developed (possibly from a *Dictyostelium*-like predecessor) system-wide defenses against most cytoplasmic and nuclear viruses. Like *Dictyostelium,* nematodes have a system of RNAi. This RNAi system confers the epigenetic capacity of dsRNA to induce gene inactivation and can be transmitted to other tissues. In *C. elegans* and hydra species, the capacity to induce gene silencing is striking in its systemic nature and can be expressed in all tissue following regional exposure. The *C. elegans* RNAi response is even transmitted to the progeny of individual RNAi-exposed worms. This systemic and transmissible response to RNAi is not found in any other animal, although regionally transmitted RNAi responses are found in plants. The systemic RNAi response is so efficient and prevalent in *C. elegans* that it was operationally used (via bacteria producing dsRNA) to genetically map the functions of most (96%) of its genes, in particular chromosomes. Such a systemic system would clearly be expected to exclude the dsRNA viruses (and ssRNA viruses via transcription and replication intermediates) that are so prevalent in Fungi. Although the evolutionary origin of the RNAi system has not been well evaluated, most of the genes involved have been characterized. Recently, the *sid-1* gene was shown to encode a critical membrane-spanning protein thought to be involved in some type of signal transduction. This protein is found in nonneuronal cells but, curiously, is not found in neuronal cells. It is interesting that neuronal cells are resistant to the systemic effects of RNAi, yet neurons are not known to support infection by any virus. However, neurons do respond to autonomous RNAi. Homologues of *sid-1* are not uniformly conserved in other species that have elements of RNAi, such as *Drosophila,* although homologues are present in mammals (mice and humans). The various genes that are related to RdRPs, such as the *ego-1* gene in *C. elegans,* appear to be of more central importance since they are needed for the germ line RNAi response. These genes also include the *rrf-1, rrf-2,* and *rrf-3* RdRP genes that are involved in somatic RNAi. All of these genes are homologues of RdRP genes found in plants. However, homologues are absent in *Drosophila* and human cells. An additional RNAi-associated gene is the Dicer gene, which encodes an RNase III endonuclease that is specific for dsRNA. In addition to its role in RNAi, the germ line *ego-1* gene is also involved in suppression of Cer retroviruses and retroposons present in the *C. elegans* genome. However, as shown by the *rde-1* mutant, RNAi can be inactivated without a corresponding increase in transposition. Thus, the relationship of this RNAi function to retroposon control is not clear. Similar to the *Dictyostelium* RNAi system discussed previously, the nematode RNAi system is also virus-like in its transmissible and amplifiable nature. Furthermore, and also like the *Dictyostelium* situation, the central role of the RdRP suggests that this defense system might also have evolved from a stable colonization by a persistent RNA viral agent. This original viral agent could have superimposed the RNAi recognition system of gene silencing (possibly for persistence) while allowing exclusion of competing genetic elements.

Nematodes also lack DNA viruses

The additional absence of nuclear or mitochondrial dsDNA viruses (not just RNA viruses) from nematodes presents another problem. Perhaps this absence is only apparent, due to the limited search for DNA viruses of nematodes. The apparent absence of DNA viruses cannot be satisfactorily explained solely by the presence of the RNAi defense system, since not all DNA viruses contain dsRNA, which is needed to induce the RNAi response. Since nematode genomes (unlike *Neurospora*) appear to tolerate repeated sequences (e.g., Cer retroposons [see below]), there is no obvious genomic barrier to prevent genomic or extragenomic colonization by DNA replicons or episomes. Currently, we lack a sensible scenario to explain the absence of nuclear DNA viruses in nematodes. It remains possible that nematodes have simply not been sufficiently well studied to have identified the genetic parasites (especially latent ones) that are prevalent in natural field populations.

Nematode genomes and genetic parasites

Overall, nematode genomes are compact and have a relatively small number of retroposons (less than 1%) compared to the much larger numbers in plant (up to 90%) and animal (about 35%) genomes. These worm retroposon numbers compare to the about 300 LTR elements in various yeast species. In *C. elegans,* the most abundant of these repeat elements are the 124 members of the Cer family of elements. These Cer elements show a clear relationship via RT coding sequences to the Ty3/gypsy and vertebrate retroviruses. Nineteen families of these elements are known and have been defined by RT and LTR similarity. A majority of these elements are located at the ends of the host chromosomes. Most interestingly, and in contrast to the cases with Fungi and *Dictyostelium,* in *C. elegans* some of the Cer elements can be nondefective, containing the full complement of retrovirus genes, including an *env* gene. The occurrence of the *env* ORFs makes these Cer elements the first example in eukaryotic evolution of complete (nondefective) endogenous retroviruses (ERVs) present in the host genomes in significant numbers. However, similar to most higher eukaryotic genomes, the defective versions of nematode retroposon elements are significantly more numerous in the host genomes than are the full-length retrovirus counterparts. These defective retroposons often lack most of the coding regions, especially the *env* sequence. Yet most interestingly, phylogenetic analysis of the

C. elegans Cer elements indicates that the full-length versions of the Cer retroviruses are basal to the larger numbers of defective versions. Most of the 19 Cer families have one corresponding full-length Cer element, which is basal to the remaining defective elements. Since the coding sequences of most of these full-length elements have been conserved, this strongly suggests that some selective pressure maintains the Cer ORFs and hence also maintains colonization of the host genome by the intact Cer viruses. The selected presence of the intact Cer sequences could also allow accumulation of the more numerous defective Cer elements. A dendrogram showing the phylogenetic relationship of Cer elements is shown in Fig. 6.2. As discussed

Figure 6.2 Phylogenetic relationship of Cer elements from the nematode *C. elegans.* Abbreviations: MULV, *Muleshoe virus;* HFV, *Human foamy virus;* MMTV, *Mouse mammary tumor virus;* HERV-K, human ERV-K; HIV-1, *Human immunodeficiency virus 1.* Based on data from E. W. Ganko, K. T. Fielman, and J. F. McDonald, *Genome Res.* **11:**2066–2074, 2001.

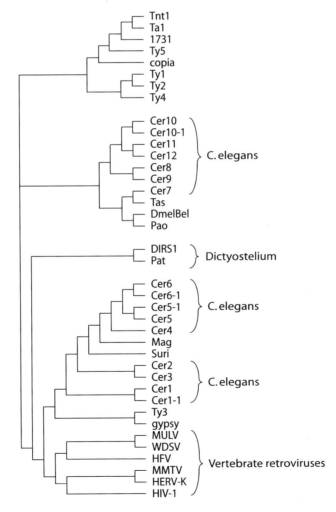

below, this pattern of a small number of conserved intact ERVs with a much larger number of related defective retroposons is seen in the genomes of various other higher animal species. Like the DIRSs of *Dictyostelium,* Cer elements also show a strong tendency to insert into other retroposons. It is interesting that the *C. elegans* Y chromosome, like that of animals, is particularly rich in retroposon sequences, such as the RT-encoding TOY element. *C. elegans* is colonized by about 30 dispersed elements related to Tc transposons (Tc1/mariner), with the exact numbers varying in a strain-specific manner. These Tc-like elements are either inhibited from activity in the germ line or defective for their corresponding transposases. However, in the presence of functional transposases, these elements undergo transposition. Unlike what is seen in larger eukaryotes and *Drosophila,* the centromere is not a site of transposon colonization in *C. elegans. C. elegans* also has an abundant but rather unusual rolling circular Helitron DNA transposon element. The miniature inverted-repeat transposable elements are also very abundant in the *C. elegans* genome. The Tourist family members, which are also present and encode a transposase, can also be found in plants, insects, and vertebrates. Overall, we see in the *C. elegans* genome the prevalent colonization of full-length retroviral elements and their more numerous defectives. We also see a striking lack of any other viral or other transmissible genetic parasites, similar to the paucity of *Dictyostelium* viruses.

Genomes of parasitic nematode species

Parasitic nematode species, such as *Panagrellus redivivus,* have genomes that are colonized by atypical DIRS-1 genetic elements, which are still recognizably similar to those of *Dictyostelium* species. Like the DIRSs of *Dictyostelium* (but unlike vertebrate retroposons), the termini of these atypical DIRS-1 elements are inverted rather than direct repeats. These elements also conserve an ORF that encodes a lambda-like recombinase, which is similar to the recombinases found in insect baculoviruses and in some plant mitochondrial genomes. Although related DIRS-1 elements have been found in other parasitic nematodes, they are not present in the sequenced genome of *C. elegans.* Thus, genome colonization by these elements appears to be host lineage specific and associated with the origin of nematode species diversification. Similar to Cer elements, DIRS-1 elements also insert preferentially into copies of themselves, often interrupting resident elements. However, about half of the DIRS-1 elements characterized so far can also insert into Cer elements, suggesting some interaction or competition between these two types of elements. Although also found in deuterostome species, such as fish and amphibians, no DIRS-1-like elements have been found in insect or plant genomes. Phylogenetic analysis indicates that these elements are basal both to the retroviruses of the Ty3/gypsy viruses and to the caulimoviruses of plants. Parasitic and free-living nematodes often have moderate to low copy numbers of the transposase-encoding mariner-like (mle-1) and Tc1-

like elements, as well as numerous defective, transposase-lacking versions of such elements (frequently found on the X chromosome). Thus, the parasitic nematodes can be differentiated from the free-living nematodes by their patterns of colonization by genetic parasites, which are more related to the parasitic elements of *Dictyostelium*.

Nematodes: basal biology

The small *C. elegans* worms very much resemble the invertebrate and insect larvae in terms of their biology, in that both worms and larvae are small, motile, wormlike forms that feed on algae. Nematodes are microorganism consumers and eat many types of algae and other cellular forms. Thus, it is felt that these simple worm forms are basal to the more complex aquatic animals.

Planarians

Freshwater planarians are small flatworms that have also been widely used to study the biology of early animals. Due to their remarkable capacity to regenerate tissue, they have received much attention. One abundant asexual North American species in particular (*Girardia tigrina*) has shown some remarkable and exceptional relationships to two genetic parasites. This species harbors a planarian extrachromosomal virus-like element that is a linear dsDNA and ssDNA that contains one ORF with clear similarity to the replicase domain of circoviruses as well as another ORF with similarity to the viral helicase domain of the human papillomavirus E1 protein (and also simian virus 40 T-antigen and adeno-associated virus 2 *ns* gene). These viral sequences are expressed in nonhead tissues. Another remarkable feature of this species is that its genome is highly colonized (8,000/haploid genome) by a dispersed mariner-like transposon (with a transposase) also found in *C. elegans* and arthropods but not found in other related planarian species. The congruence of both these extrachromosomal and chromosomal elements is remarkable, but possible interactions have not been evaluated.

Invertebrate Aquatic Animals: Deuterostome-Protostome Divergence

Aquatic animals are thought to have evolved from simple unsegmented wormlike organisms, similar in biology to *C. elegans*. Thus, the very diverse array of animal lineages that are currently found in the oceans has resulted from a major diversification of simpler wormlike forms. From the perspective of higher animal lineages, an early event in metazoan evolution was the divergence of deuterostomes and protostomes. The superphylum Deuterostomia includes the Chordata (chordates, also called vertebrates), urochordates, and Echinodermata (echinoderms). Since the deuterostomes are progenitors of bony fish (and all vertebrates) with their highly developed adaptive immune systems, they are of special interest from the perspective of evolutionary virology. The superphylum Protostomia includes the Mollusca (molluscs [bivalves and gastropods]) and the Panarthropoda (arthropods, such as crustaceans, shrimp, and insects). The crustaceans are considered the progenitors of terrestrial insects. The virology of arthropods is discussed in detail in the next chapter.

Aquatic farming and viruses

Interest in the virology of aquatic animals has been enormously stimulated in recent years by large worldwide increases in marine aquaculture. Significant losses in farmed fish and shellfish have compelled greater attention to the study of the viruses of these organisms. Yet the farming of fish is a very old practice, dating back 3,000 years before present in China. In addition, the occurrence of viral diseases on fish farms was first reported as early as 1563, at which time disease (due to carp poxvirus) was associated with enlarged livers and spleens. Bony fish and their vertebrate relatives are especially known to support the replication of large numbers of types and strains of viruses. Frequently, virally induced mortality in fish is associated with the infection of juvenile or larval forms, as some (but not all) adult fish tend to be persistently or inapparently infected. Viral infection of protostomes is also common, and here, too, virally induced mortality is frequently associated with infection of larval or juvenile forms. Correspondingly, commercially exploited shellfish populations have also suffered significant losses due to viral infections. These farmed shellfish include molluscs, echinoderms, and crustaceans.

Overall patterns of viruses in aquatic animals

Observations from mariculture have led to the identification of some overall and sometimes striking virus-host patterns. One general observation is that along with the diversification of aquatic animals, there has also occurred an associated diversification and radiation of the virus species that infect aquatic animals. This is especially apparent within the bony fish. Bony fish are known to support infection of almost all of the types of viruses that are found in mammals, plus some types of viruses that are otherwise found only in insects. These include many kinds of large DNA viruses, such as the iridoviruses (of which over 180 types are known in fish), the baculoviruses (found mostly in shrimp and insects), the herpesviruses (found in both shellfish and bony fish), poxviruses, ascoviruses, adenoviruses, and polyomaviruses, as well as other distinct classes of DNA viruses, such as white spot syndrome virus (WSSV). ssDNA viruses (gemini- and densoviruses) are also known for aquatic animals. Table 6.1 presents a summary of most of the DNA viruses found in aquatic vertebrates.

Table 6.1 DNA viruses of poikilothermic vertebrates

Virus family	Fishes		Amphibians		Reptiles			
	Sharks	Teleosts	Anurans	Salamanders	Lizards	Snakes	Turtles	Crocodiles
Parvoviridae	−	−	−	−	+	+	−	−
Iridoviridae	−	+	+	+	+	+	+	−
Poxviridae	−	−	+	−	+	−	−	+
Herpesviridae	+	+	+	−	+	+	+	−
Adenoviridae	−	+	+	−	+	+	−	+
Polyomaviridae	−	+	+	−	+	−	+	−

With respect to RNA viruses, an equally broad array of viral types is known to infect fish. These include positive-strand and ambisense RNA viruses: togaviruses, picornaviruses, birnaviruses, nodaviruses, coronaviruses, and reo-like viruses. In addition, the bony fish appear to be the first organisms in evolution to host negative-strand ssRNA viruses, including rhabdoviruses, myxoviruses, and orthomyxoviruses. In fact, within teleosts, rhabdoviruses (such as viral hemorrhagic septicemia virus [VHSV] and infectious hematopoietic necrosis virus) constitute the single largest group of isolated viruses and are responsible for epizootics and heavy losses in aquaculture. Thus, the sudden evolutionary appearance of this family of viruses in fish species is striking. Also in bony fish, for the first time in evolution an authentic, autonomous, prevalent, and disease-associated retrovirus appeared. A summary of the known RNA viruses and retroviruses of bony fish is presented in Table 6.2. Furthermore, some of the characteristic virus-tissue associations known for mammals are seen for aquatic animals and their viruses. For example, alphaherpesviruses, which are known to establish latent infections in peripheral nervous tissue, are also seen in clams and specific fish species, where they establish persistence in ganglia. It is therefore highly ironic that along with the origin of the highly complex adaptive immune system of vertebrate fish, we can also observe the surprising radiation of fish-specific virus diversity. This fish virus-host association is in striking contrast to the virus-host relationships described above for nematodes. It is also very different from virus-host associations of the prechordate deuterostomes (such as echinoderms and urochordates) and their viruses, as will now be described.

Echinoderms, the simplest deuterostomes, have a paucity of viruses

Echinoderms are deuterostomes that diverged early in the evolution of the chordates. By far, the best-studied example of a basal deuterostome is the echinoderm sea urchin *Strongylocentrotus purpuratus*. Although they are deuterostomes, sea urchins are clearly primitive compared with vertebrates, lacking even notochords. One of the more striking differences between the echinoderms and the bony fishes and amphibians is the presence of the

Table 6.2 RNA viruses of poikilothermic vertebrates

	Fishes		Amphibians		Reptiles			
Virus family	Sharks	Teleosts	Anurans	Salamanders	Lizards	Snakes	Turtles	Crocodiles
Orthomyxoviridae	–	+	–	–	–	–	–	–
Paramyxoviridae	–	+	–	–	+	+	+	–
Rhabdoviridae	–	+	–	–	+	–	–	–
Bunyaviridae	–	–	–	–	–	–	+[a]	–
Retroviridae	–	+	+	–	–	+	+	–
Coronaviridae	–	+	–	–	–	–	–	–
Calciviridae	–	+	+	–	–	+	–	–
Togaviridae	–	+	–	–	+[a]	+[a]	+[a]	–
Picornaviridae	–	+	–	–	–	+	–	–
Nodaviridae	–	+	–	–	–	–	–	–
Flaviviridae	–	–	–	–	+[a]	+[a]	+[a]	–
Reoviridae	–	+	–	–	+	+	–	–
Birnaviridae	–	+	–	–	–	–	–	–

[a] Arthropod-borne viruses termed "arboviruses."

much more complicated adaptive immune system of the vertebrates. Of special interest to virologists are the absence of a basal layer in echinoderms and their simple epithelium relative to vertebrate skin because these rapidly differentiating basal cells support the replication of so many fish viruses. Because sea urchins are commercially produced in substantial numbers in Japan and because of the intense interest sea urchins have received as an experimental model for the study of animal development, there has been ample opportunity to observe virally induced disease in these hosts. However, few, if any, autonomous viral agents have been observed for sea urchins or any other echinoderms or urochordates. For example, sea cucumbers are also cultured in Japan and have similarly not been observed to support infection by any viruses. In addition, lampreys and hagfish (urochordates) are also conspicuous in their lack of reported virus-associated diseases. More recently, however, reports of a lamprey herpesvirus have surfaced, suggesting that at least herpesviruses can infect these hosts. Both aquatic vertebrates and the aquatic superphylum Protostomia (molluscs and arthropod species) are subjected to intense mariculture and are known to be highly susceptible to acute disease caused by a broad array of viruses. Some of these viruses are able to infect both deuterostome vertebrate species and protostome species (such as fish and crabs or shrimp). Thus, the general absence of viruses in sea urchins and most urochordates presents a stark contrast with vertebrates and the echinoderms. Importantly, in this absence of viral agents, echinoderms appear to be very similar to *C. elegans*.

In the case of sea urchins, it is not fully clear how they might prevent viral infections. Systems of genome defense, such as RNAi, are known to occur in sea urchins. This might allow sea urchins to generally limit infections with

RNA viruses. Sea urchin RNAi, however, does not appear to be transmissible and systemic, as it is in *C. elegans,* so it is not clear whether sea urchin RNAi would be as effective a virus defense system as it appears to be in *C. elegans.* Furthermore, echinoderms have no specific processes (such as the sex-associated repeat-induced point mutation of *Neurospora*) that might help prevent the generation of duplicated autonomous or genomic DNA sequences that result from infections with various DNA viruses (iridoviruses, baculoviruses, herpesviruses, and parvoviruses). DNA viruses are known to frequently infect shellfish, crustaceans, and vertebrates. And in some cases, a single host is multiply infected by the same virus. The absence of DNA viruses from sea urchins is enigmatic. Furthermore, it is exceedingly ironic that the relatively simple lower animals like *Dictyostelium, C. elegans,* and sea urchins, which lack the highly sophisticated adaptive immune systems of vertebrates and maintain the seemingly simple innate immune processes, appear so inert to all (nongenomic) viral agents.

Invertebrate immunity is nonadaptive

Lower deuterostomes are considered a sister group to vertebrates but are distinct from protostomes in having a complement system. Sea urchin genomes code for an analogue of the C3 component of complement, SpC3, as well as a factor B and mannose binding protein-associated serine protease. The C3 component of complement is considered part of the "nonclassical," or alternative, pathway. However, given that it appears to be the earliest component of the complement system to have evolved, it is likely that these genes represent the origin of the complement system. Sea urchins lack many of the associated molecules and receptors involved in the induction of an adaptive immune response in vertebrates. They have no tumor necrosis factor (TNF), TNF receptor, cytotoxic cytokines, or chemokines and lack acute-phase or inflammatory cytokines. Like other lower animal forms, however, they do have several types of nonspecific cytotoxic cells.

Origin of complement

Complement attacks and destroys membranes of invader organisms and also assists in allowing phagocytosis of antibody-antigen complexes. However, the existence of this membrane attack complex needs to have a corresponding safety lock system to prevent self-killing. In mammals, this safety lock is provided by the RCA (regulators for complement activation) system of proteins, which prevents inappropriate complement activation. As the attack complex and the RCA proteins must act together and as both consist of a complex set of interacting proteins, these systems are another apparent example of components of a complex phenotype that must have evolved together. There are no examples of progenitors that have C3 analogues but lack RCA-like proteins. What might have been the origin of this complex recognition phenotype? Pore-forming protein complexes that are able to generate holes in membranes, as well as matching inhibitors of the pore formers, are well-established components of various addiction modules found in unicellular organisms (e.g., the killer phenotypes in yeasts and bacteria). Thus, C3 can also be functionally considered to be the toxic half of a two-part addiction module, since it needs the matching RCA complement-binding protein to prevent self-killing. These RCA proteins all have characteristic short consensus repeat sequences (SCR). Tunicates (urochordates) have proteins called HrSCR-1, -2, and -3 that maintain the SCR elements and appear to compose an RCA system. In the case of tunicates, it is interesting that gonad expression of these proteins is also apparent. It thus appears likely that all tunicates and lampreys maintain a C3-like complement system and corresponding RCA system of nonadaptive immunity. This is in keeping with the monophyletic character of urochordates. Members of the Echinodermata and Urochordata phyla also have nonspecific cytotoxic cells, which are discussed below.

Cytotoxic cells and tissue recognition

Sea urchins do have at least four types of cytotoxic cells (cytotoxic T lymphocytes [CTLs]) that kill target foreign cells (such as human red blood cells). Only some of these cells express C3, which is inducible. Other cytotoxic cells appear to be phagocytic amoebocytes. These cells express various markers characteristic of mammalian cytotoxic cells, such as CD14, CD56, and CD158b, but they do not express other common CTL markers, such as CD3, CD4, CD6, CD8, and CD16. These cytotoxic cells are able to lyse contacted target cells via calcium-dependent hemolysins and contain lytic granules, with acid proteases, that seem to be similar to the granules found in mammalian natural killer (NK) cells. These amoebocytes are also involved in an encapsulation response which walls off invading organisms and can involve the deposition of calcium and other substances. These cytotoxic cells can make lectins, agglutinins, and lysins. In addition, it appears that antifungal peptides can be generated from the C terminus of hemocyanin (a copper binding protein involved in O_2 transport). All of these nonadaptive immune systems evolved before the evolution of the gene rearrangement systems required for adaptive immunity. In addition to these nonadaptive immune systems, protostomes appear to have other, poorly characterized mechanisms for countering virus infections. Tissue extracts from blue crabs can be demonstrated to inhibit a broad array of viruses (including Sindbis virus, vaccinia virus, vesicular stomatitis virus, and mengoviruses) via what appears to be viral attachment inhibition. Recent evidence suggests that at least some urochordates have a cell-based system of tissue recognition. Many tunicates are able to undergo fusion with other individual organisms and form chimeric tissues, in which one partner often reabsorbs most of the cells of the other fusion partner but also maintains some of the partner's cells in its germ line and stem cells. This fusion is precluded for genetically distinct partners by a nonadaptive compatibility system (Fu/Hc) that involves one highly polymorphic gene, which has up to 500 alleles. The origin of this compatibility system and its possible relationship to viruses are not known. However, it seems likely that such a recognition system was present in urochordates before the origination of the adaptive immune response.

Sea urchin genomes and ERV genetic parasites

Sea urchin genomes and those of other urochordates (tunicates and lampreys) are relatively simple compared to the genomes of vertebrates. The biggest difference is that the vertebrate genomes appear to have undergone large-scale sequence duplication in comparison to the genomes of urochordates. The sea urchin genome sequencing project is still under way, so we cannot yet evaluate all the specifics of these differences. Likewise, the various genetic elements that have colonized the sea urchin DNA have not been determined, as has been done for *C. elegans* (described previously). However, it is still clear that the urochordate genomes are colonized by both defective and full retroviruses, and the level of this colonization (well below 1%) is much less than that of vertebrates. For example, the genome of the sea urchin is colonized by retroposons such as sea urchin retrovirus-like elements (SURLs). Based on the genome project, sea urchins are predicted to have 27,350 major ORFs. Within this genome, 315 copies of SURLs have already been identified. Most of these elements conserve the RT coding domain, although 46 are also known to be interrupted within the RT ORF (which is related to Ty3/gypsy elements). The mutation rate of the RT ORFs is similar to that of single copies of other genes, which indicates that they are not hypervariable. Furthermore, analysis of rates of synonymous substitutions in RT ORFs indicates that these ORFs are under strong positive selective pressure to be maintained in the genome. In other echinoderm species, SURL-like elements are frequently interrupted, although most species also conserve RT elements with an intact RT ORF. Of particular interest is the observation that *env*-like ORFs also exist for at least a small number of these SURLs in sea urchins. However, unlike with *C. elegans,* it has yet to be determined if these complete retroviral sea urchin elements are basal to the more numerous defective copies. The SURLs exist in distinct families, and each specific family of SURL shows a clear link to its host species. SURL clones from the same species are very similar to each other but are distinct from the SURL families of other echinoderm species. Thus, it appears that large-scale colonization of echinoderm genomes by specific families of retrovirus-like elements occurred at the divergence of these species from one another, and these species-specific SURLs have been maintained under positive selection. This pattern of lineage-specific retroposon colonization not only applies to sea urchin species but also applies to tunicates, starfish, and herring, in that the specific retroposons are clearly conserved within a lineage but distinct from those of other lineages. Many of these SURL sequences are also known to be highly expressed in sea urchin eggs prior to fertilization, suggesting a large-scale ERV reactivation early in development. In sea urchins, very high levels of transcription of poly(A)-containing retroposons occur in prefertilized eggs prior to gastrulation. This constitutes up to 50% of the total RNA at this point. The purpose of such intense retroposon transcription is not clear, and it has been proposed to be a system for the storage of nucleotides needed during early development. However, other rapidly developing eggs, such as frog eggs, do not accumulate retrotranscripts to such high levels. The sea urchin SURLs clearly resemble the ERVs of mammalian genomes in two ways: both types of organisms are colonized by lineage-specific ERV families, and both types of ERVs become highly activated for expression in the early embryo. As mentioned above, the function of this SURL reactivation in sea urchins remains unknown. The possible effect of these elements on other, potentially competing genetic parasites or viruses has not been evaluated, but it is conceivable that these elements could interfere with other genetic parasites.

Retroposon DNA methylation and the genome defense theory

The general linkage and phylogenetic congruence of retroposon colonization to a specific host species suggest that retroposons play some role in the process of host speciation. But how ERVs might contribute to host speciation is not clear. It might be expected that the patterns of ERV activation or transcription could relate to any putative role they might have had in host speciation. However, ERV expression patterns are distinctly different in sea urchins and mammals. The differences in the patterns of retroposon transcription in urochordates are most apparent at the early embryo stage. In sea urchins, high-level ERV transcription occurs in the blastocyst, after fertilization. In most mammalian embryos (but not avian or marsupial embryos), high-level ERV expression is also observed and is possible due to the unmethylated state of the genomic DNA, which later becomes methylated during blastula formation. This DNA methylation in mammalian embryos suppresses retroposon transcription in most of the somatic cells. Somatic cells, for the most part, are characterized by high-level gene expression, as observed in most tissues. Thus, global DNA methylation of retroposons but not of expressed genes is typical of mammalian genomes. The suppression of somatic retroposon expression has led to theories that this methylation and the consequent ERV suppression have evolved as a genome defense system to counter the colonization of the germ line by retroviruses. This concept is similar to the repeat-induced point mutation system described in chapter 5 for *N. crassa*. Methylation of retroposons in mammalian genomes would thus be a genome defense system. However, DNA methylation in urochordates differs significantly from this mammalian pattern. In urochordates (e.g., *Ciona intestinalis*), the DNA methylation pattern is the inverse of the global methylation pattern seen in mammals, in that retroposon or repeat DNA is not methylated whereas gene-constituting DNA is methylated. This appears to be the opposite of what would be expected from the genome defense theory. Moreover, both the mammalian and tunicate patterns of DNA methylation are in fact in direct conflict with the theory that methylation is a genome defense system selected to suppress the colonization of germ lines by retroviruses. In the case of the urochordate genome, the ERVs are in the unmethylated fraction of embryonic DNA, the opposite of what would be expected for ERV suppression based on the genome defense theory. In the case of the early mammalian embryo, the genome is specifically susceptible to retrovirus expression and integration prior to the differentiation of germ line cells (in the blastula), and it is the somatic cell lineages that suppress high-level retroposon expression through methylation, after germ line commitment. Ergo, the genomes of germ line cells of the early mammalian embryo are open to retrovirus colonization, as they are not protected against ERVs by DNA methylation. Thus, the viral-defense theory fails to explain the well-conserved patterns of ERV expression and suppression. Yet the patterns of retroviral colonization and expression remain major differences between the genomes of urochordates and vertebrates. This issue is further discussed below.

The aquatic protostomes

Protostomes, as a group, are not as well studied as deuterostomes, and no member has had its genome sequenced. Like echinoderms, protostomes lack adaptive immunity systems, and in this feature they resemble the early deuterostomes. Chelicerata, a basal protostome subphylum, appear to have conserved the RNAi system and represent a progenitor of the class Arachnida. For example, horseshoe crabs are chelicerate species that are basal to arachnids. Protostomes include the Mollusca, a very diverse phylum with more than 110,000 species, many of which have mineralized shells. Molluscs support the replication of many various virus types, including iridoviruses, baculoviruses, and herpesviruses. The sea slug represents a simple mollusc. Although not well studied from a viral perspective, one species of sea slug (*Elysia chlorotica*) is especially noteworthy. This sea slug is often mentioned in the evolutionary biology literature as a prominent example of symbiosis, since it eats filamentous algae, but can capture and maintain the alga-derived chloroplast in functional states for extended periods (a green photosynthetic slug). However, these slugs undergo phased seasonal die-offs. During this mass mortality, all slugs produce large quantities of nuclear and cytoplasmic retroviral particles, apparently from ERV. The biological significance of this virus induction has not been well evaluated, but it is likely to be involved in the phased mass mortality of these slugs. The most evolved protostome is the octopus, a mollusc which is also known to be susceptible to virus infections, although these agents are for the most part poorly studied. The Crustacea, a subphylum of the protostomes, are characterized by chitinous mineral exoskeletons and include crabs, lobsters, barnacles, and shrimp. It is worth recalling that some chlorella viruses also code for chitin production. This protostome subphylum also supports many types of viruses. Crustacea are progenitors of the arthropods, which include the terrestrial insect genera that are considered further in chapter 7. It is curious that although crustaceans are predominantly water-dwelling organisms, they are the predecessors of insects, and yet almost no insects dwell in the oceans. However, viruses that infect both insects and crustaceans, such as baculoviruses and nodaviruses, are often rather similar to each other, but they can be quite distinct from the viruses that infect vertebrates.

Virus-host ecology

The protostomes all have some general similarities in their life strategies. All are hatched from eggs, which are generally produced in large numbers. All give rise to small larval and juvenile forms, which tend to feed on plankton and algae. Overall, all of these orders appear to host viruses of various types, often lots of them. These viruses have a tendency to infect and cause pathology in younger or juvenile forms and either do not infect or inapparently persist in adult forms. However, it often appears that there are some specific examples in which specific viruses are able to persist and inapparently infect younger forms of specific hosts. In these cases, the persistently infected hosts often function as reservoirs for viruses that cause disease in other, sometimes related species (specific examples are discussed below). This situation suggests that virus or host ecology could have a major impact on the host population dynamics and probably accounts for the numerous examples of population crashes in farmed populations of protostomes. Unfortunately, this type of host-dependent persistent or acute virus ecology has not, for the most part, been well studied, making it difficult to generalize this issue. Thus, the ecologies of virus and host are linked and distinct for persistent and acute virus life strategies.

Protostome immunity

Although not well studied, the immune systems of protostomes are clearly nonadaptive. In addition, protostomes appear to lack the C3 complement system found in deuterostomes, as described above. Nor do they encode any of the other receptors and signal molecules associated with the adaptive immune system, such as TNF, TNF receptors, inflammatory cytokines, acute-phase proteins, or CTL receptors. All protostomes lack hemoglobin and instead use hemocyanins, copper-containing globin family proteins, for O_2 delivery to tissues. Chelicerates, crustaceans, myriapods, and a few insect species also use hemolymph (the circulatory fluid of organisms having open circulatory systems) as an element of a defense system that results in coagulation, which walls off parasites. This process is also clearly apparent in the horseshoe crab. This coagulation is not like the fibrinogen-mediated clotting of vertebrates, as it uses no proteins common to those of vertebrates. Protostomes do appear to have an RNAi defense system. This RNAi system is a basal adaptation since it is found in the chelicerates, which are basal to the phylum Arthropoda. (RNAi is also found in paraphyletic spiders.) However, this protostome RNAi system is not fully preserved in all descendants, such as *Drosophila* spp. Protostomes have cytotoxic cells, although the specific features and functions of these cells are not well characterized. However, it still appears to be clear that these cytotoxic cells resemble NK cells and may even kill contact target cells by a Fas/FasL-like process. One common response seen in most shellfish (and conserved in insects) is that foreign cells or organisms and irritating materials tend to be walled off by host responses. The cells that do this are amoeboid cells from the hemolymph. The material that walls off the invader or irritant can be either a chitinous material, a mineral (most often calcium), or a lectin-like coagulated matrix. In farmed pearl oysters, this process of mineralization is exploited by the introduction of a foreign irritant, leading to the deposition of calcium carbonate pearls. It seems possible that these walling-off processes used to protect the host from predation and other assaults could have evolved from innate immune reactions. It is curious that except for RNAi, none of these defense systems clearly suggest processes that would be expected to be effective against viral infections.

Viruses, farming, and shellfish with herpesviruses

The problems posed by virus infection of shellfish became very apparent with the expansion of mariculture. Mass mortalities were especially noted in 1994, when the French oyster aquaculture industry crashed, as well as in 1999, when a similar crash was seen in Japan. Other major crashes in commercial shellfish populations have also been experienced, such as in the shrimp aquaculture in China. These crashes have most often been due to infections with various types of viruses. Nodaviruses, which have two linear, positive-sense ssRNA gene segments, are an especially big problem in the cultivation of clams in the Mediterranean Sea. Herpesviruses in clams are also a major impediment to commercialization. For example, ostreid herpesvirus 1 (OsHV-1) is generally highly lethal in juveniles and larvae of various oyster species (marine bivalves) at 4 to 5 days postfertilization. OsHV-1 is expressed mainly in connective tissue, but surprisingly, it is also specifically found in the oyster nervous system, such as visceral ganglia. High-level viral expression in hemocytes is also seen. It is most interesting that the tendency of alphaherpesviruses to persist in ganglia is a unique biological characteristic that is maintained in the alphaherpesviruses of vertebrates. OsHV-1 represents an early occurrence of this virus-ganglion biology during the evolution of a true herpesvirus. OsHV-1 is a member of a large family of multimembraned, large nuclear DNA viruses with a well-defined genome organization and replication strategy. Additionally, and like the alphaherpesviruses of mammals, OsHV-1 is able to efficiently establish latent or persistent infections. Furthermore, the persistence of OsHV is common, as the virus is known to asymptomatically infect 80 to 90% of all adult Pacific oysters. Some of these OsHVs may also replicate acutely in other hosts. There appear to exist many versions of OsHV-1-related herpesviruses, which might explain why OsHV-1 cross-reacts immunologically with channel catfish herpesvirus. It would be interesting to know the relationship of OsHV-1 to the herpesviruses reported for lampreys and sharks.

Oyster farms and other viruses

Besides herpesviruses, shellfish are susceptible to other types of virus. Japanese pearl oysters can be infected by a marine birnavirus (for which seven strains are known). The virus infects many deuterostome fish species and provides another example of the tendency for aquatic viruses to cause disease in highly divergent species (often inclusive of prostostomes and deuterostomes). Infections are especially prevalent in summer, possibly due to higher water temperatures. This virus caused the commercial oyster population crash in 1997 and 1998. Mortality rates for infection by marine birnavirus are high in many, but not all, susceptible species. In those species in which persistent infections are established, the virus can be reisolated, suggesting that persistence creates a stable reservoir of virus that can spread to other hosts. Virus replication is especially high in hemocytes, where persistence also appears to be established. Inapparent persistence in wild-caught shellfish thus poses a major threat to the commercial farming of shellfish.

Shrimp farms and baculovirus and the origins of WSSV

In the shrimp industry, infections by baculoviruses, such as baculovirus penaei, have been a major cause of problems, resulting in massive farm mortalities. WSSV (a DNA virus in the family *Nimaviridae*) has been an especially important pathogen. As with the oyster situation described above, baculovirus infection of shrimp tends to be pathogenic in larval and juvenile forms but inapparent and persistent in adults. WSSV is probably the best studied of the DNA viruses that infect shrimp and has a wide host range among the various shrimp species. In general, infection with WSSV results in disease. WSSV is a rod-shaped enveloped virus and is the only large DNA virus of shellfish that has been sequenced. At 292,967 bp, it represents one of the largest animal DNA viruses yet characterized, surpassed only by eukaryotic phycodnaviruses and mimivirus of amoebae in terms of genome size. The WSSV genome is a circular DNA with 184 ORFs. Interestingly, only 6% of these ORFs are related to any sequence in the GenBank database, and these related sequences are mainly for proteins involved in virus DNA replication. Thus, the WSSV genome represents a large amount of genetic novelty. One of the WSSV helicase/nuclease genes, however, clearly appears to be similar to those found in arthropods, but it may be basal to these host genes. One very interesting WSSV gene is that for collagen, in that genes for such extracellular matrix proteins are not usually observed in virus genomes. This is reminiscent of the hyaluronic acid synthase gene found in chlorella virus. Another very interesting and curious gene of WSSV appears to regulate transforming growth factor beta, which controls inflammatory reactions in vertebrates but has not yet been established to function in shrimp or any protostome (which lack inflammatory cytokines).

As these examples demonstrate, WSSV has a large number of relatively unusual virus-specific genes of unknown function that are not found in the host or in other viruses. Therefore, the origin of this virus and these genes is obscure. Although WSSV resembles a rod-shaped version of a baculovirus in several morphological characteristics, it shares no sequence homology to that group of viruses. The closest virus genes by sequence similarity are the DNA polymerase genes of the animal herpesviruses. However, the significant differences between WSSV and the herpesviruses (e.g., a much bigger genome, a distinct virion structure, a distinct genome organization, no multimembranes, no nuclear virus assembly, no genes for transcription, and no other gene similarity) indicate that WSSV represents a family of DNA viruses distinct from the herpesviruses. This, along with the high number of virus-unique genes in WSSV, suggests that WSSV represents a new and distinct clade from some distant progenitor DNA virus or polyphyletic mixture of DNA viral progenitors.

Unlike many of the viruses described above, WSSV strains tend to be highly species specific with respect to host mortality. Yet WSSVs are found in a broad array of species. This may indicate the existence of many species-adapted versions of WSSV. Virus replication is especially high in the hemolymph. In fish hosts, WSSV shows high mortality only in young fish and has been isolated from fish kidney tumor cells. In wild-caught specimens of shrimp, WSSV can be found in 60% of larvae; thus, it is ubiquitous in nature and not simply a product of aquaculture. It has been isolated from five shrimp species, two freshwater prawns, four crab species, and three lobster species—all of which were susceptible to some level of disease. In some shrimp species, mortality can be as high as 100%. WSSV can also infect blue crabs, as well as other crab species. Interestingly, however, in two species of mud crabs, WSSV infection results in no symptoms, yet the virus can be reisolated from these previously infected hosts. This is especially the situation with the benthic larvae of mud crabs, which can frequently be found to harbor virus without virally induced disease. In addition, wild-caught *Metapenaeus dobsoni* (a penaeid shrimp) also carries the virus with no symptoms of disease. Others have reported that in addition to crabs, prawns and lobsters can be asymptomatic carriers of WSSV. The WSSV pattern of virus ecology thus seems to be that persistence is asymptomatic and highly host specific, while acute replication and virally induced disease are seen in related (but distinct) host species.

There is one interesting exception to the normal larval versus adult biology of shellfish virus infection. The usual pattern whereby juvenile crustaceans are more susceptible to viral disease than adults is not true for the baculovirus midgut gland necrosis virus, which infects larval stages but induces disease only in late, postlarval stages. In this regard, this shrimp baculovirus is more similar to the insect baculoviruses, which also often infect gut tissue and cause disease in postlarval forms.

Polyomaviruses

Polyomaviruses are a family of small nuclear dsDNA viruses with circular genomes that commonly establish persistent infections in many mammals (including the majority of the human population) and can cause acute disease in avian hosts. Polyomaviruses have not been seen in plants, insects, or lower eukaryotes, but it appears that clams may harbor this virus family. For example, 86% of the *Tapes semidecussata* clam populations from the northern Mediterranean coast of Spain are infected with a virus that shares the ultrastructural, morphological, and cytopathological characteristics of polyomaviruses. Viral particles with icosahedral symmetry were found in both the cytoplasms and the nuclei of numerous cell types. Virus-infected cells showed severe alterations, including hypertrophy, reduction of the intracellular compartments, and extrusion of the nuclear envelope. But this may have been due to a superimposed bacterial infection. Nonetheless, gill epithelial cells showed disorganization and swelling in the apical region, which affected the ciliary structures. These cellular alterations are very similar to the effects that mouse polyomavirus has on newborn mouse lung epithelia. Thus, it seems that the simplest metazoans, such as clams, are able to support replication of polyomaviruses and that there has been a strong conservation of the polyomavirus and host biology. It is worth repeating the observation that some planarian species (e.g., *G. tigrina*) harbor extrachromosomal virus-like elements that conserve the helicase domain found in polyomavirus T antigen, suggesting a possible progenitor for this clam virus.

RNA viruses of protostomes

Several types of RNA viruses are known to infect protostome species, including the reo-like virus that infects Chinese mitten crabs. Various rhabdoviruses specific for protostome species, such as yellowhead virus of shrimp, have also been identified. These rhabdoviruses are one of the few characterized types of negative-strand ssRNA viruses of protostomes. Of some note are the picorna-like viruses present in the oceans and known to infect oceanic animals. Throughout this book, I will use the term "picorna-like viruses" loosely to mean positive-strand ssRNA viruses that include the families *Picornaviridae, Caliciviridae, Comoviridae, Sequiviridae, Dicistroviridae,* and *Potyviri-*

dae. It should be noted that in the context of plant RNA viruses, this usage is often controversial because many feel that members within these viral groups share little genetic homology. However, since it is the intent of this book to consider the possible evolutionary history of viruses from the perspective of the host, this use of the term "picorna-like virus" can be well justified. Recent surveys of ocean samples based on consensus PCR primers to RdRPs have identified the existence of large and diverse populations of marine picorna-like viruses that can be continuously isolated from all oceanic sources. Phylogenetic analysis of these amplified sequences has established the existence of two entirely new families of viruses. A dendrogram of this

Figure 6.3 Phylogenetic relationship of picorna-like viruses isolated from the oceans to known picornaviruses. Abbreviations: HaRNAV, *Heterosigma akashiwo* RNA virus; CrPV, *Cricket paralysis virus*; DCV, *Drosophila C virus*; HRV, *Human rhinovirus*; PV, *Poliovirus*; ERBV, *Equine rhinitis B virus*; FMDV, *Foot-and-mouth disease virus*; AiV, *Aichi virus*; EMCV, *Encephalomyocarditis virus*; PTV, *Porcine teschovirus*; SV, *Sapporo virus*; VESV, *Vesicular exanthema of swine virus*; RHDV, *Rabbit hemorrhagic disease virus*; MCDV, *Maize chlorotic dwarf virus*; RTSV, *Rice tungro spherical virus*; PYFV, *Parsnip yellow fleck virus*; BBWV2, *Broad bean wilt virus 2*; CPMV, *Cowpea mosaic virus*; ToRSV, *Tomato ringspot virus*; PVY, *Potato virus Y*; RGMV, *Ryegrass mosaic virus*; SPMMV, *Sweet potato mild mottle virus*; WSMV, *Wheat streak mosaic virus*. Reprinted from A. I. Culley, A. S. Lang, and C. A. Suttle, *Nature* **424**:1054–1057, 2003, with permission.

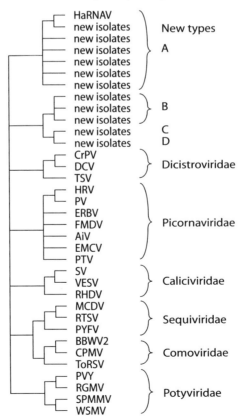

analysis is shown in Fig. 6.3. One of these families includes a new virus that is lytic to *Heterosigma akashiwo,* a toxic red tide alga responsible for fish deaths. Members of this family were found to be the most numerous in the oceans and the most diverse and distant from the animal and plant picorna-like viruses, relative to the other marine picorna-like viruses. It therefore seems likely that this family may be the oldest and most basal family of the picorna-like viruses. Marine viruses that belong to the family *Dicistroviridae,* such as Taura syndrome virus (TSV), have received much attention because they cause large-scale disease on shrimp farms. This positive-strand ssRNA virus shows clear relationships to both viruses of plants and the cricket paralysis-like viruses of insects. Similarities include the use of an alternate to AUG initiation. Thus, TSV may represent an evolutionary intermediate towards the evolution of terrestrial picorna-like viruses. It is interesting that seagulls, which often feed on shrimp, can excrete TSV in their droppings, suggesting some interesting interspecies virus ecology as well as a system for terrestrial virus transport.

Mixed viruses

The abundance and diversity of marine viruses that acutely or persistently infect shellfish species make it appear likely that virus-virus interactions occur in these hosts. If so, it is expected that persistent viruses, in particular, would often have genes that affect the ability of other viruses to colonize the same host. The resulting interactions could be either interfering or complementing (as seen with satellites and their helper viruses). In shrimp, support for complementing mixed virus interactions is seen in at least one situation: the shrimp mid-crop mortality syndrome, which has caused major losses in the shrimp farming industry. Virus analysis has shown that this syndrome is multifactorial, as it is known to involve simultaneous infection with four distinct viral types. These four viruses include a reo-like virus, a parvo-like virus, and two rhabdo-like viruses. Although it is felt that only two of these agents directly contribute to disease, the occurrence of such a mixed state suggests that virus-virus interactions can indeed have a large impact on infected hosts. However, the inherent complexity of mixed infections makes it difficult to experimentally evaluate the evolutionary pressures exerted by such situations, either on the host or on the virus genomes. Prevalent mixed infections would appear to entail a selective pressure for the functioning of a virus's genes in the context of all the infecting viruses and the host, in contrast to the situation of single infections, where the virus's genes must function in the context of only the specific host. This mixed-virus scenario is not unique to aquatic animals, and as will be seen below, pathology due to mixed virus infections is also well established in many higher plants.

Protostome genomes: a paucity of data

Although it would be interesting to compare the pattern of acute viral infections in protostomes to any genomic colonization by virus-like genetic parasites, this is not possible because these protostome genomes have not been well studied. It is known that internal DNA deletion occurs during development in some crustaceans, reminiscent of the ciliate genomes and of a related process that occurs in the parasitic nematodes. However, details of these genomic rearrangements are lacking. It is also clear that retroposons are present in protostome genomes, but until more sequence data are available, we cannot evaluate if, like for sea urchin, there exist intact basal and species-specific versions of retroviruses associated with a set of more numerous, defective ERV derivatives. It is known that mollusc genomes do have non-LTR RtE-1-like elements, as found in *C. elegans,* which contains an ORF corresponding to a 1,200 amino-acid protein. As mentioned previously, this type of retroelement is distinct from other retroposons and involves a distinct mechanism of integration. The presence of such elements appears to link the evolution of protostome genomes to the genome of *C. elegans.* Similar elements are also found in flatworms. However, these RtE-1-like elements are absent from the genomes of vertebrates, suggesting a boundary between the genomes of protostomes and those of deuterostomes and vertebrates.

Jawed Vertebrate Fish: Another Major Evolutionary Discontinuity in Host and Virus

Genomic changes from urochordates to vertebrates

The urochordates, which include jawless fish such as lampreys, have most of the types of tissue present in vertebrates. They are characterized by the presence of a notochord, which is a collagen-containing structure that follows the neural crest and provides a main structural element for the urochordate body plan. In addition, there are many differences between the urochordates and the vertebrates that establish an evolutionary discontinuity between these groups. For example, genomes of vertebrate species are expanded, especially through gene duplication events, relative to the smaller and simpler genomes of urochordate species. Along with this vertebrate genome expansion, there is also a significant increase in the numbers of retroposons colonizing all vertebrate genomes, which have also maintained most (but not all) previously existing retroposons, transposons, and repeat families observed in urochordate genomes.

Virus types

Bony fishes, which are the earliest divergent vertebrate lineage, have both innate and acquired systems of immunity. This implies that all of the gene complexity required for these complex immune systems was acquired at the origin of all vertebrate lineages. In bony fish, we also see the evolution of jaws, spinal chords, vertebrae, crania, and bones. Thus, there are numerous distinctions between bony fishes and the urochordate predecessors, and these define a major evolutionary discontinuity. However, along with this discontinuity (at the origin of aquatic vertebrate species) there also exists a parallel, although more obscure, evolutionary discontinuity: a matching and major radiation of virus families that infect vertebrates also occurred. There is evidence that at the origination of the vertebrates, many viruses also evolved. This includes several classes of virus families that are not hosted by any of the lower life forms that have so far been considered. These new viruses fall into four families of negative-strand ssRNA viruses (rhabdoviruses, bunyaviruses, paramyxoviruses, and orthomyxoviruses) and also include the first example of nondefective, autonomous, and prevalent retroviruses. Some new DNA viruses are also seen, such as papillomaviruses (distantly related to polyomaviruses), as well as the first example of autonomous, or extracellular, adenoviruses. In addition, other DNA virus lineages became established, apparently derived from previously established DNA viruses. These include the herpesviruses, baculoviruses, and poxviruses, all of which appear to have descended from distinct but apparently older large DNA viruses related to phycodnaviruses, mimiviruses, and T-even phage. We also see the evolutionary reentry of some viruses (reoviruses, parvoviruses, and birnaviruses) which were present in ancestral lineages (such as Fungi) but had been absent from less distant basal metazoan predecessors (worms and basal deuterostomes). This aquatic virus expansion mirrors the Cambrian host radiation and accounts for a large fraction of the virus families we currently see in mammals, plants, and insects, which all evolved near the origination of aquatic vertebrates. Many of these viruses were also seen to infect the protostomes as described above, but with a notable reduction in virus types.

Bones

Characteristic differences between urochordates and vertebrates include the development of the central and peripheral nervous systems, which are surrounded by bone (cranium and vertebrae, respectively), as well as the association of lymphoid and hemopoietic cells with bone (bone marrow). What was the origin of these bony structures, and why are they associated with nervous and immune tissues? Although the shells of protostomes also use calcium carbonate deposits to form structural elements, the phosphate-containing hydroxyapatite used to form bone is distinct. The origin of bone calcification is unknown. However, this type of calcification resembles the nonspecific cellular immune responses of early metazoans, which wall off parasitic agents via calcification mediated by surrounding lymphoid cells. This type of response is still characteristic in vertebrates as a chronic inflammatory response.

Skin and viruses

In addition to bone, the architecture of various other tissues of vertebrates is also different from that of the urochordates. One notable example is that vertebrate epithelia (skin and other tissues, such as the gut lining) have a basal cell layer that undergoes continuous terminal differentiation. Along with this capacity for continuous differentiation, the first evolutionary occurrence of high rates of virus-induced skin growth anomalies (hyperplastic dermal lesions or benign tumors) can be seen. No lower metazoan and few protostomes show any such related growth effects of virus infection. Both large DNA viruses (e.g., iridoviruses, WSSV-related viruses, herpesviruses, and adenoviruses) and retroviruses (e.g., walleye dermal sarcoma virus [WDSV]) show a clear tendency to cause growth abnormalities in fish and amphibian skin (but not in shrimp, crabs, or oysters). Some of these viruses can also cause occasional growth abnormalities in other vertebrate tissues (e.g., lymphatic tissue, connective tissue, and kidneys). In many cases, the altered skin growth is the main consequence of virus infection, as mortality is not necessarily associated with these vertebrate virus infections. In these cases, the viruses tend to have specific genes that can control vertebrate cellular division or differentiation, as discussed below. Furthermore, these skin infections and growth anomalies are often prevalent in wild fish and amphibian populations, so they are not products of aquaculture, with its attendant large and crowded fish populations.

Adaptive immunity

Possibly the most striking difference between urochordates and vertebrates is the presence of a full adaptive immune system in the latter. The adaptive immune system uses elements from the innate immune system to both initiate and control the acquired immune response. The basal phylogenetic position of bony fish makes these hosts of special interest for understanding the origins of immune systems. Adaptive immunity is a highly complex phenotype that appears to have been acquired in its functional totality, and thus it is monophyletic. The adaptive immune response is a linked interaction between inflammatory and acute-phase proteins—cytokines (interleukins 1 and 8, interferon, transforming growth factor beta, and TNF-α), and chemokines (CC and CXC), their receptors (various CD types), and acute-phase proteins (serum amyloid A and serum amyloid P component)—and the signal transduction systems that initiate both the humoral and cellular antigen-specific immune responses. These signal transduction systems activate the generation and selection of sequence diversity in antigen-reactive immune proteins (e.g., antibodies and T-cell receptors). The adaptive immune response also involves complement. As described above, complement protein C3 and its RCA control proteins are clearly present in urochordates. In addition, a nonadaptive major histocompatibility complex (MHC)-like system is present in tunicates. The complement system in higher vertebrates is composed of about 30 proteins. Many of these seem to have evolved from duplication of earlier genes that were present at the protostome-deuterostome boundary. For example, the C4 and C5 genes appear to have been generated through duplications of the C3 gene. It appears that complement proteins C3 to C9, mannose-binding-lectin-associated serine protease, mannose-binding lectin, and Bf factor B are all present in bony fish. C3 is encoded by a single-copy gene in all tetrapods. However, in fish species, multiple copies of the C3 gene are present, with up to 11 copies in some species. Usually, such increases in gene number are considered to have occurred through gene duplication events. Bony fish also have antigen-specific CTLs that have the characteristics of the CTLs from higher vertebrates. For example, the CTLs of channel catfish appear to kill target cells via a Fas/FasL-like process. It also appears that they may use a perforin/granzyme response for killing. These CTLs have antibody receptors for Fc and immunoglobulin M. Thus, bony fish appear to essentially have all the elements of the humoral and cellular adaptive immune responses. All of these complex characteristics appear to have evolved together, seemingly at the same time, since they can all be found in all bony fishes examined (e.g., catfish and salmonids). Aside from C3-like complement and nonspecific cytotoxic cells, none of these elements are present in urochordates (or protostomes).

The currently accepted view concerning the function of the adaptive immune response is that this surprisingly complex system has evolved to control parasitic invasion, especially by viruses. However, the extreme irony, noted above, is the significant expansion of the virus families that can infect aquatic vertebrate hosts. In other words, it is not apparent that adaptive immunity is more protective against virus infections than the previously existing innate immune systems. However, it is possible that this virus expansion followed an evolutionary period during which virus infection of jawed fish was rather restricted. The early vertebrates may have indeed been rather resistant to most virus infections, whereas vertebrates that evolved later became more susceptible to viruses. One reason for thinking this is that the most primitive of the jawed fishes, represented by the shark family, are susceptible to very few viruses. Sharks appear to have essentially all the basic elements (albeit simplified) of the adaptive immune system, including an antigen-specific humoral response and an allogeneic cellular response (in which CTLs have clear gene families of T-cell receptor homologues). Sharks are also notable for several other features. Aside from their jaws, they use cartilage, not bone, for their skeletal systems. Various species of sharks are viviparous, giving birth to live young. This raises the immunological dilemma of how the adaptive immune system of the mother fails to recognize an allogeneic embryo, a characteristic dilemma shared with the placental mammals (considered in detail in chapter 8). Little is known concerning the details of this immunological dilemma in sharks, except that the placental yolk sac is a site of high-level cytokine expression. With respect to virus susceptibility and virally induced tumors, sharks differ noticeably from the bony fishes in that almost no virally associated pathology has been described for sharks. Although sharks cannot be reared in aquaculture, and thus viral diseases that become more apparent in such a farmed setting may not yet have been recognized, they are fished in large numbers. Thus, at the very least virally induced skin growth anomalies occurring in nature should have been identified. The one exception to the lack of shark viruses is the 1985 report of a herpesvirus infecting wild and captive smooth dogfish, which showed necrotic skin lesions on fins and trunks. Thus, herpesviruses may be one of the only viruses found in both lampreys and sharks, organisms that span the development of the adaptive immune system. Sharks are also remarkable for their extremely low tumor incidence relative to that of bony fish (although it should be noted that sharks have a simpler skin tissue architecture than bony fish). If sharks are representative of the very first organisms with an adaptive immune system, then it seems very possible that these organisms were in fact successful at preventing infection and colonization by many virus types and that the expansion of virus susceptibility in bony fish was a later development in evolution.

Fish retroviruses

In fish, we see the first clear example of extracellular, autonomous, and acute retroviruses, best exemplified by WDSV. WDSV is the best-studied example of a fish retrovirus and is well established for being able to induce skin tumors. This virus is most related to the vertebrate Moloney murine leukemia virus (MoMLV) family, which is basal to a large family of mammalian retroviruses. Because WDSV represents the first example of an autonomous retrovirus and shows a predilection for replication in differentiating cells that were not represented in predecessor organisms, it seems likely that these fish retroviruses may have been the predecessors of the autonomous vertebrate retroviruses. This view is consistent with phylogenetic analysis of the retroviruses, which supports the basal positioning of the fish retroviruses.

The three known types of WDSV are unusual relative to other retroviruses in several respects. For one thing, they share a surprising level of sequence similarity among themselves and also conserve a very unusual protein cleavage site. In addition, these viruses are highly atypical of mammalian retroviruses in that they all encode a D-type cyclin, which is distinct from the host cyclins. No other retrovirus has these two characteristics. The regulatory proteins (such as Ras and Myc) found in mammalian retroviruses (such as the transforming retrovirus Rous sarcoma virus) are often considered to have been acquired as accessory proteins from host genomes because in the absence of such proteins, retroviruses can often still replicate in dividing cells.

A most intriguing biological characteristic of these fish retroviruses is that they are biologically prevalent in natural settings. The prevalence of WDSV in fish of Oneida Lake in New York has been studied and shown to cause a seasonal pattern of a proliferative skin disease in fish. This study suggested that the seasonality resulted from the coincidence of fish migration (for spawning) and acquisition of WDSV infection (from unknown sources). Fish that returned from migration to the lake showed skin growth disease for only a short period and then recovered. WDSV production did not persist in these fish, and thus the infections appeared to be acute. However, it was also observed that essentially all fish in the lake were eventually infected with WDSV. These observations support the existence of an exogenous source of WDSV that fish migrating from Oneida Lake encounter. The oceans present the most probable source of this exogenous virus. It is suspected that WDSV may persist in another host, possibly as an expressed ERV. However, there is currently no information concerning what other organism(s) might harbor WDSV. Other aquatic retroviruses are also known to be prevalent in specific hosts: retroviruses have been reported for eel, snakehead fish, sea bass (associated with erythrocyte growth), and salmon (associated with plasmacytoid leukemia). These viruses are rather distinct from one another. For example, snakehead retrovirus has a unique map location for its *env* gene and shows little sequence similarity to WDSV. In general, however, these viruses are not well studied, although there is some evidence that these viruses can suppress host immune responses.

Retroviruses of amphibians are also known and show clear relationships to the viruses of fish and reptiles. Many of these amphibian viruses, however, are endogenous in their respective hosts and are only distantly related to the currently accepted seven retroviral genera. Retroviruses related to these amphibian viruses are not widespread in other vertebrates, so they appear to be restricted to fish and amphibian lineages. It is interesting that a recently described, complete ERV of the python *Python molurus* is highly expressed in all *P. molurus* pythons but absent from all other python species (although *Python curtus* does have a related ERV). This *P. molurus* virus has an additional and unknown ORF within its polymerase gene sequence. The python viruses show little similarity to the retroviruses of higher vertebrates and cannot be classified with them. Similarly, retroviruses of fish do not group with true retroviruses of mammals or birds. The latter viruses fall into five genera: lentiviruses, human T-cell leukemia viruses, avian leukemia viruses, type D retroviruses, and mammalian type B retroviruses. In other words, the fish and reptile retroviruses represent essentially novel retroviral genera, which show interesting links to each other and their hosts but not to mammalian and avian host lineages and their corresponding retroviruses.

Retroposons and fish genomes

In the genomes of fish, we see a widespread colonization by retroposon elements derived from retroviruses. However, as shown by the genome colonization of many other vertebrates, the retrovirus-derived sequences present in fish genomes represent distinct virus groups that are specific to and highly reiterated in the fish genomes but are not part of any other major retrovirus group. Fish genomes have vertebrate short interspersed repetitive elements and related, RT-encoding long interspersed repetitive elements in relatively large numbers. Both of these elements appear to have evolved from common fish-specific retroviruses. For example, fish have Poseidon and Neptune elements, which are highly reiterated, fish-specific retroposons related via RT similarity to the Penelope element of *Drosophila virilis*. In *Fugu* fish genomes, we can find the Xena element, an example of a lineage-associated retrovirus. Also in fish genomes, we can find the Jule element with *gag* and *pol* ORFs but no *env* ORF. Jule is a member of the MAG (Ty3/gypsy) retroelement family found in *C. elegans* and silkworms (described above), a classification consistent with deuterostome evolution from worm forms. Furthermore, the Jule element is related to the sea urchin SURL element. However, Jule is only present in three or four copies in the zebra fish genome and not in several hundred copies per genome, as is SURL in sea urchin genomes. A Ty1/copia element is also found in fish, amphibians, and reptile genomes, establishing a link between these organisms. The host lineage-specific nature of these fish ERVs and retroelements and the unique class of retroviruses involved suggest that the ERV colonization of the fish genomes occurred early in the origination of these species and that further horizontal transmission of retroviruses between vertebrate classes occurs relatively infrequently. As mentioned previously, the ERVs of amphibians and fish are clearly related to each other but distinct from those found in mammals and avians. A dendrogram of some of the major families of retroposon LTRs is shown in Fig. 6.4.

Fish genomes, adaptive immunity, and addiction modules

The acquisition of the adaptive immune system represents a punctuated and transforming event in the evolution of animals that involved the acquisition of a highly complex phenotype. This phenotype includes an evolving and dynamically adapting genetic system which is able to recognize and attack nonself while preventing the development of self-recognition and self-attack. The adaptive immune system was assembled by acquisition of a new and complex set of communicating molecules (e.g., cytokines), receptors (e.g., of the T-cell receptor family), and transducers, which all lack clear evolutionary predecessors, for the most part. This acquired gene system was adapted (or designed) to detect new nonself agents in the context of a polymorphic self-detection system (i.e., involving the MHC receptors). After the detection of nonself agents, the system generates a response that creates and utilizes a new molecular process (genome level gene recombination, not present in predecessors) to generate genetic diversity and stimulate the clonal growth of specific cells that recognize these nonself agents. These resulting cells then either secrete a novel class of molecules that bind and inhibit nonself agents or allow amoeboid cytotoxic cells to find, contact, and destroy cells harboring these agents. Most of the features of this adaptive immune system were not present prior to the evolution of vertebrates. We know that predecessor urochordate tunicates had a polymorphic MHC-like system linked to nonadaptive amoeboid hemolymph cell-induced killing. But this tunicate system

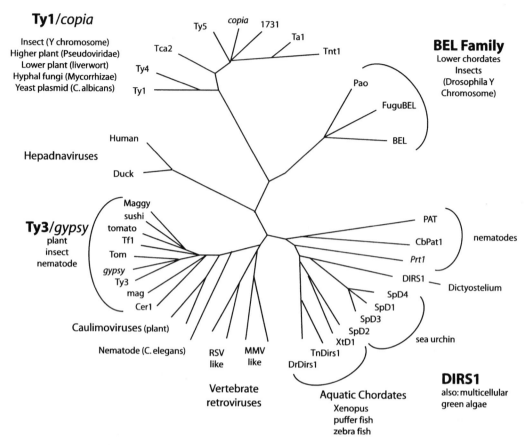

Figure 6.4 Dendrogram of LTR regions of DIRS-1-, Ty1-, and Ty3-related elements. Based on data from T. J. Goodwin and R. T. Poulter, *Mol. Biol. Evol.* 18:2067–2082, 2001.

lacks molecular similarity to the vertebrate MHC system. In order to create the adaptive immune system of vertebrates, a tunicate-like system must have acquired an adaptive component linked to nonself-recognition that prevented self-recognition. That is, the adaptive immune system is destructive (of nonself agents) and protective (against destruction of self cells) at the same time. These linked features are characteristics of an addiction module. Thus, it can be proposed that this adaptive immune system should be considered an elaborate addiction module. Hosts that have acquired adaptive immunity have also acquired a very destructive system (equivalent to a powerful toxin) that can kill essentially any cell it contacts. The hosts must, however, also be protected from this destructive power by the simultaneous acquisition of a self-recognition system which prevents self-killing (equivalent to the antitoxin). As in other addiction modules, the killing (toxic) capacity of adaptive immunity is stable and long-lived, but the antitoxic capacity (self-recognition) is transient and occurs mainly during the development of the immune cells.

Hence, adaptive immunity displays both parts of an addiction module (toxin and antitoxin) and the differential stability of the toxic component relative to the protective antitoxin.

A scenario for the viral origins of adaptive immunity

Many of the individual elements needed for an adaptive immune system appear to have arisen individually during biological evolution, but not in pre-existing cellular organisms (as noted above). Rather, these strategies and systems can be found in various types of viruses and are used by them to colonize their hosts. For example, the capacity of virus (a bacterial prophage) to recognize and alter host bacterial surface receptors is the well-known phenomenon of phage conversion and represents the most dynamic genetic feature of phage-host systems. This conversion is related to both the success of phage colonization and bacterial virulence (and its resulting host phenotype). Such phage-mediated receptor conversion is still used to type specific bacterial strains. With respect to the origin of the adaptive immune system, we can start by posing the questions, What is the most basal gene function in the adaptive immune system required for the generation of diversity (which is needed for non-self recognition), and from what source did this gene likely evolve? The RAG proteins are responsible for the DNA rearrangements and recombination that generate the genetic diversity, which results in surface receptor diversity. Thus, the genes encoding RAG proteins are the starting material for selection of adaptive immunity. Phylogenetic analysis of these RAG proteins suggests that they have no predecessor genes in early eukaryotic or prokaryotic genomes. Instead, RAG genes are more closely related in both sequence and function to the integrase genes of various retrovirus genomes. Can a virus employ such an integrase-RT protein for the purpose of generating diverse surface receptors, as is seen in the adaptive immune system? Along these lines, it has recently been reported that a prophage of *Bordetella* bacteria is responsible for tropism switching by altering cell surface receptor expression. This phage-directed tropism switching results from a template-dependent process that uses reverse transcription of phage RNAs (encoding a surface receptor protein) to introduce nucleotide alterations and genetic diversity into the host genome. Thus, it generates a vast repertoire of possible ligand-receptor interactions. This process is the only currently known prokaryotic example of a genetic cassette that uses reverse transcription and can adaptively generate large pools of genetically diverse receptors. The RT of this phage most closely resembles that of MoMLV. MoMLV is a retrovirus that is basally related to many other retroviruses and retroposons of vertebrate organisms, including the ERVs found in fish genomes. Thus, it can be seen that the molecular machinery needed to generate the adaptive component of vertebrate immunity can be found in various persistent viral, not cellular, genomes.

Other novel components of the adaptive immune system are also found in viruses. For example, the abilities of viruses to express cytokines and cytokine receptors and to alter host cell signal transduction and growth are well-established characteristics of most bony fish DNA viruses and retroviruses. These molecular strategies are frequently used by the numerous prevalent viruses in fish species. However, such virus strategies appear to have been absent prior to the evolution of bony fish. Viruses, such as the poxviruses and herpesviruses, that encode viral versions of cytokines and cytokine receptor molecules are often believed to have acquired such "host-like" immune regulatory genes by "stealing" these genes from their hosts during evolution in order to suppress host immunity and allow active virus replication. However, phylogenetic analysis of these viral immune and growth regulatory genes does not generally support this hypothesis because the viral versions of such genes are usually either unrelated at the sequence level to host homologues or basal to those homologues that do show sequence similarity. In general, viral regulatory genes appear to encode more primitive or basal proteins (e.g., using single protein domains) relative to the host homologues. In this light, it is highly interesting that herpesviruses which infect both lampreys and sharks are known, implying the existence of a persistent virus in predecessor hosts (which predated this evolutionary transition to vertebrates and provided the source of these signaling systems). Also relevant to the possibility that viruses can provide genetic novelty for the origins of host regulatory molecules are observations of fish retroviruses. The fish retroviruses have virus-specific versions of growth regulatory genes, such as the D-cyclin genes, which are absent from the hosts. These genes are clearly of viral origin. In addition, retroviruses themselves have a highly distinctive characteristic relative to all other mammalian viruses: they use the greatest diversity of viral and cellular receptors to infect cells. In particular, retroviruses are noted for using receptors associated with the hemopoietic and the adaptive immune systems. Vertebrate retroviruses have long been noted to have an inherent propensity to infect cells of the immune system.

What is the viral ancestor of adaptive immunity? One seemingly major problem with the viral-origin hypothesis is that we cannot currently identify a specific candidate virus family that might have been the progenitor of the adaptive immune system. No one virus family has all these needed functions. Beyond the basal requirement for the acquisition of RAG function and its control, which clearly could have come from a retrovirus, the extensive complexity of the adaptive immune system suggests that a specific or single viral agent would be unlikely to be able to provide all the required gene functions. This point, along with the high-level ERV colonization that now characterizes all vertebrate genomes, suggests that the acquisition of adaptive immunity was a complex and punctuated genetic event. That event was not the product of a single genetic parasite or virus, but rather it resembles a stable colonization by sets or swarms of complementing and defective genetic agents.

These agents must have superimposed onto the notochord host a most complex addiction module which was able to compel stable colonization but also able to exclude competing genetic parasites. The predecessor host most likely already had an MHC-like recognition system, a system of cytotoxic cells, and a complement system that were parasitized and regulated by the newly colonized agents, resulting in the creation and evolution of the adaptive immune system.

Fish iridoviruses

Iridoviruses, large dsDNA viruses, are one of the most abundant types of viruses that infect fish. Over 100 types of fish iridoviruses are known; some of them can also infect amphibians and reptiles, including turtles. These viruses have genomes that range from 100 to 200 kbp, with linear DNA, and encode well-conserved capsid proteins. The virion membrane is noncellular, and the virus is very stable, allowing for aquatic persistence. Two large groups of these viruses are known: the *Lymphocystis disease virus* (LCDV) group (lymphocystiviruses) and the *Tipula* iridescent virus group (iridoviruses). Mandarin fish infectious spleen and kidney necrosis viruses are some of the better-studied LCDV members. The LCDV group virus has a 111,362-bp dsDNA genome with 124 identified ORFs. Several of these viral genes, especially those encoding replication proteins, are related to genes of other DNA viruses. All iridovirus genomes code for a DNA polymerase and two subunits of a DNA-dependent RNA polymerase. Also, they show some similarity (via capsid genes) to *African swine fever virus*. The LCDV DNA *pol* gene is most similar to the phycodnavirus DNA *pol* gene; thus, iridoviruses likely evolved from these older aquatic DNA viruses of algae and other marine protists. The iridescent viruses (iridoviruses) of marine animals encode a complex DNA-dependent RNA polymerase that consists of between 8 and 14 subunits, with 2 large subunits, the larger of which (RP01) contains the conserved universal hexapeptide sequence found in all RNA polymerases. This large subunit is more similar to the RNA polymerase of insect viruses than to those of other cytoplasmic DNA viruses. The fish iridoviruses are clearly most related to insect iridoviruses and were their likely ancestors. As an aside, it is curious that there are no mammalian or warm-blooded host versions of iridoviruses. Iridoviruses have a nuclear phase but, unlike herpesviruses and more like poxviruses, assemble nucleocapsids in cytoplasm. This results in the characteristic iridescent inclusion bodies observed in infected insect cells. The early nuclear phase requires host RNA polymerase activity, leading to initial nuclear viral DNA synthesis, followed by a cytoplasmic phase of viral DNA replication via concatenated DNA intermediates. The DNA appears to be packaged by a headful DNA packaging process. The DNA is also circularly permuted and terminally redundant, and thus it resembles phages P22 and T4. The *Iridoviridae* are the only other virus family (in addition to the P22 and T4 phages) with this type of replication strategy. The presence of this unusual process of DNA replication suggests that iridoviruses may have evolved from a mixture of several DNA virus lineages. It should also be noted that the insect and vertebrate iridoviruses differ from each other in some general characteristics. For example, insect iridoviruses have methylated DNA, whereas the vertebrate viruses do not, suggesting some host-dependent link to viral DNA methylation.

Iridoviruses induce cellular growth in fish, octopi, and amphibians

The iridoviruses are generally ubiquitous in their hosts. These viruses tend to have many genes that affect the host cell cycle, which appears to account for virally induced growth alterations in infected cells. These growing cells are not invasive. For example, fish LCDV induces no substantial pathology in infected flounder, as virus replication results only in transient and benign surface lesions on the skin that eventually disappear. Goldfish iridovirus GFV has little sequence similarity to LCDV but is more closely related to frog iridoviruses, such as *Frog virus 3*. Like with LCDV, no virally induced disease is observed in infected goldfish. The iridoviruses thus tend to show persistent, unapparent life strategies in their natural hosts. However, some viruses can cause severe acute disease, as observed for Pacific herring virus and viral hemorrhagic septicemia virus, which causes a disease that poses a significant commercial problem. It is interesting that fish pathology can involve myocardial mineralization and hepatocellular necrosis. Infected fish can show chronic inflammation and severe focal skin reddening. In some cases, a virus that is inapparent in fish (e.g., LCDV) appears to be lethal in marine bivalves, establishing that these viruses can also jump species and infect protostomes. A related iridovirus is also known to infect *Octopus vulgaris*. This virus is also associated with tumors but otherwise causes little disease. Additional related viruses can also be found in amphibians and reptiles, and such viruses compose a large amphibian virus family. Frog virus 3 and 23 related viruses were isolated from renal tumors of field frogs and toads. Thus, the amphibian viruses are highly prevalent in natural populations. These amphibian virus families, like all iridoviruses, are limited to poikilothermic (cold-blooded) animals. Also, all these hosts have an aquatic phase in their life cycle. No warm-blooded-animal versions of iridoviruses are known.

Fish herpesviruses

With the evolution of fish, the evolutionary introduction and expansion of true herpesvirus members can also be seen. Channel catfish herpesvirus (CCV) (*Ictalurid herpesvirus 1* in the new nomenclature) is one of the best-studied fish herpesviruses, due to its ability to induce acute disease with high mortality and large population losses in farmed catfish. The virus tends to infect gills of juvenile fish and establishes persistence and latency in adult fish. The capsid gene and the DNA polymerase gene of CCV have discernible homology to those of herpes simplex virus. No other sequence homology to additional herpesviruses is apparent. By virion morphology, morphogenesis, and genome replication, CCV is clearly identifiable as a herpesvirus. These fish viruses have multimembranes with nuclear replication, assembly, and budding, as do other herpesviruses. Conserved genetic maps and morphogenesis are characteristic of these herpesviruses. These fish viruses show a tendency to cause epithelial tumors, especially at higher temperatures, but all tumors are benign. Viral epidermal hyperplasia (also known as walleye epidermal hyperplasia) is caused by a herpesvirus infection that results in mortality of hatchlings, where affected fish show epidermal hyperplasia in fins and skin. Herpesviruses are also known for salmonid species, such as the white bream and rainbow trout, the latter of which is infected by salmonid herpesvirus 1. Like CCV, this acutely replicating virus results in high mortality rates after infection of gills. Frog versions of herpesviruses are also known, such as Lucké frog tumor herpesvirus (ranid herpesvirus 1), which can induce renal adenocarcinoma in the American leopard frog (*Rana pipiens*). This virus is clearly related to fish herpesviruses but is different from mammalian and avian herpesviruses. The herpesviruses of fish and amphibians encode growth factors, some of which are clearly non-host-like. Although latency seems to occur frequently, especially in adults, it is not understood in any great detail. However, in the case of cyprinid herpesvirus, virus DNA is found latently in spinal nerves even after virally induced papillomas have regressed, suggesting a tendency to establish latency in nervous tissue, similar to the alphaherpesviruses.

Herpesviruses: inapparent persistence and acute disease are host species dependent

In the above examples it can be seen that some fish herpesviruses appear to induce acute disease in juvenile fish. Other disease-inducing herpesviruses are also known, such as the lethal koi herpesvirus of carp, which is distinct from other fish herpesvirus. However, not all herpesvirus-induced disease in fish or amphibian hosts is restricted to the juvenile stage. A herpesvirus of the green turtle is known to induce fibropapillomas even in adults. Interestingly, the same turtle virus can be found in some fishes (e.g., the saddleback wrasse), which are asymptomatic carriers. This suggests that persistence of this virus may be restricted to specific fish hosts. In fact, the ability of some herpesviruses to persist in specific hosts but induce acute disease in other (sometimes very related) hosts appears to be a rather general characteristic of the entire herpesvirus family. For example, in the European eel, anguillid herpesvirus establishes an inapparent persistent infection. Infected eels are prevalent and remain healthy. Even when these eels are treated with dexamethasone, which can induce herpesvirus reactivation and production, the eels remain healthy. Yet this virus is lethal to Japanese eels, causing gill disease and high levels of virus replication in fibrocytes. Because the natural prevalence of this herpesvirus is high in European eels, any contact between Japanese and European eels seems destined to eventually expose the Japanese eels to lethal anguillid herpesvirus infections. This host-dependent persistence and host-dependent lethality would seem to predict major consequences with respect to interspecies competition and the natural selection of these populations. Along these lines, it is interesting to consider that all herpesviruses are monophyletic. Yet, most individual types of herpesviruses are also, by and large, phylogenetically congruent with the hosts they persistently infect, but not the hosts they acutely infect. This suggests that the long-term evolutionary stability of herpesviruses is maintained in the persistently infected hosts, not the acutely infected hosts. Although the herpesvirus families conserve replication proteins, genetic organization, and morphology, they also tend to show large-scale gene rearrangements (i.e., different gene orders), especially for genes that regulate host functions.

Other fish DNA viruses and altered cell growth

Other DNA viruses are also known to infect fish, and for the most part these virus families are also absent in urochordates. One such virus is papillomavirus. Atlantic salmon, brown bullhead, and winter flounder are all known to support the replication of fish papillomaviruses and to show associated alterations in cellular growth. This altered growth characteristic is also apparent during infection (of fish) by adenoviruses. For example, infection of Atlantic cod by an adenovirus is associated with hyperplastic dermal lesions. Like those other DNA viruses, adenoviruses show a pattern of virus-host coevolution as shown in Fig. 6.5. Thus, these "new" fish-specific DNA viruses are very much like the fish iridoviruses, fish herpesviruses, and fish retroviruses described above, in that all are associated with cellular (mainly epithelial) growth abnormalities. For viruses of bony fish (but not of sharks), there is thus a surprisingly general pattern: these diverse families of fish viruses all depend (to a variable but vital degree) on the host nuclear machinery and induce cellular growth. Additionally, although the ability to alter cell growth by mammalian members of these virus families has long been recognized (hence the name *tumor viruses*), it is striking that for the mammalian versions of these viruses, altered cell growth is a biological abnormality and not a normal outcome of virus reproduction. In natural infections of mammals, few herpesvirus, adenovirus, or retrovirus infections commonly result in cell growth alterations. Of these mammalian viruses, papillomaviruses seem the most prone to induce warts and other epithelial growth alterations. But even here it is well established that the vast majority of human papillomavirus infections of human cervical epithelia and skin are silent and persistent, showing normal epithelial growth. It seems likely that this difference of fish viruses from the mammalian viruses relates to a distinct cellular or nuclear habitat (specific to fish cells) for these fish viruses.

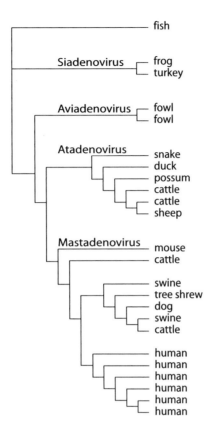

Figure 6.5 Phylogenetic relationship of adenoviruses based on their DNA polymerase regions. Redrawn from M. Benkő, P. Élő, K. Ursu, W. Ahne, S. E. LaPatra, D. Thomson, and B. Harrach, *J. Virol.* **76:** 10056–10059, 2002, with permission.

Also, it is clear that the cellular growth associated with infection by these fish viruses does not suggest a general host response to any virus infection, since such growth abnormalities are not characteristic of the numerous RNA viruses that infect fish, as described below.

Fish RNA viruses

In bony fish, a large array of RNA virus species, many of which were not present in the predecessors of the vertebrates, can be observed. These RNA viruses include fish rhabdoviruses, paramyxoviruses, orthomyxoviruses, picornaviruses, coronaviruses, caliciviruses, and birnaviruses. Of special note from the perspective of virus evolution is that these RNA viruses now include several families of negative-strand ssRNA viruses (rhabdo-, paramyxo-, and orthomyxoviruses), which where essentially absent from all the organisms that would represent predecessors of the bony fish. It is striking that rhabdoviruses, in particular, are a common source of fish infections. Currently, we have no coherent explanation for this observation, especially considering that these viruses seem to be less dependent (relative to other types of viruses) on the host machinery for their replication. Yet surveys of natural epidemics show important and devastating infections with VHSV, infectious hematopoietic necrosis virus, and spring viremia of carp virus, indicating that rhabdoviral infections can occur in nature and are not restricted to mariculture. However, as was the case with the fish DNA viruses and retroviruses, many of these viral agents have come to the attention of fish biologists due to their strong impacts on fish farms. For example, Japanese flounder and yellowtail are two of the most heavily farmed fish species. They are both prone to infections and disease with various fish RNA viruses. The most problematic and numerous of these viruses are the birnaviruses (causing viral ascites) and the nodaviruses (causing viral nervous necrosis). Birnaviruses, which are dsRNA viruses, are abundant in fish and some shellfish. Currently, 231 strains (in six genogroups) of birnaviruses have been isolated from various fish species. For the most part, infections with these strains are disease associated and appear to be acute. In some instances, however, birnaviruses can also be isolated from healthy fish, which establishes that inapparent persistent infections also occur. Birnavirus persistence may also occur in some shellfish. Sixty percent of wild mollusc shellfish species have been reported to be positive for viral RNA by PCR-based assays. Because these shellfish show no disease, this suggests that the viruses may be in persistent states. However, it is important to note that with filter feeders, the presence of viral RNA does not necessarily indicate that an infectious process is ongoing. Nonetheless, the RNA sequences of these shellfish birnaviruses appear to be distinct from those of fish birnaviruses, suggesting that shellfish are in fact persistently infected by birnaviruses. A specific example of this relationship can be seen with fish infectious pancreatic necrosis virus, which causes serious disease. This virus is also found in healthy shellfish species. In general, birnaviruses clearly resemble arthropod-borne viruses (e.g., arboviruses), Venezuelan equine encephalitis virus, and Sindbis virus, and they are situated between these insect arbovirus groups by phylogenetic analysis. In terrestrial hosts, these arboviruses are of interest because they cycle between (and replicate in) two widely separated taxonomic host groups, the vertebrates and the invertebrates. Many marine birnaviruses also appear to cycle between different host orders and are able to replicate in bony fishes and protostome species.

Nodaviruses (which have positive-strand ssRNA genomes) are also a major source of fish mortality on farms. These viruses are ubiquitous. The diseases induced by them include viral encephalopathy and retinopathy in sea bass. As noted above, in contrast to the case with fish DNA and retroviruses, virally induced growth abnormalities are not characteristic of nodavirus or birnavirus infections.

Host and virus diversity and host evolution

From the perspective of virology, vertebrate fish constitute a truly important biological transition, including the large-scale diversification of host species, the origination of the adaptive immune system, and the origination of the predecessors of terrestrial vertebrates. The most striking virus-related change among these is the creation of the adaptive immune system, with its breathtaking complexity and adaptability. Such a highly sophisticated system would clearly be expected to limit growth of viral parasites in these hosts. So it is most ironic that instead there is a large-scale radiation of virus species infecting bony fish, corresponding to the radiation of these host species. This may identify a broad pattern that will be seen with other hosts and their viruses. On a large scale of evolution, there appears to exist a discernible virus-host connection. Host species diversity appears to be associated with corresponding host-specific virus diversity. This host-virus pattern is not restricted to the oceanic organisms mentioned in this chapter, and similar host and virus diversity-linked patterns are presented in the next two chapters, which cover the origination of land plants, land animals, and their corresponding viruses. However, at the origin or base of these important radiations in host species, one can frequently observe an order of host species (such as sharks, sea urchins, or nematodes) that appear to remain relatively devoid of viral parasites. It appears that it is the descendants of these early virus-poor organisms that not only develop much greater species diversity (such as molluscs and bony fish) but also develop a correspondingly large diversity of viruses. It is possible that this perceived broad pattern is artificial, due perhaps to experimental bias in the study of mainly those viral parasites of economically important host organisms. However, in several specific examples this does not appear to be the case, as noted above. If this broad host-virus pattern is indeed real, how can it be explained? How does the existence or absence of large numbers of viral parasites affect host evolution and species formation? In bacteria, there is strong evidence that viral parasites do indeed sculpt the basic nucleotide word bias and evolutionary capacity of host genomes. What might be the evolutionary consequence to the eukaryotic host genomes in a situation of either prevalent viruses or a viral paucity, and how might this relate to host speciation or virus colonization of those host genomes? Perhaps by simply posing this question, we can now begin to evaluate the contributions of virus-associated forces to the evolution of host species. The lungfish genome, for example, has 40 times the DNA content of the human genome. Why is it so heavily colonized by repeat sequences (genetic parasites)? The lungfish acquired numerous important biological characteristics that were basal to the evolution of many diverse terrestrial vertebrate species. Yet as a species, lungfish are not diverse. We know that some representatives of early vertebrate fish, such as the puffer fish, compose much more diverse orders and have highly compact genomes, devoid of many of these genomic parasites. We do not known how or if a lack of these genetic colonizers changes or affects the interaction of these species with their viruses. It might be proposed that the evaluation of such a question would help us to better understand the selective pressures that created the host genomes.

Recommended Reading

Fish Viruses

Ahne, W. 1993. Viruses of *Chelonia*. *J. Vet. Med. Ser. B* **40**:35–45.

Essbauer, S., and W. Ahne. 2001. Viruses of lower vertebrates. *J. Vet. Med. Ser. B* **48**:403–475.

Muroga, K. 2001. Viral and bacterial diseases of marine fish and shellfish in Japanese hatcheries. *Aquaculture* **202**:23–44.

van Hulten, M. C., J. Witteveldt, S. Peters, N. Kloosterboer, R. Tarchini, M. Fiers, H. Sandbrink, R. K. Lankhorst, and J. M. Vlak. 2001. The white spot syndrome virus DNA genome sequence. *Virology* **286**:7–22.

ERVs

Ganko, E. W., K. T. Fielman, and J. F. McDonald. 2001. Evolutionary history of Cer elements and their impact on the *C. elegans* genome. *Genome Res.* **11**:2066–2074.

Gonzalez, P., and H. A. Lessios. 1999. Evolution of sea urchin retroviral-like (SURL) elements: evidence from 40 echinoid species. *Mol. Biol. Evol.* **16**:938–952.

Goodwin, T. J., and R. T. Poulter. 2001. The DIRS1 group of retrotransposons. *Mol. Biol. Evol.* **18**:2067–2082.

Herniou, E., J. Martin, K. Miller, J. Cook, M. Wilkinson, and M. Tristem. 1998. Retroviral diversity and distribution in vertebrates. *J. Virol.* **72**:5955–5966.

Leaver, M. J. 2001. A family of Tc1-like transposons from the genomes of fishes and frogs: evidence for horizontal transmission. *Gene* **271**:203–214.

RNAi and DNA Methylation

Barstead, R. 2001. Genome-wide RNAi. *Curr. Opin. Chem. Biol.* **5**:63–66.

Tweedie, S., J. Charlton, V. Clark, and A. Bird. 1997. Methylation of genomes and genes at the invertebrate-vertebrate boundary. *Mol. Cell. Biol.* **17**:1469–1475.

Deuterostome Immunology

Gross, P. S., W. Z. Al-Sharif, L. A. Clow, and L. C. Smith. 1999. Echinoderm immunity and the evolution of the complement system. *Dev. Comp. Immunol.* **23**:429–442.

Lin, W., H. Zhang, and G. Beck. 2001. Phylogeny of natural cytotoxicity: cytotoxic activity of coelomocytes of the purple sea urchin, *Arbacia punctulata*. *J. Exp. Zool.* **290**:741–750.

Magor, B. G., and K. E. Magor. 2001. Evolution of effectors and receptors of innate immunity. *Dev. Comp. Immunol.* **25**:651–682.

Miyazawa, S., K. Azumi, and M. Nonaka. 2001. Cloning and characterization of integrin alpha subunits from the solitary ascidian, *Halocynthia roretzi*. *J. Immunol.* **166**:1710–1715.

Nonaka, M., and S. Miyazawa. 2002. Evolution of the initiating enzymes of the complement system. *Genome Biol.* **3**:REVIEWS1001.

Genomes

Bowen, N. J., and J. F. McDonald. 1999. Genomic analysis of *Caenorhabditis elegans* reveals ancient families of retroviral-like elements. *Genome Res.* 9:924–935.

Cameron, R. A., G. Mahairas, J. P. Rast, P. Martinez, T. R. Biondi, S. Swartzell, J. C. Wallace, A. J. Poustka, B. T. Livingston, G. A. Wray, C. A. Ettensohn, H. Lehrach, R. J. Britten, E. H. Davidson, and L. Hood. 2000. A sea urchin genome project: sequence scan, virtual map, and additional resources. *Proc. Natl. Acad. Sci. USA* 97:9514–9518.

Viruses, Land Plants, and Insects: a Trinity of Virus, Host, and Vector

Introduction

Rationale for the trinity

In this chapter, I examine the deep issues concerning the origin and evolution of land plants, insects, and their viruses together. Earlier chapters examined the relationships between viruses and their hosts from the perspective of host evolution. The prior chapters, however, have all been able to consider one specific host lineage and to overlay the known relationships of that host lineage with its persistent and acute viruses. Accordingly, in chapter 5 I examined the oceanic species of green microalgae, red algae, and filamentous brown algae. In chapter 6, I examined the evolution of aquatic animals and their viruses. In this chapter, however, I examine the evolution of higher plants and insects together, along with their viruses. The reasons for combining these hosts are addressed in more detail below, but suffice it to say that plants, insects, and their viruses all show intimate linkages which indicate that they often coevolve. As it is the premise of this book to evaluate situations of virus-host coevolution for the potential role played by viruses, this chapter seeks to link viruses to plants and insects.

The earliest insects date to the Devonian period (400 million years before present [ybp]). The evolution of land plants and insects corresponds to a major evolutionary explosion (evident in fossil data) at the start of the Cretaceous period (about 135 ybp). I begin this chapter by examining the relationship of algae to the evolution of higher green plants. As the green algae are the accepted progenitors of land plants, they are worth reexamination at this time. In this chapter, I also consider the oceanic crustaceans along with their viruses and the evolution of terrestrial insects. As the crustaceans are the accepted progenitors of land insects, it is hoped that the consideration of these oceanic crustaceans and their viruses may help us better understand insect origins. It was noted in the last chapter that viruses infecting crustaceans are surprisingly diverse relative to the viruses of more primitive hosts (e.g., nematodes and *Dictyostelium* spp.). Viruses infecting shrimp, in

223

particular, have been well studied, and a shrimplike organism is considered the most likely ancestor of insects.

With respect to land plant evolution, there is reason to believe that symbioses with various mycorrhizal fungal species may have been important for the evolution of the land plants' root systems. In addition, fungal species also function as important vectors for plant viruses. Thus, I briefly examine (in the context of plant evolution) the filamentous Fungi and their ubiquitous infection with RNA hypoviruses. Finally, it must not be forgotten that understanding virus-host evolutionary dynamics also requires a consideration of the various virus life strategies, including the evaluation of virus-virus interactions. It is thus important to consider the relationships between persistent and acute viral agents of plant and insect hosts. This consideration will include those persistent viral agents (and their defective derivatives) that have colonized the host genome in lineage-specific ways. As will be made clear, virus-virus interactions seem to be especially prevalent for the RNA viruses of land plants, and it is common to find mixed virus infections in natural settings.

Conditions and limitations of the trinity

I seek to integrate two major lineages of host evolution along with their viruses and the vectors that transmit them. This is a daunting task which would seem to pose the risk of adding an unnecessary layer of complexity to an already complex issue. There is little to no fossil record of viruses to use to calibrate and understand the origin of various virus-host relationships, many of which have existed since early in the evolution of these hosts. This leaves mainly phylogenetic analysis data from which to extract inferences concerning viruses and host evolution. There are clear limitations to such an analysis. Furthermore, knowledge about persistent plant and insect viral agents is most often incomplete relative to our understanding of viruses that cause acute diseases in these hosts. Thus, it is generally more difficult to evaluate the contributions of persistent viruses to host evolution. In addition, the literature on viruses that infect plants and insects is highly biased towards the study of viruses infecting crop species of plants or viruses that can be used as biological control agents to infect insect pests. This literature bias must always be kept in mind in an evaluation of what is known about broad evolutionary patterns. Finally, human activity, especially agricultural activity in the last 10,000 years, has generated large, closely spaced, and genetically homogeneous plant populations, which have frequently been introduced into new habitats. This human agriculture has almost certainly affected the ecologies of the viruses and insects that are now prevalent.

Paucity of natural biology

As a rule, we have little knowledge concerning the natural biology of most plant viruses and their hosts. For example, *Tobacco mosaic virus* (TMV) was the very first viral agent to be discovered and isolated, and it is easily the best studied of all plant viruses. However, our understanding of the possible natural origin of TMV and its relationship with its natural host is surprisingly incomplete and has only recently received attention (discussed below). Fifty-five virus types are currently known to be capable of infecting tobacco crop plants. Thus, with this most important cash crop a very large diversity of viral parasites can be seen. However, for the most part we know little concerning the natural biology or origin of these viral agents. Given the overall high diversity of viruses of higher plants, especially positive-strand single-stranded RNA (ssRNA) viruses, we must seek explanations for such distinct virus-host patterns and appreciate how little we actually know of the natural forces that have led to such broad relationships. In natural, nonagricultural settings rich in plant life, such as in a tropical rain forest, virus persistence appears to be the common situation, and little virus-mediated disease is evident. However, there are few systematic studies on this topic. In this chapter, I seek to provide an overview of this issue with what can at best be considered incomplete and at worst possibly misleading information. To address this lack of balance in our studies, at times it will be necessary to draw strong inferences from a relative few, better-studied examples of natural virus-host relationships. However, such inferences from few examples have the potential to be misleading when generalized. Land plant evolution and insect evolution are highly linked, especially for the angiosperm plants that depend on insects for pollination. That linkage of evolution, however, must occur on a fitness landscape in which competing and interacting viral agents represent very important and ubiquitous factors that influence the host landscape's shape. The role that viruses have played in the evolution of their hosts has seldom been addressed in the context of either plant or insect evolution. In this chapter, I first present the overall patterns of host plant evolution and then address the evolution of host insects. Next, I consider the plant and insect viruses. Finally, I attempt to integrate these issues by considering how host insect and insect virus evolution intersects with the evolution of plants.

The Coevolved Trinity of Plant, Insect, and Virus

As presented in earlier chapters, early plant and animal life forms both initially evolved in the oceans. From the perspective of a virus, the oceans represent a very supportive habitat, in that the aquatic medium conducts viruses to all nearby hosts, avoiding desiccation and the explicit need to use vectors. In the oceans, viruses are susceptible to UV light-mediated damage and inactivation, which accounts for the majority of virus killing in this habitat. At depths of a few meters, filtration of damaging UV light would aid virus survival. Thus, the interaction of viral systems with light is a crucial aspect of the oceanic habitat. As plant and insect hosts adapted to a terrestrial habitat, the viruses of these hosts would also have been under selective pressure, not only to survive the more direct and intense irradiation from sunlight but also to withstand the much greater rate of desiccation. In addition, the absence of an aqueous medium would have required viruses to develop much less passive, non-diffusion-based systems for transmission to new hosts, such as the use of mobile vector hosts. Thus, major problems confronted the evolution of viruses as they moved to the terrestrial habitat and encountered major shifts in habitat characteristics and the characteristics of their land-adapted hosts. Except for hosts that still required water for egg and larval development (such as amphibians), terrestrial viruses now required specific adaptations that would allow interhost transmission in the absence of passive diffusion. In many cases, this necessitates a motile vector (e.g., an insect) to infect an immobile host (e.g., a plant). Thus, we can see how viruses may provide a direct linkage between the evolution of insects and that of plants.

Early plants and insects

It is now accepted that a progenitor of land plants was the green microalgae, based on phylogenetic analysis of chloroplastic and mitochondrial DNA. The earliest land-based plant descendants of these oceanic algal species are now best represented by the bryophytes, which are simple, mosslike plants. The progenitor organisms of terrestrial insects appear to have been some type of shrimp species resembling the extant fairy shrimp (order Anostraca). The first land-based insect descendant is best represented by species of Blattaria (which include modern cockroaches). Of these two groups (insect versus plant progenitors), clearly the insect progenitors are the more complex organisms, as they have already developed complete organ systems. Nonetheless, virus infections appear to be prevalent in the extant representatives of both early plant and insect progenitors. As already mentioned in chapter 4, algae are highly prone to infections with large double-stranded DNA (dsDNA) viruses of the phycodnavirus-like family. Both lytic, acute infections and persistent, inapparent infections (transmitted through gametes) are known. Other virus types, such as RNA viruses, are also known but much less common in algae (discussed below). Shrimp species are also known to be infected with an array of viruses, which includes almost all families of DNA and RNA viruses. As mentioned in chapter 6, the picorna-like Taura syndrome virus has especially been a problem for shrimp farming. Since most of these viruses appear to be highly specific to and congruent with their hosts, it is inferred that similar virus relationships prevailed in the oceans at the originations of land plants and insects. Thus, it can be fully expected that at the originations of these two groups of land species, the viruses that infected them were prevalent in the oceans.

At the evolutionary junction corresponding to the initial land colonization by life, there is a corresponding sharp shift in virus-host relationships for both plants and insects. Plants underwent adaptation to land, creating bryophyte-like organisms that acquired root systems and then vascular

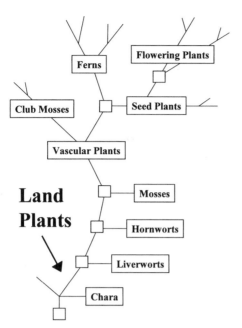

Figure 7.1 Phylogenetic relationships of early plants. Redrawn from http://faculty.washington.edu/mandoli with permission.

structures. This led to the emergence of the ferns and later to the origination of gymnosperms. A phylogenetic dendrogram that outlines this early plant evolution is shown in Fig. 7.1. Along with these plant adaptations there is what appears to be another transitional virus paucity in evolutionary biology, similar to the "virus void" situation seen in the evolution of *Dictyostelium* and nematodes (chapter 5). Although viruses of modern plants are very numerous and prevalent, few viruses appear to be able to infect the extant representatives of early terrestrial plant hosts (based on the general lack of reports in the literature of such viruses). This pattern of virus paucity also includes the ferns, which represent a very successful plant type that was an early colonizer of land. It must be stated, however, that such an absence of data is always suspect and possibly distorted due to insufficient or uneven analysis. However, I can state with some confidence that the occurrence of acute viruses infecting such plants is exceedingly rare. Currently, there are few data concerning the possibility of persistent or inapparent virus infections in these hosts. Nonetheless, it is known that, unlike the situation often observed for other persistently infected hosts, such as the filamentous algae or filamentous Fungi, virus reactivation in reproductive tissues and gametes has not been reported for the representatives of the early land plants.

The genomes of flowering plants (angiosperms) are known to be highly colonized by various genetic parasites (mainly retroposon-derived elements; see Fig. 6.4). However, the evolutionary pattern of the colonization of early plant genomes by genetic parasites has not been fully studied. The main limitation to this analysis is that the genomes of early plant species have not been sequenced. Nevertheless, the little information that has been collected supports the idea that high-level colonization by retroposons, retroviruses, viruses, and their derivative elements is much less prevalent in these early plant species than in the genomes of ferns and flowering plants and gymnosperms (e.g., conifers, cycads, and ginkgos). For example, fern genomes are estimated to be about 160 trillion base pairs in length, compared to about 3 trillion base pairs in the human genome. Fern genomes are surprisingly huge, surpassed only by the genomes of some amoeba species, which can contain from 290 trillion to 670 trillion base pairs of DNA. Clearly, fern genomes underwent some type of massive colonization by genetic parasites, but the exact makeup of these colonizers has yet to be determined. Lilies, which appeared early in the evolution of the flowering plants, also have very large genomes, with about 36 trillion base pairs. It is not known, however, if such genome colonization by genetic parasites has any effects on virus-host relationships. Thus, based on current data, it can only be concluded that during the evolution of higher plants, there was a great accumulation of genetic parasites.

Angiosperms and virus-linked species radiation

Although the evolution of gymnosperms (nonflowering plants bearing naked seeds) is associated with the appearance of some viruses able to infect these species, it is the evolution of angiosperms (flowering plants bearing fruit-enclosed seeds) that exhibits a very large radiation in both host species and virus types. This radiation of so many angiosperm species was noted by Charles Darwin to pose a puzzling and significant problem for evolutionary biology, as there seemed to be no apparent driving force for such large-scale speciation. Figure 7.2 shows a phylogenetic dendrogram with the currently accepted overall pattern of green plant evolution. With respect to the corresponding radiation of viruses, the great majority of viruses that infect angiosperms are various types of positive-strand ssRNA viruses, which appear to have several distinct origins. Some of these virus families are unique to plant species. As is discussed below, the confusing complexity of the nomenclature of these viruses poses a significant problem for non-plant virologists, similar to the problem with endogenous and autonomous retrovirus nomenclature that is discussed in chapter 8. A most difficult issue concerning the extant plant viruses is evaluating the possible evolutionary origins of all of these virus families. As is discussed below, there is reason to think that many of the plant virus families are polyphyletic, as they may not share common origins. For example, positive-strand ssRNA virus families exist in both rod and isometric forms, which are thought to have distinct lineages. In addition, some of these virus families have segmented genomes, which are also considered to be distinct. Yet it would be expected that all of these extant virus families would most likely trace their origins to some oceanic virus, similar to those found in oceanic algae, animals, and Fungi. The most promi-

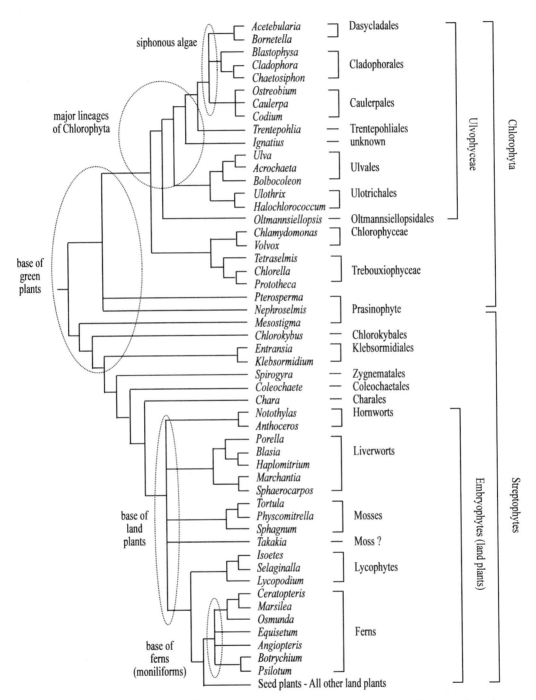

Figure 7.2 Overall evolutionary relationship of green plants. Redrawn from http://faculty.washington.edu/mandoli with permission.

nent and diverse of all virus families are positive-strand ssRNA plant viruses. As noted in chapter 6, if the "picorna-like" viruses are loosely defined as positive-strand ssRNA viruses, then the oceans currently contain six known and four previously unknown families of such viruses. In plants, two of these known ocean families, the *Potyviridae* and *Comoviridae*, are present and are among the most numerous plant viruses, in terms of species numbers. Thus, some linkage between the "picorna-like" viruses found in plants and the oceans seems clear. The *Tombusviridae*, *Bromoviridae*, and the rod-shaped *Closteroviridae* are the next most species-rich plant viruses, but these are not considered to be picorna-like viruses. However, as discussed below, it may still be possible to link some of these virus families to those that infect likely predecessor hosts (such as fungal *Barnaviridae*). Of special interest are the plant tobamoviruses (a floating genus) and the family *Luteoviridae*, whose taxonomy is tightly linked to that of its angiosperm hosts. These two virus families appear to have distinct relationships to their hosts, as discussed below. Some virus families are associated with high mortality of their hosts, whereas others are not associated with much host disease. In terms of early evolution, it might be inferred that the various plant positive-strand ssRNA virus families most likely evolved from viruses that are found in oceanic hosts. Although some positive-strand ssRNA viruses infect algae, such viruses are especially common in shrimp species. Thus, it is important to consider the possible interactions between insect and plant orders mediated by viruses. It is curious that although green algae are prone to infections with large nuclear DNA viruses, no large or intermediate DNA viruses are known for any land plant species. Thus, there is a clear and sharp demarcation of the virus-plant relationship.

Insect-plant-virus radiation

Along with the origin and radiation of angiosperm species, one can see a corresponding radiation in insect species. Similar to the flowering plants relative to other plant orders, insects show the greatest diversity relative to all other animal orders. That is, insects are a highly diverse and successful group. Flowering plants are, for the most part, dependent on insects for their pollination. Conversely, many insects depend on plants as food sources. This has the practical consequence that in commercial terms, insect predations account for the majority of agricultural losses. Pollinators are mostly flying insects (Hymenoptera, Diptera, and Lepidoptera), which evolved in parallel with angiosperms. Most flying insects undergo metamorphosis from larval wormlike forms into adult winged flying forms. The origination of insect flight and the evolution of metamorphosis both appear to be linked to plant evolution. As an example, a linkage of flying insect evolution to the origination of vascular plants with tall stature has been proposed. In addition, many larval forms of flying insects are responsible for much of the foraging on and consumption of plants, but these larvae may themselves be subjected to parasitization by parasitoid wasps. As discussed below, endogenous DNA viruses of these wasps can be essential for the parasitoid-host relationship. Furthermore, an argument can be made, based on the parasitoid wasp-flower relationships seen in some lilies, that insects may have played a role in the origination of flowers. Intriguingly, it appears that plants may also be able to manipulate the parasitoid-host insect relationship by producing sex attractant signal molecules that recruit the parasitoid, following predation by a larval host of either the parasitoid or the flowering of the plant. Finally, it needs to be emphasized that by far the greatest number of insect species that are vectors for plant viruses fall within the order Hemiptera (true bugs). This very diverse order includes aphids and leafhoppers. For the most part, plant viruses transmitted by these insects are specific for the insect vector, although they seldom replicate in the vector. In the case of aphids, this virus vector specificity can be very tight. The unusual biology of aphids (e.g., aphids are often parthenogenetic, haploid, and clonal) and their link to viruses make an important point about the intimate relationships that exist among viruses, hosts, and vectors. Several explanations account for the specificity of a plant virus for its insect vector. Often, this specificity relates to the specific binding of the virus to surfaces in the insect mouthparts. In the case of some aphids, however, it has been established that virus specificity relates to effective virus persistence within aphid cells, which is mediated by virus-specific binding to heat shock proteins (GroEL) expressed by endosymbiotic bacteria (*Buchnera* spp.) of the aphid. Like the various orders of plants, the insect orders have broad patterns of virus-host relationships that are well maintained.

The earliest land insects appear to be of the Blattaria order, which includes modern cockroaches. A dendrogram, based on comparison of *cox1* sequences, depicting the overall pattern of insect evolution is shown in Fig. 7.3. Although some positive-strand ssRNA viruses have been reported for cockroach hosts, they are more commonly found in other species, such as bee and cricket species. Overall, however, there appears to be a clear paucity of positive-strand ssRNA viruses in insects, not only representatives of early insect orders but also those insects, such as aphids, that function as vectors for plant RNA viruses. Other major virus families currently seen in insects in general appear to be absent from these early insect orders (i.e., baculoviruses, entomopoxviruses, negative-strand ssRNA viruses, etc.). This contrasts with the case of Orthoptera (crickets and grasshoppers), which are known to support infections with ascoviruses and entomopoxviruses. Overall, there appears to be a clear and general trend in the insect-virus relationships. The majority of insect viruses are DNA viruses of moderate to large genome size. Mostly, these viruses (such as baculoviruses) appear to have lytic, acute relationships with their hosts and also tend to infect and cause acute disease in larval forms of insects. Unlike for bacteria and algae, however, there are currently no reports of insect dsDNA viruses (nongeminiviruses) that are able to integrate and persist in host genomes (as do the

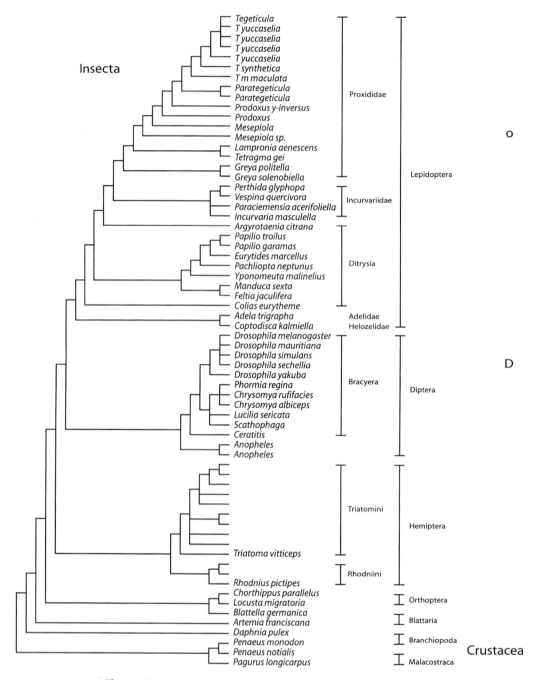

Figure 7.3 Overall evolutionary relationships of insects. Adapted from M. W. Gaunt and M. A. Miles, *Mol. Biol. Evol.* **19**:748–761, 2002, with permission.

phaeoviruses of filamentous algae or lysogenic phage of bacteria). However, as discussed below, the polydnaviruses (PDVs) are integrated genomic viruses of parasitoid Hymenoptera.

Thus, there are many observations that suggest strong linkages between plants, insects, and their viruses. In considering the combined evolutionary relationships between land plants and insects and the viruses that infect them, it seems clear that they will not be represented by a simple treelike topology with a common trunk. Instead, some type of overlapping reticulated or superimposed network will be required. This superimposed network will need to show clear patterns of coevolution between these distinct orders of life. Along these lines, one might envision two major lineages (plants and insects) that are mostly congruent with each other, but symbiotic with Fungi at their origin. Viruses will have both host-congruent and host-incongruent patterns of evolution. Typically, a viral agent is not considered a basic element of the evolutionary tree of its host. However, as presented below, viruses may well provide the thread that binds the fabric of coevolution between plants and insects.

Plants and Their Viruses

All green plants can be considered to belong to two major phyla: the Streptophyta (which includes all land plants and their charophyte relatives) and the Chlorophyta (which includes the rest of the green algae) (Fig. 7.2). All land plants appear to have evolved from one common progenitor which underwent diversification to currently existing families and some extinct plant families. A common origin is inferred from various features shared among all land plants, such as the characteristics of chloroplasts and RNA editing in plastids. The overall path of land plant evolution began with green microalgae in the oceans, which were associated with a particular type of chloroplast. Plant adaptation to land is now accepted to have occurred along with the acquisition of a root system via endosymbiosis with filamentous Fungi. The origination of leaves and the origination of vascular systems were subsequent major developments. The origination of seeds in gymnosperms and the origination of flowers in angiosperms then followed; these acquisitions are associated with major radiation of plant species. The extant representatives of simple plants (e.g., mosses) are considered paraphyletic to the lineages that led to angiosperms. The major taxonomic clades of extant plants are Gnetales, sphenopsids, ferns, gymnosperms (e.g., conifers, cycads, ginkgos, and ephedras), and angiosperms (monocotyledons, eudicotyledons, and magnolias).

Early plants and viruses

Land plants (embryophytes) are considered to have evolved from charophycean green algae since both of these orders have lignin, which is absent from other green algae. In addition, the two orders have similarities in their chloroplasts, such as the presence of group II introns in chloroplast tRNA (a conserved characteristic of all land plants). Bryophytes represent a lineage that diverged early from the land plants and is now considered paraphyletic to them. Bryophyte species resemble some algae in that they produce diploid motile spores but no seeds. The evolution of early land plants is also associated with the development of a root system, and it is now generally accepted that a symbiotic relationship with filamentous Fungi, with their capacity to make penetrating hyphae, was the likely origin of the land plant root system. A symbiosis between plant root systems and separate fungal organisms is especially evident for tree species, such as pines, which generally have a mantle of mycorrhiza covering their root systems. Fungi can also be found to be associated with lower plants and can produce structures analogous to the vesicular arbuscular structures of higher plants. Various extant symbiotic Fungi (such as species of the *Zygomycota, Ascomycota,* and *Basidiomycota*) show such associations with lower plants. These associations are similar to that of the root fungus mycorrhiza and are consistent with the hypothesis that symbiosis between Fungi and plants was important for land plant evolution.

However, this hypothesis of symbiosis with Fungi poses an interesting situation with respect to possible virus relationships. As essentially all extant filamentous Fungi appear to be persistently infected with various types of RNA viruses (mainly hypoviruses and other RNA viruses; see chapter 5), it is fully expected that the progenitor of the land plant root systems would also have been persistently infected with RNA viruses. If so, this symbiosis would have provided a good opportunity for the evolution of some land plant RNA viruses through adaptation to the new plant host. Given that *Mushroom bacilliform virus* (MBV) (positive-strand ssRNA) is the only member of the family *Barnaviridae* and also that it has clear similarity to some plant RNA viruses (discussed below), it seems quite possible that this virus could represent the viruses of Fungi that were present early in the evolution of land plants. Thus, some virus lineages might date back to the very origins of land plants and their associations with Fungi.

RNA viruses of Fungi

MBV is the sole member of its virus family and is unique to Fungi. Its genome consists of one RNA segment of 4 kb that encodes a replicase and a coat protein and contains a 72-kbp open reading frame (ORF) which appears to encode a helicase. It has no movement proteins. Thus, MBV represents a relatively simple RNA virus. MBV bears a striking resemblance to *Alfalfa mosaic virus,* which is one of the better-studied potyvirus members (of which there are many). Within the potyvirus group, *Spanish latent virus* appears to be the basal member, based on phylogenetic sequence analysis. *Alfalfa mosaic virus* has three RNAs, one of which encodes a movement protein; thus, it is a more complex virus than MBV. It has been suggested that MBV may represent a progenitor virus with one RNA segment that evolved to the multiple RNA segments of potyviruses. This idea may be general in that most RNA viruses with segmented genomes may have evolved from progenitor viruses with single RNA segments. However, the possibility that multiple virus lineages may have converged to generate multisegmented viruses is also a viable hypothesis. Bromoviruses also have multipartite RNA genomes (typically three). Their RNA replicases show some similarity to that of TMV. However, in other genes bromoviruses are more similar to cucumoviruses, so it is possible that the bromoviruses had multiple viral progenitors.

RNA viruses of algae

Although certain DNA virus families are known to be prevalent in algae (large DNA phycodnaviruses and phaeoviruses), these viruses have no counterparts in land plants and clearly could not have been direct progenitors of the viruses of modern plants. However, some RNA viruses of algae are also known, and it is these that may represent plant progenitor viruses. For example, of the rod-shaped RNA viruses that infect algae, the best characterized and most relevant is *Chara* alga virus (CAV). This virus has several characteristics which suggest that it could represent an ancestor of various extant plant viruses. A relationship between CAV and TMV can be demonstrated by cross-reactivity of antibodies to capsid proteins of TMV with proteins of CAV. In other words, CAV is a tombusvirus-like agent that resembles viruses known to infect a wide range of angiosperm species. Furthermore, phylogenetic analysis supports the idea that CAV might represent the oldest and most basal member of this virus clade. The implications of these results are that CAV does not seem to represent a more recent adaptation of a modern plant tombusvirus to algae, but rather that CAV may be the oldest member of this virus family, again suggesting that such viruses may have been prevalent prior to the origination of angiosperms.

Fungal vectors and zoospores

In addition to directly supporting the replication of fungal viruses, Fungi are also known to transmit RNA viruses (especially the furoviruses) to plants. Of these, *Cucumber green mottle mosaic virus* (CGMMV) (a tobamovirus) is a well-studied example. CGMMV also has the curious ability to parasitize algae. Partitiviruses are also known to infect Fungi as well as two genera of plants. In plants, partitivirus infection is typically cryptic or inapparent, with no or few symptoms. Thus, partitivirus appear to establish a persistent virus-host relationship. Partitivirus infections tend to be transmitted through seeds. The biological relationship between these plant viruses and their fungal vectors is striking. Fungus-mediated virus transmission often involves the motile fungal zoospores, which are able swim to roots of host plants and carry the virus with them. In a sense, the motile zoospore provides the same function as an aphid does for many angiosperm-infecting viruses—motility. Partitiviruses are mostly rod-shaped, positive-strand ssRNA-containing viruses which become encysted within zoospores. In most cases, the virus is partitioned in the spore while infecting the fungal host, establishing a tight biological relationship between virus and fungus. Once encysted within these zoospores, the virus is protected and can survive over 20 years in some cases.

Spores, seeds, and early plants

It is likely that both algae and early land plants had motile spores before the evolution of seeds. Most algae are haploids that occasionally become diploid during sexual reproduction and zoospore generation. Interestingly, in filamentous brown Fungi and other higher Fungi, zoospore formation is frequently associated with reactivation of species-specific persistent virus replication (see chapter 4). Along with the evolution of seeds in plants, there must have been a shift from the diploid alga-like spores to haploid seeds, as well as a shift from the alga-like haploid to diploid soma cells (seen in many extant plants). The forces that led to this shift in life strategy are unknown. However, it seems likely that examining extant representatives of early land plants might yield some information on this issue. Identification of the likely early ancestors of land plants is based mainly on the analysis of mitochondrial DNA (mtDNA) and chloroplastic DNA (cpDNA). Such analysis suggests that the Prasinophyceae (algae) appear to most closely represent the early ancestors of green plants.

Plastid DNA and early plants

Prasinophyceae, such as *Mesostigma viride,* appear to be basal to both the Streptophyta (bryophytes, ferns, and all other multicellular land plants) and the Chlorophyta (all green algae). This alga has a circular mtDNA that resembles those of other green algae in that it is small (about 42,000 bp), with high gene density (87% coding sequence, 65 genes). This mtDNA has four group I introns (in *cox1*) and three group II introns (in *cox2*), which is characteristic of all plants. *M. viride* is a scaly green biflagellate with very large cpDNA (135 genes), which distinguishes it from other plant species. This alga appears paraphyletic to higher plants and may be the earliest green plant to have diverged from the lineage that led to land plants. The biological and reproductive characteristics of *M. viride,* however, resemble those of algae more than those of land plants, strongly suggesting that early land plants were very alga-like.

In bryophytes, such as *Codium fragile,* the cpDNA is circular. At 89 kb, it is the smallest cpDNA known, lacking the large repeat elements present in other plant cpDNAs. Curiously, within these hosts, the cpDNA has a low evolutionary rate (i.e., a slow accumulation of nucleotide substitutions). This is in stark contrast to the plastid genomes of other plants, which show much higher evolutionary rates. It is interesting that the bryophyte cpDNA codes for an RNA polymerase that resembles the RNA polymerases of bacterial phages. However, in mosses such as *Physcomitrella patens* (which is a sister group to all land and flowering plants), the plastid RNA polymerase has become nuclearly encoded by the *PpRPOT1* and -2 genes. These are clearly phage-type RNA polymerases and appear to be a relatively recent acquisition in the genomes of these land plants. These observations suggest that virus-like agents participated in the early evolution of cpDNA in these descendants of early land plants. It is interesting that the phage-like (T7-like) RNA polymerase that copies the avocado sunblotch viroid is plastid encoded. And this virus is known to infect chloroplasts, an otherwise an unusual situation for the viruses of higher plants.

Characteristics of early land plants

The origination of vascular plants and the development of seeds represent two linked early and major developments in land plant evolution. Vascular systems appear to have evolved in bryophyte-like ancestors, which tended to be parasitic sporophyte species whose spores were diploid. Vascular plants comprise two large classes—Lycopodiophyta (club mosses) and Euphyllophyta (seed plants, ferns, and horsetails). Lycopodiophytes have leaflike structures but have short plant stature and do not make seeds. Euphyllophytes are vascular plants that can be of tall stature and do produce seeds. These two major characteristics of euphyllophytes (tall stature and seed production) allowed these plants to adapt to new, previously unavailable habitats. The capacity for increased plant stature resulted from vascularization (such as with trees) and also required a corresponding increase in root penetration, suggesting that there must have been some coordination in the development of multiple characteristics. The creation of seeds clearly enhanced habitat availability to plants and allowed plants to colonize drier land. This high-land colonization most likely led to enhanced soil production and subsequently nutrient-rich runoff. It has been suggested that this increased, nutrient-rich runoff would affect the oceans, increasing the frequency of algal blooms. These algal blooms could themselves result in widespread anoxia and lowered CO_2 production and consequently could induce glacial periods (i.e., ice ages). In addition, as discussed in chapter 4, most algal blooms are associated with or terminated by corresponding increases in lytic virus production. This suggests that a major shift in global virus-host dynamics was likely to have occurred at the origination of vascular land plants.

Virus paucity in early land plants

The situation concerning viruses and early land plants is curious. Although these plant species are relatively well studied (with approximately 6,000 scientific papers published in the last 10 years) and the viruses of the extant land plant predecessor representatives (algae and Fungi) are prevalent, there are no reports of viruses infecting any of the representative early land plant species. No virus, for example, has been reported for any bryophyte. This paucity of virus reports also applies to ferns, which are also sporeformers (like bryophytes). Very few ferns have been observed to support the replication of any virus, although tobra-like virus particles have been found in the hart's-tongue fern. And yet, ferns are sufficiently well studied to have made likely the identification of prominent virus infections. It is therefore clear that lytic virus infection of these plant species is either rare or nonexistent. However, as has been argued in earlier chapters, the lack of virulent virus infection does not preclude the possibility of prevalent infections with persistent, nonlytic viruses, as has been described for higher Fungi and algae. The observation of tobra-like virus particles in one fern might be such a situation. Yet persistent viruses are often identified by the presence of their nucleic acids in host cells or the transient production of virus particles (such as in zoospores of brown algae and higher Fungi). No such observations have been made for ferns, suggesting that persistent virus infections are also not common in these plants. Given the technical difficulty of observing persistent viruses, however, it remains very possible that such persistent viruses are prevalent and have simply been overlooked in these plant species. Still, relative to higher plants, there is clearly a paucity of virus diseases for lower land plants. As mentioned, this putative virus void is most reminiscent of a similar void described for the early aquatic animal life forms (such as *Caenorhabditis elegans* and sea urchins) in chapter 6. In that situation, it was noted that these hosts had also acquired the RNA interference (RNAi) system, which in these early animals has the added characteristic of being a systemic and transmissible response that could preclude most virus infections. However, in the case of early land plants, we do not currently know of the existence of a general host antiviral defense system that might be able to preclude infections by various viruses. It is curious that ferns have such huge genomes relative to those of most species, indicating massive colonization by genetic parasites. Since the genomes of these plants are poorly studied, however, there may well exist some yet-to-be-discovered system that can preclude infections by numerous viral families.

Gymnosperms

Gymnosperms (gym = naked, sperm = seed) represent a major group of land plants characterized by seeds not enclosed in an ovary (fruit) and include the cycad and conifer families and various palmlike plants. Extant gymnosperms are considered to be a sister group to angiosperms and diverged early on from other higher plants. In other words, gymnosperms do not appear to have been direct predecessors of angiosperms. Gymnosperms are monophyletic. Within the gymnosperms, the Gnetales appear to be the closest relatives to the conifers. Based on fossil evidence, cycads appear to have been the first gymnosperm plant lineage to diverge, followed by ginkgos, then the

Gnetales, and then the Pinaceae (a sister group to the monophyletic conifer families). Extant cycads are often viewed as living fossils. Some viruses of gymnosperms are known, but their numbers appear to be very much lower than those of angiosperm viruses (discussed below). Some reports of nepoviruses can be noted, such as for *Cycas revoluta*, which has been reported to harbor a nepovirus-like virion particle. It seems likely that such an observation could represent a persistent infection. However, as is often the case concerning persistence, this issue has not been evaluated carefully. Prevalent infections with acute viruses, however, are clearly not observed in most gymnosperms.

Gymnosperm genomes

Especially noteworthy are the genomes of gymnosperms, which are often highly colonized by various specific families of retroposons. These retroposons can constitute the majority of gymnosperm genomes. Curiously, authentic autonomous retroviruses for gymnosperms, or any other plant for that matter, have not been observed. This poses the question of how these retroagents gained initial access to plant genomes. Even more curious, and in apparent distinction from most other eukaryotic genomes, plant genomes appear to lack any full-length versions of endogenous retroviruses (ERVs) containing *env* sequences, although a few copies of *gag* and reverse transcriptase (RT) gene-containing plant retroposons have been identified in rice (*Oryza sativa*) genomes. Pararetroviruses or badnaviruses (which replicate via extranuclear DNA genomes) are known for many plant species and are also represented in relatively small numbers within the genomes of gymnosperms. However, as these agents lack integrases, their integration is likely dependent on other sources of integrase to allow for genome colonization. The presumed mediator of such integration events would likely be activated RT from endogenous plant retroposons.

The analysis of gymnosperm retroelements was initially accomplished using PCR consensus primers to amplify and sequence the various families of retroposons. From these data, it was observed that gypsy-like elements, their non-long terminal repeat (LTR) long interspersed repetitive element (LINE) derivatives, and copia-like elements were the most widely distributed. Most of these element families appeared to be monophyletic and well conserved within a specific plant lineage, even though the different retroposon groups (e.g., gypsy, copia, and LINEs) were clearly distinct and widely separated from each other within one plant lineage (see Fig. 6.4). The great majority of these elements lack intact *gag* and *pol* ORFs, so no functional proteins can be expressed from these elements. The gypsy-like elements are rather similar to the retroviruses found in animal genomes. However, in no situation has a transmissible version of a plant retroposon been reported to have been propagated, so there seems to be no example of reactivation of a plant ERV. Badnavirus-related elements formed clades distinct from those of the retrovirus-related elements. These badnavirus-related elements appear to be basal. Curiously, the host clades as defined by the various retroelements did not clearly define major taxonomic plant clades (conifers, ferns, gymnosperms, etc.). This is rather different from the situation for ERVs and mammalian genomes (chapter 8). Within the various families of gymnosperm trees, the gypsy and copia elements are both monophyletic. This monophyletic pattern is evidence against a recent horizontal transfer of these elements to these plant lineages and suggests that they were acquired at the origination of these lineages. Thus, we are left with a pattern of retroposon colonization that is rather reminiscent of that seen in the early animals (e.g., *C. elegans* and sea urchin). For unknown reasons, gymnosperm lineages, like the simple animals, are associated with early events of widespread colonization by lineage-specific retroelements. The great majority of these elements are defective, especially for the *env* sequences. Although some intact copies appear to have been conserved, discovery of the possible significance of such conservation awaits comparative full-genome analysis of various gymnosperms.

Angiosperms

Angiosperms are land plant species characterized by ovules enclosed in an ovary or the flower of flowering plants. Angiosperms have the most complex reproductive organs of land plants. Furthermore, flowering plants represent by far the largest and most diverse of all land plants, with over 250,000 species. Among the various types of angiosperms, the orchids are most diverse and also an old lineage. The enormous angiosperm diversity was an enigma to Darwin and was referred to as "Darwin's abominable mystery." This outstanding mystery remains in that there is still a need to explain the forces that might have led to such a diversification. A linkage between flowering plants and insects seems clear. The large majority of flowering plants rely on pollination by insects for their reproduction; thus, plant reproductive success is partially determined by insect behavior. However, less well appreciated is the large diversity of plant viruses also associated with angiosperms, as discussed below. Flowering plants comprise two major groups: monocotyledons and dicotyledons. Monocotyledons are flowering plants with a single cotyledon (leaf seed and embryo) and include lilies, orchids, grasses, and cereals. Monocotyledons constitute about one-quarter of all flowering plants. Dicotyledons (flowering plants with two cotyledons) include mustards, legumes, cacti, and most fruit trees.

The origination of angiosperms can be traced to the early Cretaceous period. For flowering plants, there is strong evidence of repeated, recent, and diverse transfers of mtDNA genes to the nucleus. Therefore, some system or process must exist that facilitates nuclear gene migration. In addition, angiosperms have experienced a large amount of horizontal gene transfer, especially into the mtDNA. For example, the invasive homing group I introns of the *cox* gene have been acquired over 1,000 times in various angiosperm species. The mechanisms that allow these sequences to move, seemingly from one species into another, are obscure. However, the possibility that these "horizontal transfers" might actually reflect independent colonization events mediated by related genetic parasites, such as viruses, cannot be ruled out and actually seems probable. Thus, unlike the early land plants, the plastids of angiosperms are rapidly evolving and under constant invasion by genetic parasites. Of the 14 extant monocot lineages, all appear to have diverged from each other during the early Cretaceous period (~100 million ybp).

Viruses and angiosperms: overall patterns

Like the angiosperms themselves, the viruses that infect these plants are by far the most diverse and numerous for the entire plant kingdom. Because of this diversity, most published considerations of viruses of plants are actually focused almost exclusively on viruses that infect angiosperms. Consequently, in this chapter, there is also a more detailed examination of angiosperm viruses and their possible origins relative to discussions of other plant viruses. Overall, several patterns of these viruses and their angiosperm hosts can be noted. As mentioned repeatedly, most viruses of angiosperms are RNA viruses of positive-strand polarity. An additional distinction from other host orders is that angiosperms are also frequently infected with mixtures of viral agents, including satellite viruses. This type of mixed infection is common in field isolates. As previously noted, angiosperms lack the larger dsDNA viruses, other than the geminiviruses. Unlike for insects and algae, no phycodna-like viruses, pox-like viruses, baculoviruses, or herpesviruses are known for any plant species. A prevalent explanation for this lack of DNA viruses has been that plants, with their very thick and relatively impervious cellulose cell walls, present an impenetrable barrier to all moderately sized DNA viruses (although *Agrobacterium* can penetrate plant cells, bringing along tumor/transforming DNA [T-DNA] plasmids). Communication between plant cells occurs via plasmodesmata. Plasmodesmata have relatively small openings that normally restrict movement of macromolecules between cells and could preclude DNA virus transmission. These structures are targets of many RNA virus proteins and can be enlarged by the movement proteins of various positive-strand ssRNA viruses. That only plant RNA viruses have such movement proteins and that these proteins are known to be crucial for plant virus viability strongly support the idea that virus spread is a special problem in plants. And yet, the DNA viruses of algae and other aquatic cells are clearly able to penetrate the tough cell walls by hydrolyzing an opening and injecting DNA from the outside of the cell. Some phycodnaviruses actually have ORFs that appear to code for various hydrolytic enzymes and even cellulose synthase, suggesting that they encode enzymes that could be used to breach plant cell walls. The absence of any such plant DNA viruses might indicate that the thickness of plant cell walls is simply beyond the capacity of such external virus injection mechanisms to overcome. Another very curious but general issue of angiosperm-virus relationships concerns the retroviruses. Angiosperms lack any true retrovirus that is able to integrate into host genomes following reverse transcription of the RNA virus genomes. Yet angiosperms do support various pararetroviruses, such as badnaviruses, that replicate using RT via cytoplasmic DNA forms. Thus, it is difficult to explain the absence of true retroviruses in plants. This observation is especially puzzling when it is noted that the genomes of most plant species show high-level colonization by various retroposons, which are clearly derived from retroviruses and contain integrase ORFs. It seems that retroviruses may have been highly active early in the evolution of plant genomes but are no longer supported by these host orders.

In terms of presenting a habitat for viruses, plants have several features worth noting. Plants lack adaptive immune systems but clearly retain an array of posttranscriptional control and other defense systems. Although plants do not retain the fully transmissible RNAi system, they do show strong regional responses to dsRNA and regional RNAi induction, which affect the possible expression of virus genomes.

Because of the preponderance of positive-strand plant RNA viruses, at this time it is worth reconsidering in greater detail the likely origins of these virus families. Currently, 14 families of plant viruses are known, encompassing 70 genera. Mostly, these virus families are distinct. As previously mentioned, overall there are two clearly recognizable physical types of plant RNA viruses: rod shaped and isometric. In addition, within these broad morphologies, there exist numerous virus families with variations in the following characteristics: single or multiple gene segments, membrane coverings, distinct 5′ and 3′ RNA ends [with or without 5′-terminal proteins or 3′ poly(A)], gene order, and expression of polyproteins or use of internal ribosome entry sites for translation. All of these features, along with methods of transmission (plant host, insect vectors, etc.) are used to define specific virus families, resulting in a large set of angiosperm-specific viruses.

Such a large array of virus types immediately poses the question of whether there can be a common lineage for these viruses or whether they must be polyphyletic. The most common feature that seems to link this entire set of viruses is the need to replicate RNA using an RNA-dependent RNA polymerase (RdRP). Thus, much attention has been directed towards the analysis of RdRPs to determine whether it can be used to trace virus origins. However, there clearly appears to be a fundamental problem with this approach that concerns the rate of sequence change in RNA viruses and the appropriate use of this information to infer origins. It was argued by E. C. Holmes in 2003 that essentially all extant RNA viruses appear to have evolved as recently as 50,000 ybp. This result is based on molecular clock estimates from various RNA virus genomes. Because the rate of error in replicating RNA virus templates is high ($\sim 10^3$ substitutions per site per year), virus RNA genomes can essentially erase their ancestral or historical sequence information rapidly. This high rate of change poses a significant practical problem with regard to establishing the phylogenetic relationships of RNA viruses. Yet it is equally clear that various RNA viruses are phylogenetically congruent with their hosts, strongly implying coevolution with hosts on the much longer timescale of host evolution. The conservation of molecular characteristics (active sites, capsid types, molecular strategies, etc.) might, then, reflect the need for the virus to retain limited sequence solutions to meet functional constraints on a background of very high rates of sequence change in the rest of the genome. Such an overall high rate of change would result in phylogenetic branch lengths that are too long to quantify and hence too long to provide meaningful lineage identification. Virus-host coevolution would then be visible due to the maintenance of host-restricted molecular constraints on otherwise high rates of virus evolution. In addition to the problem with high clock rates, RNA viruses appear to have undergone frequent recombination events that resulted in new virus lineages. Such a situation would clearly imply that these virus lineages have mixed ancestry, thereby further complicating phylogenetic analysis. Thus, the existence of numerous positive-strand ssRNA virus families that infect angiosperms does not really raise a theoretical problem in terms of understanding how such viral variation might have come about, but it does pose a practical problem in that our methods of analysis seem to be unable to deal with this high level of diversity.

Several early reports, such as those from P. Argos et al. in 1984 and 1988 and A. E. Gorbalenya et al. in 1995, argued that all positive-strand ssRNA viruses evolved from a common ancestor. A catalytic GDD sequence motif was identified that was essential for the RNA-polymerizing activity of the enzyme. Thus, this motif appeared to link most viral RdRPs. However, P. M. Zanotto and colleagues argued later (1996) that phylogenetic analysis does not support the inclusion of all the positive-strand RNA viruses in a single lineage. Nonetheless, an analysis by A. Gibbs et al. in 2000 did support a monophyletic relationship for these viral RdRPs and postulated that an alphavirus-like supergroup represented the predecessor of the positive-strand ssRNA viruses. Such a supergroup might conceivably have evolved from the single-segment isometric *Leviviridae* family of phages. In 1993, E. V. Koonin and V. V. Dolja and also V. K. Ward suggested that dsRNA viruses may have diverged multiple times from positive-strand ssRNA virus predecessors. The RNA viruses have been classified into six supergroups: carmo-like, sobemo-like, picorna-like, flavi-like, alpha-like, and corona-like viruses. Interestingly, most of these supergroups appear to be represented in ocean isolates. Clearly, positive-strand ssRNA viruses can be split into major families based on RdRP similarities. These families are distinct from those based on plant movement protein similarities (which are polyphyletic). The replicases for all positive-strand ssRNA viruses can also be sorted into three distinct supergroups. Group II includes RdRPs from dianthoviruses, tobamoviruses, carmoviruses, and a subset of luteoviruses (described below). Also within this group are the replicases of RNA phage Qβ, flaviviruses, and yeast dsRNA elements. The implication of this inclusion of RdRPs from viruses of lower organisms is that this lineage of RNA viruses predates the origination of green algae, the predecessors of plants, and includes major families of all the positive-strand ssRNA plant viruses.

Additional results appear to support long-range evolutionary linkages between positive-strand ssRNA viruses. The capsid of *Tobacco ringspot virus* shows an antigenic link to the picornavirus superfamily. This virus is a member of the nepoviruses, which are known for causing disease in fruit crops. In fact, nepoviruses, comoviruses, and picornaviruses appear to be in one superfamily because of the similarity of their capsid structures determined by crystallography. Yet the *Comoviridae* have two RNA segments expressed as polyproteins, and thus they appear to be distinct. In contrast, the *Bromoviridae* have three positive-strand ssRNA segments. Multisegmented virus families are expected to have evolved from single-segmented ancestors. One member of the *Bromoviridae* family that may represent a basal version is *Olive latent virus 2* (OLV-2). The virus has no vector and is not associated with disease, which suggests a persistent life strategy. In addition, and also consistent with a persistent life strategy, OLV-2 has a narrow host range.

The virus-plant dynamic: recently altered by human activity

As I discuss the relationship of plant viruses to their angiosperm hosts, I need to raise a note of caution. There are an estimated 250,000 species of angiosperms, which are by far the most diverse of all plant groups. Given that most angiosperms are susceptible to virus infection and that the number of viruses infecting these hosts is great, it is estimated that there may be up to 26×10^6 possible virus-angiosperm combinations in the world. This is a staggering number that has yet to be sampled sufficiently, and our understanding of these possible virus-host combinations is narrow and distorted. For the most part, plant viruses have been studied due to their ability to cause acute disease in crop species. Besides the fact that this focus tends to ignore nonpathogenic or persistent virus-plant relationships, crop disease is strongly influenced by human activity. For example, humans have been known to introduce viruses and vectors into new areas; to create large, dense monocultures that are highly susceptible to viral infection; and to prolong crop seasons, thereby facilitating virus transmission. All of these situations tend to promote viruses as agents of disease in crop species. It thus seems possible that human agricultural activity has distorted and is distorting the natural relationships between plant viruses and their hosts that have existed for much of plant evolution.

Despite the clear influence of human activities on virus-plant relationships, there are examples that demonstrate human-independent evolutionary links between viruses and their plant hosts. As an example, consider the Solanaceae (nightshade) family of angiosperms. Although angiosperms in general appear to have arisen about 120 million ybp, the Solanaceae are the earliest angiosperms that can be identified from the fossil record of the Cretaceous period (65 million ybp). Within the Solanaceae, *Nicotiana* species have been most studied as models of virus-plant relationships. Currently, the major center of *Nicotiana* species is South and Central America. This area also corresponds approximately to the center of native plant resistance to TMV (see below). It has been proposed that the current distribution of *Nicotiana* species may relate to the early continental breakup. A tobamovirus group clusters with these *Nicotiana* hosts, and this clustering suggests coevolution of these viruses and hosts. Such a suggestion would be consistent with the idea that these hosts and the viruses that infect them have been biologically linked since the earliest period of angiosperm evolution.

Models of Acute Plant Viruses

Tobamoviruses

Tobamovirus, a floating virus genus, is a well-studied example of rod-shaped, positive-strand ssRNA viruses of plants. Consequently, we can examine tobamoviruses as models for virus-host relationships and possible origins. In most host populations, tobamovirus populations are both stable and broadly congruent with their host angiosperms. Furthermore, tobamovirus taxonomy mostly correlates with the taxonomy of the host plants. In fact, the majority of tobamoviruses appear to have codiverged with their hosts. This includes the tobamoviruses that infect Solanaceae (nightshades), *Brassica* (e.g., mustards and cabbages), and Leguminosae (legumes). The majority of these infections appear to be acute, as they are disease associated, and thus do not appear to offer an obvious explanation for the virus-host congruence. However, virus-induced plant diseases tend to develop slowly and to be less pathogenic than virus-induced diseases of animal hosts. Perhaps it is also the narrow host restriction that maintains the virus-host congruence. Yet in this situation the maintenance of genetic stability would pose a problem that requires explanation. In some cases, tobamoviruses have been isolated from unexpected sources. And in a few of these unusual hosts (such as orchids infected by *Odontoglossum* ringspot virus), the viruses are known to establish persist infections, resulting in inapparent maintenance within their hosts. Sequence analysis suggests that the most basal members of the tobamoviruses are *Sunn-hemp mosaic virus* and CGMMV. CGMMV is of special interest, as Fungi also transmit this virus (as described above). Thus, assuming that these two viruses represent early tobamoviruses, the possible virus progenitors of modern tobamoviruses may have been the filamentous viruses of Charophyceae (algae) (and also possibly plasmodiophoromycete Fungi).

Other tobamoviruses have also received much attention. *Tomato mosaic virus* is of interest because it has been isolated from glacial ice cores, suggesting that these isolates correspond to viruses prevalent about 10,000 ybp.

Such isolates are identical to contemporary *Tomato mosaic virus* isolates, supporting the view that tobamoviruses tend to be very stable genetically. Assuming that these observations are correct, this result also indicates that molecular clock estimates based on high rates of virus replication error (noted above) must be highly inaccurate when applied to such stable virus populations. The most studied tobamovirus is clearly TMV, the first virus ever characterized. Despite this distinction, the natural biology of TMV and TMV-host relationships are not well understood. Recent studies now suggest that the original habitat for TMV was centered near Peru, Bolivia, and Brazil (see the article by E. C. Holmes in Recommended Reading). This area is inhabited by three species of *Nicotiana* (*Nicotiana glauca*, *N. raimondii*, and *N. wigandiodes*) that tolerate TMV infections. These *Nicotiana* species show few or no symptoms when infected by TMV and thus appear to establish persistent relationships with these viruses, providing a long-term niche for TMV. TMV was intially studied due to its ability to cause severe disease in the crop species *Nicotiana tabacum*. In this host, TMV is only an acute agent. Since *N. tabacum* is an amphidiploid (an organism having a diploid set of chromosomes from each parent), and thus is a species found only as a crop, the TMV infecting *N. tabacum* is most likely a "crop fugitive." In other words, this virus-host relationship represents an unlikely origin for this virus. This point emphasizes the consequences that human activity can have on virus-host relationships.

Carlaviruses

Other rod-shaped plant viruses show distinct host relationships. The carlaviruses (a floating genus) are distinguished by their long filaments and single segment 8-kb positive-strand ssRNA genome, with a poly(A) 3′ end. Diseases caused by carlaviruses are rare; thus, this virus family has been poorly studied. However, persistence with these viruses can be highly prevalent. Carlaviruses tend to be found wherever host plants are found. Carnation latent virus is possibly the best-studied example of the carlaviruses and tends to induce either mild disease or asymptomatic infections. Another well-studied member is potato virus S, which causes a persistent or latent infection in most potato plants. In the case of some carlaviruses, transmission occurs via aphid vectors. Other carlaviruses have no known vectors, nor are they known to be seed transmitted. Examples of other viruses that cause latent infections include mulberry latent virus, passiflora latent virus, and cowpea mild mottle virus. Carlaviruses are not known to infect cereal crops. These observations strongly suggest that for the most part, carlaviruses have a stable persistent life strategy within their hosts.

In contrast to the carlaviruses are the *Closteroviridae*, another family of filamentous positive-strand ssRNA. These viruses can have either mono- or bipartite RNA genomes. In addition, these viruses are often associated with crop disease.

Luteoviruses

Luteoviruses, a diverse group, can be considered a well-studied model of an isometric, positive-strand ssRNA virus supergroup. In contrast to tobamovirus phylogeny, luteovirus phylogeny does not correlate with the host phylogeny. In addition, luteoviruses are associated with widespread disease in their crop hosts. There seem to be conflicting evolutionary pressures on luteoviruses and their hosts, in contrast to the tobamoviruses and their hosts. For the most part, luteoviruses establish an acute life strategy within their hosts. It seems likely that this difference in life strategy is the important distinction between luteoviruses and tobamoviruses that determines the differential evolutionary pressures faced by these supergroups of viruses. In addition, luteoviruses are frequently observed to assist in transmission of other viruses. Thus, there also appear to be many virus-virus interactions associated with this supergroup that could affect evolutionary pressures.

Many more luteovirus lineages have evolved through interspecies recombination than have tobamovirus lineages. Thus, in contrast to the genetic stability of tobamoviruses, luteoviruses appear to have adapted a life strategy characterized by high rates of genetic change and exchange. In this light, a major event in luteovirus evolution appears to have been the replacement of the RdRP: luteovirus RdRPs exist in two distinct sets that have little or no homology to each other. One RdRP set is closely related to the carmovirus RdRP, while the other RdRP set is related to the sobemovirus RdRP. Recombination may have also been important for the evolution of *Odontoglossum ringspot virus* (ORSV). ORSV appears to be an atypical tobamovirus in that it is a virus of orchids (a monocot). Sequence analysis suggests that ORSV may have originated from a recombination of parental viruses that had unrelated dicot hosts. However, in spite of these recombinational events that appear to have resulted in a new order of RdRP and ORSV lineage, the virion structural proteins of all luteoviruses are clearly members of a single family based on structural considerations. This indicates that at least some luteovirus descendants should have mixed lineages. As for the origins of luteoviruses, it has been suggested that luteoviruses may have arisen from an RNA virus of an animal or insect host that adapted to plants. However, a fungal origin also seems possible, given the similarity of fungal viruses to animal viruses.

Other viruses have also been considered possible progenitors of the luteoviruses. For example, the necroviruses, such as tobacco necrosis virus, have a single-segment, positive-strand ssRNA genome, without a capped 5′ end. They are associated with a limited amount of disease and are host restricted. They are found to naturally persist in roots of host plants worldwide. In addition, they are commonly associated with the smaller satellite virus *Satellite tobacco necrosis virus 1* (STNV-1). Another possible origin of luteoviruses is the carmoviruses, as they also seem to be a source of RdRPs in some luteoviruses. The carmoviruses are small icosahedral positive-strand ssRNA viruses with genomes of about 4 kb in length. Carmoviruses tend to cause mild or asymptomatic infections in their hosts. *Turnip crinkle virus* is a well-studied member with a very restricted host range. Carmoviruses, such as carnation mottle virus, can be found in ornamental host plants. The fact that they can be transmitted by fungal zoospores (and beetles) suggests that these viruses could have originated from fungal sources.

Bromoviruses

Bromoviridae, which are also positive-strand ssRNA viruses, are among the most common plant viruses known, infecting over 1,000 host species, with 85 plant families represented. As a group, these viruses have the broadest host range of all plant viruses, although individual members can have a narrow host range. Curiously, all appear to be transmitted via beetle vectors, emphasizing the importance of virus vector evolution. *Cucumber mosaic virus* and *Peanut stunt virus* are the best-studied members of this family. One genus of virus with distinct biology is the oleavirus OLV-2. This virus has no known vector and induces no known diseases. Because it is also a member of the alphavirus supergroup, which includes animal viruses, it may be representative of a progenitor virus, possibly of fungal origin.

Vectors and plant viruses

I discuss insect vectors below, but some comments on insect vectors are appropriate here. Up to 30% of all plant viruses belong to the family *Potyviridae*. These are among the most destructive of crop viruses and are generally transmitted by aphids. Each potyvirus genus appears to have adapted to a specific vector organism. Thus, virus biology and vector biology are tightly linked, and this linkage can also involve aphid symbionts, as discussed in detail below. Other insects seem to seldom transmit plant viruses. Orthoptera (crickets, katydids, grasshoppers, etc.) and Dermaptera (earwigs), for example, include few vectors for plant viruses, whereas Coleoptera (beetles; 55,000 species) include numerous vectors for tymo-, como-, and bromoviruses. Some pollinating insects can also transmit virus, such as blueberry leaf mottle virus via honeybees, but given the numbers of pollinating insects, this is not very common. For the most part, these striking patterns of virus vector biology are not well understood.

Other viruses

There are many individual families of plant viruses that could be considered here in greater detail. From the perspective of virus or host evolution, however, these other virus families are not very informative on the whole and therefore are not discussed specifically. However, several additional points are nonetheless worth noting. The geminiviruses of plants are ssDNA viruses that replicate via rolling circular mechanisms. They have some similarities to nanoviruses (multipartite ssDNA viruses). Viruses with related replication mechanisms are found in most living orders. For the geminiviruses, however, virus DNA replication is supported by both plant and bacterial replication machineries. For example, tomato leaf curl virus can also replicate its DNA in *Agrobacterium tumefaciens*, suggesting an origin from ssDNA phages or Ti plasmids that are injected into plant cells from bacterial parasites. Furthermore, sequences related to geminivirus DNA can be found in the genomes of some plants, such as *Nicotiana* spp., suggesting some involvement of geminiviruses at the origination of those plant lineages. This result has been used to argue that geminiviruses evolved from prokaryotic replicons and have invaded plant genomes. Nepoviruses also add an interesting note. These viruses are isometric RNA viruses that are exclusively transmitted via nematodes. They show a wide host range throughout North America and Europe but not worldwide. They are frequently associated with satellites and other RNAs. It is curious that these viruses can be highly specific to their nematode hosts, yet as discussed in chapter 6, nematodes are not known to support the replication of any virus. Plant rhabdoviruses are often able to replicate in insects and are highly specific for their insect vectors. However, plant rhabdoviruses can also be adapted for plant-only replication, being maintained by serial passage in plants. This phenomenon suggests that plant rhabdoviruses likely evolved from insect viruses that acquired an additional plant movement protein. Thus, overall, there are several instances in which it appears that an animal virus may have adapted to infect plants.

Mixed autonomous and satellite viruses

It was noted above that, overall, plants appear to have a strong tendency to support mixed infections with RNA viruses. Numerous specific examples exist in the plant virus literature that establish this situation. Less clear, however, are the evolutionary consequences (to both the viruses and the hosts) of such mixtures. In some cases, interactions between two different viruses are able to affect virus host range. For example, TMV does not normally infect wheat. But if TMV coinfects wheat with barley stripe mosaic virus (which supplies needed movement proteins), TMV will be complemented and can replicate. Another example of mixed virus disease is groundnut rosette disease. This is caused by a complex of two autonomous viruses plus a satellite virus: groundnut rosette assistor virus, an aphid-transmitted virus that provides the coat protein; groundnut rosette virus (GRV); and the RNA satellite GRV. Some umbraviruses can persist in their host plants without a helper virus. During such persistence, they make significant amounts of virus-specific RNA but produce little progeny virus or disease. These viruses have no coat proteins, and they seem to be fully adapted to use the coat proteins of helper luteoviruses. The umbraviruses are related to tombusviruses, and their infections can be highly host specific. In addition, luteovirus-infected plants can also support the dsRNA satellite GRV. Satellite GRV is so prevalent that it can be found anywhere worldwide that the host crop is grown. Thus, it is very clear that an essentially defective and persistent life strategy of satellite GRV can be highly successful in nature. It is not clear, however, what effect this would have on the fitness and reproductive success of the helper virus or the host, as discussed below.

Tight linkages between a helper virus and a satellite can also be observed. For example, TNV can be classified on the basis of the specific satellite virus it activates. This link of helper (host) to satellite (parasite) is reminiscent of typing bacteria based on phage sensitivity and appears to support the concept that virus-virus relationships can be crucial for success in the field. STNV is a well-studied satellite of TNV. This was the first plant satellite to be discovered (in 1962). STNV has an uncapped mRNA that functions in both prokaryotic and eukaryotic translation systems. Also, the RNA lacks a 5'-terminal viral protein, and the 3' end resembles a tRNA in structure and lacks a poly(A) tail. These features make STNV very useful for efficient in vitro translation from wheat germ extracts. STNV contains only one gene that encodes a coat protein. Helper-satellite virus activation is strain specific. There is also high specificity for the fungal vector involved and some specificity for the host plant (this being a common situation for satellites). The STNV coat protein shows some relationship to phage proteins, suggesting a phage-based origin. There also appear to be relationships between this and various other satellite viruses.

Parasitic RNAs

In addition to satellite viruses, there are also many small satellite RNAs found in plants. These small RNAs tend to be highly structured, with RNA base-pairing-stabilized secondary structures, and they do not code for gene products. They can be either linear or circular. In being noncoding parasitic replicators of helper viruses, these small RNAs clearly resemble fully defective viruses. And in some cases (e.g., *Cucumber mosaic virus*), they can attenuate symptoms induced by the helper virus and can also interfere with other viruses. Other satellite RNAs, however, are associated with enhanced virulence during helper virus infection. Thus, satellite RNA relationships with the hosts are complex. In addition to satellite RNAs, plants can also harbor endogenous dsRNAs. These tend to be found in various cultivated crop species, such as rice, bean, and barley. These dsRNAs resemble hypoviruses in that they are cryptic and code only for replicase and coat proteins. They do not code for movement proteins. Also like hypoviruses, they show no clear association with virus particles and are often prevalent and persistent. However, unlike the fungal hyphae that transmit hypoviruses, plants do not tend to fuse in the field and thereby transmit these dsRNAs. Currently, their mechanism of movement between plants is unclear. It is also unclear what effect they have on either the host plant or the ability of the plant to support other virus infections, although it seems reasonable to predict that there is some interaction with other infectious agents. It has been suggested that these dsRNAs may have originated from defective ssRNA viruses.

Seeds, angiosperms, and viruses

In filamentous algae, zoospore-associated transmission of virus is commonly observed. It might thus be anticipated that a similar type of virus transmission might be associated with the seeds of angiosperms. Such seed-associated transmission could allow persistent viruses to attain a high probability of infecting subsequent host generations. The invention of the seed allowed land plants to colonize drier lands away from shore, and thus it was a major determinant of plant reproductive success. Seed plants appear to be monophyletic, having evolved from one apparent common predecessor. Four major groups of seed plants are currently found: cycads/ginkgos, Gnetales, conifers, and angiosperms. Ferns appear as a sister group to these clades. Seeds are notable for the large quantity of stored proteins they can contain. Tonoplast proteins, for example, are abundant and highly expressed in the seeds of many monocots, dicots, and gymnosperms. In fact, plant seeds and embryos (e.g., wheat germ) are excellent sources of translation system extracts, and they also contain ribosome inhibitors that self-regulate translation, as described below. Seeds can also express self-incompatibility functions. With respect to virus infections, seeds appear to offer some very specific advantages.

Efficient seed-based transmission would reduce the need for vector-dependent virus transmission. Virus persistence in infected plants seems necessary to allow efficient host reproduction. Thus, the fitness of persistently infected hosts would need to be maintained. One way persistent infections could affect the fitness of infected host plants would be to interact with acute viruses or prevent their colonization of the host. This idea, however, has not been evaluated. Tymoviruses, which are mostly seed transmitted, appear to have originated from a common source about 200 million to 240 million ybp. This example suggests that virus-seed association may date to the origination of angiosperms. Some potyviruses and partitiviruses can also be seed transmitted. As noted above, many Fungi appear to be involved in zoospore-based virus transmission. One example of such a bromovirus is OLV-2 (which has three ssRNA segments).

The expectation that seed-transmitted viruses might frequently be persistent suggests that such relationships might be difficult to observe in the field. There is evidence that this may indeed be so. A case in point relates to the plant pararetrovirus tobacco vein clearing virus (TVCV) of *Nicotiana* species. This virus is transmitted 100% vertically via seeds. It is possibly the only example of a seed-transmitted pararetrovirus studied so far. Prior to this report, caulimoviruses were unknown in *Nicotiana*. It is interesting that TVCV came from a hybrid plant—*Nicotiana edwardsonii*—which is used as an indicator plant for various viruses. The parents of this hybrid were a male *Nicotiana glutinosa* plant and female *Nicotiana clevelandii* plant. TVCV was not observed in either parent. Nor was TVCV transmission able to be demonstrated by mechanical methods or by grafting infected plants to seven other *Nicotiana* species. Neither can TVCV transmission be accomplished by aphid vectors. TVCV DNA hybridizes to the genomic DNA of *N. edwardsonii* and the male parent, *N. glutinosa*. TVCV DNA does not hybridize to female parental DNA. Interestingly, TVCV DNA shows a 78% sequence identity to some pararetrovirus-like sequences present in high copy number in the *N. tabacum* (and *Nicotiana rustia*) genome. TVCV is unrelated to other caulimoviruses with related biology, such as *Banana streak virus*, which can also be integrated. Caulimoviruses (circular dsDNA retroviruses), such as *Cauliflower mosaic virus*, exist in four known groups and have very narrow host ranges. There is no sequence relationship of the caulimovirus Gag protein to that of any animal retroviruses, although its RT has a clear relationship to those of the animal retroviruses. Cauliflower mosaic virus is mainly aphid transmitted and not typically seed or pollen transmitted. However unusual it may seem, TVCV makes an important point. The hybrid-based reactivation of TVCV appears to identify a situation in which endogenous viruses are reactivated from the male host genome following mating to an uninfected female host. This link of endogenous virus reactivation to sexual reproduction very much resembles the situation of hybrid dysgenesis for *Drosophila melanogaster,* described in detail below. Such sex-based virus reactivation has major implications for the sexual isolation of related species and the separation of interbreeding populations. These implications are also considered in "The *Drosophila* Genome, Retroviruses, and Speciation" below.

Antiviral and self-responses

Plants have developed an array of antiviral response systems encoded by their genomes. First, however, I describe extranuclear plant systems that can affect virus infection. One such process has been named "pathogen-derived resistance." This type of protection consists of a virus-mediated cross-protection that can occur during mixed virus infection. In 1929, an early example of this was first reported by H. H. McKinney for tobacco plants infected with a green mosaic virus strain (TMV). These infected plants developed no further symptoms when coinfected with a yellow mosaic virus strain. This type of interference is very common in related strains of virus. Tungro virus is found in Southeast Asia, where it causes asymptomatic infections of its natural host and provides cross-protection in that host. Another process of virus protection was discovered when transgenic plants were first generated. Tobacco plants expressing TMV coat protein were shown to resist infection with TMV. Such experiments led to the identification of various systems by which plants can posttranscriptionally repress expression of foreign genes. One such response system was described in detail in chapter 6, dealing with *C. elegans*. Like the nematodes, plants also have an RNAi system. This consists of an RdRP along with other genes (for nucleases) that are induced to degrade and silence foreign genes. In *Arabidopsis*, the relevant genes are the RdRP *SDE1* and *SGS2* genes. Homologues of these genes are not found in *Drosophila* or humans, but they are clearly related to genes found in *C. elegans*. However, unlike those of *C. elegans* and other simpler animals, the RNAi response of plants is not systemic or transmissible. It is restricted to a more local response. It should be remembered that no known RNA viruses have been observed for those early animal systems, yet RNA viruses are highly prevalent in angiosperms. It would therefore seem likely that the extant plant RNAi response is not sufficiently strong to prevent RNA virus parasitization of plants.

Plants have additional systems that can recognize self and differentiate self from nonself. For example, three distantly related families of flowering plants, the Solanaceae, Scrophulariaceae, and Rosaceae, all have self-incompatibility RNases that limit self-fertilization. Like RNAi, these systems also depend on RNases from monophyletic clades. These RNases are reminiscent of the SK18 system of yeasts for controlling dsRNA or Dicer of *C. elegans*. The possible relationships of these systems to virus colonization have not been evaluated. Although many other plants also limit self-fertilization, the mechanisms employed vary.

Another plant regulatory system can be found in seeds and embryos. As mentioned above, plant embryos (e.g., wheat germ) are excellent sources for translation system extracts. However, most of these seeds have potent protein inhibitors of translation systems called ribosome-inactivating proteins (RIPs). RIPs are often localized within the endosperm, which must be removed to yield functional extracts capable of stable and ongoing in vitro translation. RIPs can constitute a major fraction of the proteins in seeds and can depurinate even a single adenine from the stem-loop of 28S rRNA. Ricin is perhaps the most potent example of such a RIP. Both single- and double-chain versions of RIPs are known, but often RIPs are stored in nontoxic forms as polymers or bound to RIP inhibitors (resembling a toxin-antitoxin addiction module). Tritin is the major RIP found in wheat seeds, and RA39 is found in rice. Pokeweed seeds, melon seeds, kernels of camphor trees, maize, barley seed, and iris bulbs also have RIPs. When they were first discovered, it was thought that the main function of these proteins was to inhibit pathogenic Fungi, since several RIPs were observed to be antifungal. However, it is now known that some Fungi have their own RIPs. Fungal RIPs are found in fruiting bodies. In addition, it has been established that not all plant RIPs are antifungal, as they fail to inhibit important fungal pathogens. Other plant examples include RIPs expressed in roots: *Agrobacterium rhizogenes*-transformed hairy roots make RIPs, as do *Mirabilis jalapa* roots. Another function of RIPs appears to be antiviral: several RIPs are known to have potent antiviral activities. RIPs with established antiviral activities include the pokeweed antiviral factors, PAP and PAP II, which are both highly active against potato virus X, potato virus Y, and potato leafroll virus. These RIPs appear to operate by preventing ribosomal protein EF2 binding, and they inhibit virus translation. Yet RIPs are most frequently associated with seeds and embryos, which suggests that these may not be RNA virus-friendly tissues. However, the consequences of abundant RIP production for seed-mediated virus transmission have not been evaluated.

Nuclear and plastid genome colonization

With the sequencing of entire plant genomes we can start to get a clear picture of the effects that viruses and virus-like transposons have had on the evolution of plant genomic DNA. Overall, retroelements are found in high numbers in most, but not all, plant chromosomal DNAs. The two major groups of retroelements found in plants, copia-like and gypsy-like elements, are both better studied and more widespread than are other types of elements. Copia and gypsy elements differ from each other in the order of RT and integrase domains. Related fungus-derived retroposon sequences appear to be basal to those of plants and *Drosophila*. This supports the view of fungus involvement in early plant and animal evolution (or possibly colonization of Fungi, animals, and plants by the same elements; see Fig. 6.4). Curiously, these fungal elements have not been found in algae. Also curious, phylogenetic analysis indicates that the *Drosophila* gypsy element is basal to those gypsy-like elements found in plants, suggesting the possibility that an animal species was the original source of the elements currently found in land plants. As for the copia-like elements, in most plants these elements are monophyletic. This strongly suggests that retroposon colonization is associated with the origins of the various plant lineages and that subsequent selection has not eliminated these agents or contributed to additional large-scale colonizations.

In plants, the large majority of retroposon copies are inactive, defective, and unable to transpose or make virus. Very few ORFs for intact *gag* or *pol* genes exist, and no copies of *env* have yet been sequenced. However, a small number of retroposons do appear to have retained Gag and RT activity. For example, the BARE-1 copia-like element found in large numbers in the barley genome has maintained the *gag* ORF. Curiously, this retroposon also appears to retain transposition activity. Overall, the degree of plant genome colonization by retroposons varies significantly. Some plants, such as *Lilium henryi*, have many fewer gypsy-like elements than usual. It appears that increases in genome colonization by retroposons occurred during the evolution of higher plants. In the genomes of cereals, members of the family Gramineae (grasses), there is an especially large fraction of retroelements. For example, retroposons are present in high numbers in barley (BARE-1 copia-like element, ~50,000 copies), wheat (BIS-1 constitutes 5% of the genome), and maize (retroposons constitute 50% of the genome). The maize situation is of special interest, as it has been reported that the maize genome only recently (i.e., 3 million ybp) underwent large-scale retroposon amplification. A similar situation applies to the BARE-1 element of barley, but evidence suggests that habitat-specific amplification of BARE elements in wild barley species has occurred. Gymnosperms (e.g., conifers) are similar to wheat with regard to retroposon content. In the very large genomes of gymnosperms, which are not frequently polyploid, retroelements tend to make up about 50% of the DNA. One explanation for this high-level, retroposon-based genome plasticity is that amplifying retroposons may provide the genomic rearrangements required during periods of rapid adaptation. A puzzle that remains, however, is how each plant lineage initially became highly colonized by particular families of retroposons. It seems likely that some exogenous source of ERV was involved at the origination of these plant species. For example, it is known that each major clade of rice retroposon element is more related to elements in other species than to other elements in the rice genome. This suggests that the original source of these amplifiable retroposons was other organisms' genomes, not other retroposons already in the rice genome. In fact, grasses (rice) in particular appear to have high numbers of retroposons. Two complete plant genomes are now known. The first sequenced was the *Arabidopsis thaliana* genome, which is a 460-kbp nuclear genome. AtC2 is a retroposon found in *A. thaliana* that is present as a full-length copia-like element. Three hundred AtC2-related elements can be observed in the genome, constituting about 1% of the total genomic DNA. Of these 300 copies, 23 are full-length copies with putative ORFs but without *env* sequences. Interestingly, all of these full-length copies appear to represent distinct retroviral families, similar to the ERV patterns observed in *C. elegans*. In *A. thaliana*, the gypsy-like elements are monophyletic and appear to have been early colonizers of genomes in this lineage. The subsequent gypsy element sequence diversification appears to have occurred mainly during vertical, not horizontal, transmission. A similar conclusion with respect to gypsy element colonization applies to the rice genome.

All three of these major plant retroposon families (copia-like, gypsy-like, and AtC2-like) are abundant in animal lineages. LINEs are related to these families and can be found in both plant and animal genomes, although generally at very different frequencies. LINEs can be defined as retroposon-derived sequences that lack LTRs or 3′ poly(A) sequences. In placental vertebrate animal genomes, LINEs are present in exceedingly high numbers (10^6/genome). Generally, LINEs contain degenerate *gag* and RT ORFs. Some of these retroposon families are not well represented in plants. For example, plants may have many fewer or no Ty3/gypsy elements or Pao-like elements. In *Arabidopsis*, Tal-1 and Tal-17 LINEs have been identified. However, Tal-1 is present at a very modest level—between one and three copies per genome. A related sequence found in potato is the Tst1 element, which is also found at between one and three copies per genome. In plants, Ty1/copia-like LINE-related elements are also known, but these have been given the unfortunate (and obscuring) name of *Pseudoviridae*. The 276 pseudovirus elements are simply the LTR-deficient Ty1/copia-like (LINE-like) elements present in *Arabidopsis* DNA.

The rice genome is similar in many respects to that of *A. thaliana*. Rice (*O. sativa*) has a 466-Mbp genome, of which ~42% is repetitive DNA. Miniature inverted-repeat transposable elements are present at about 33,000 to 50,000 copies per genome. Most of these repeats can be found in fungal genomes, but one-third are not found in any fungal species' DNA. Rice has about 98,000 transposable elements, 18% of which are retroposons. Thus, the ratio of DNA to retroposons in plant genomes is very different from that in vertebrate genomes and much more similar to that in the *C. elegans* genome. Of the rice retroposons, 80% are similar to those of *Arabidopsis*. Rice has 20 copia-like element families; for comparison, maize has 10 such families. In comparisons of the rice genome sequence to the partial sequence of a second, Japanese rice species (both studied by the International Rice Genome Project), little difference was seen between the coding sequences of the two species. However, a major difference between these Japanese species was found in their transposable elements—of which 63% were retroposons. The different retroposons were mainly of the Ty1/copia type, plus some Ty3/gypsy types.

Wild rice has similarly acquired the RIRE1 copia-like retroposon element. The RIRE1 *pol* sequence has conserved sequences similar to those found in *Drosophila* retroposon element *pol* ORFs. This observation again raises several major questions concerning what forces drive plant speciation. Two big questions can be posed. First, why is speciation associated with genome colonization by specific retrovirally derived agents? And second, how do these new elements enter the rice genome, especially considering that plants are not known to support autonomous retroviruses, nor are insect vectors known for plant retroviruses?

In contrast to plants, insects (such as *Drosophila*) can express ERVs as infectious and transmissible viruses, especially from reproductive and embryonic tissues (discussed below). Also, phylogenetic analysis supports the idea that the fungal and insect versions of these genomic retroviruses are more basal than those found in plants. This leads to the startling possibility that insects or early animal predecessors of insects may have hosted and transmitted the retroviruses that, at the very originations of land plant lineages, colonized plant genomes in significant numbers and developed into the large numbers of ERV defectives. In other words, the counterintuitive idea is that, in a sense, insects begat the higher plants, leading to the creation of the flowers that now feed them. Another observation that may be consistent with such a counterintuitive idea is that the DNA replication genes of rice are phylogenetically incongruent. These rice genes are more similar to the human genes than to those of *Drosophila*. Some important animallike replication genes appear to have entered the rice lineage from unknown sources. As strange as this idea may seem, it is not without precedent. Others have previously suggested that the evolution of plant flower structures may have been driven by insect parasitization, in which insect embryos were deposited in plant tissue, induced growths (galls) in plant tissue, and led to the development of the protective flower structures of angiosperms. Such a scenario, however, would also require the permanent colonization of the plant genome by insect-derived genes from the insect embryos. If this idea is correct, it would suggest a truly amazing evolutionary cycle in which insects were involved in the origination of flowering plants, which then contributed to the evolution and diversification of insects! Some calla lilies and water lilies, which are among the most basal and ancient flowering plants, have interesting and intimate relationships with insects that could relate to this idea of an insect origin. Some lilies and other flowers, such as dead-horse arum, generate "body heat," which can volatilize the scent molecules (stench odors) that attract female blowflies. These blowflies lay their eggs in the flowers and in the process pollinate the flowers. The amount of heat generated by these flowers is very high, exceeded only by the heat of insect flight muscles, and has no other apparent function for the plant physiology. However, this heat aids and supports insect activity. Other heat-generating flowers also attract specific insects for reproductive purposes. In addition, as discussed below in the section on parasitoid wasp evolution, sexual relationships between insects and flowers may lend further support to the idea of an insect-based flower origin. Although this idea may account for many of the currently observed genetic characteristics of plant and insect genomes, more comprehensive sequence analysis of both putative insect and plant progenitors is needed to support or dispel such a hypothesis.

Plastid DNA colonization

Unlike the situation observed in lower plants, the plastid DNA genomes of higher plants show much evidence of evolution via repeated colonization and invasions with genetic parasites. The most remarkable of these is the observation that in angiosperm mtDNA, homing group I introns within the *cox* gene have been acquired over 1,000 times during evolution of various angiosperms. The open question, then, is how these homing introns might have been transmitted to so many angiosperm mtDNAs. The insertion process very much resembles an insertion sequence-like insertion event, such as that used by ssDNA plectoviruses of bacteria. It would thus seem reasonable to hypothesize that some infectious agent must have existed that allowed the transmission of these introns to all of their hosts, but this putative virus is not currently known. Lacking a system of packaging and transmission, how introns move remains a mystery. Mechanical insect-mediated transmission has been proposed. One other example of plasmid DNA transmission in plants should be mentioned. That involves the T-DNA that is transmitted from *Agrobacterium* symbionts to plant root cells, inducing the cells to form N_2-fixing nodules. Bacterially derived plasmid DNA is able to enter the plant cells, pilot to the plant nucleus, and integrate into host cellular DNA. This bacterial plasmid clearly shows a "virus-like" behavior. Yet it does not replicate in the host plant cell but simply integrates, expresses some genes, and persists, very much like a lysogenic phage. No movement or transfer genes are expressed by the T-DNA in the plant cell. The plasmid does express plant-specific growth factors that allow the nodules to support the symbiotic bacteria. It is remarkable to consider that this T-DNA shows an ability that is absent from any plant DNA virus. It can penetrate host cells, move to the nucleus via the well-conserved virD2basic protein, enter open chromatin, and express proteins, leading scientists to wonder why no plant virus can do these things. It seems possible that all the functions needed to penetrate plant cells might be beyond what could be expressed on the surface of a virion. It is simply too complex a task. If so, then DNA plasmid-"virus" transmission in plants appears to require a complete *Agrobacterium* cell.

To conclude, one can see clear overall patterns among plants, viruses, and insects when they are examined from the perspective of plant evolution. Next I consider the perspective of insect evolution.

Viruses and the Origin of Insects

I previously mentioned some overall relationships that link plant evolution and insect evolution. The most prominent link is the congruent explosive radiation that resulted in the coincident diversification of both angiosperm and insect species. I now seek to include the patterns of virus infection and transmission in insects as I consider insect evolution. In some cases, very tight links between plant viruses and the insects that transmit them are known. Some of these viruses are transmitted exclusively by one insect vector. Plant virology literature frequently points out such relationships. However, that literature does not generally discuss the viruses that infect insects themselves, which I do below. When this is done, it becomes apparent that the insect orders and families show a very uneven pattern of virus susceptibility and vector function. For example, tetraviruses appear to exclusively infect lepidopteran hosts. Such issues have not previously received much attention.

Arthropods and arachnids

There is little doubt that the first insects evolved from arthropods in the oceans prior to angiosperm evolution. We can use fossil evidence to trace the evolution of arthropods to arachnids and hexapod ancestors of insects. Arthropods comprise the hexapods (insects), myriapods (centipedes and millipedes), crustaceans (e.g., shrimp and crabs) and chelicerates (e.g., sea spiders, horseshoe crabs, and arachnids), all of which are characterized by exoskeletons made of chitin, segmented body plans, and paired appendages. These represent the most successful of all animal species, numbering about one million species in total and constituting the majority of animal biomass. The smallest of the arthropods, mites and parasitic wasps, measure less than 1 mm in length. Fossil evidence clearly suggests that early arachnids occupied seaside habitats. These eight-legged arachnids included trilobites, the horseshoe crabs, and the sea scorpions. To this day, however, we know little about viruses that infect these species. In horseshoe crabs, viruses do not seem to be prevalent, but the topic is poorly studied. However, it is clear that the arachnid class was not nearly as successful on land as the insect orders became. Furthermore, with respect to plant virus interaction, only one of the twelve terrestrial arachnid orders is involved in virus transmission and infection. This order consists of the mites and ticks. Arachnids such as the eriophyids (spider mites) have mostly been studied as mediators of fungal pathogen transmission to plants. Viral diseases of mites themselves are known, but they are much less common than the corresponding diseases of insects. Two known virus infections of arachnids includes those of the citrus red mite and European red mite, which can be infected with nonoccluded, but poorly characterized, DNA viruses. Although not sufficiently well studied, these viruses may be related to iridoviruses, such as the viruses reported to infect the *Varroa* mites associated with moribund honeybee colonies. However, mites are often parasitic to insects and directly associated with transmitting insect viruses, especially to bees. Generally, these viruses do not infect their mite vectors but can show some vector specificity.

Hexapoda (insects)

Because of their enormous numbers, hexapods will be the main focus of this discussion of viruses and the origin of insects. Insects are by far the most diverse of all animals, with hundreds of thousands of extant species. Thirty orders are recognized, of which nine (including mostly chewing insects) are known to feed on green plants. (Insects are especially diverse in their mouthpart structures.) Although the insect class is older than many modern lineages, insects did undergo an explosive radiation, which was coincident with angiosperm radiation as observed in megafossils from the Cambrian explosion. Also at the time of the Cambrian explosion (about 500 million ybp) was the continental breakup of Gondwanaland, resulting in the separation and isolation of South America. Fossils of the Hexapoda (insects) became abundant at this time, as did the first plant fossils (about 428 million ybp). (See Fig. 7.3 for the overall pattern of insect evolution.)

Basal hexapods

The basal hexapods that first appeared on land were related to species of Blattaria as well as other wingless organisms resembling species of Protura and Collembola. Paired wings, common to so many insects, arose later, around the time vascular plants developed. The earliest insect fossils correspond to those for Blattaria (cockroaches). Blattaria are estimated to have evolved around 380 million ybp; Orthoptera and Hemiptera (true bugs, leafhoppers, and aphids) originated much later, around 95 million ybp. Early Blattaria now appear to be the common ancestors of all insects, as inferred from limited sequence analysis of the mtDNA *cox* sequence. However, it now also appears that the oceanic ancestors of these early insects were the Anostraca (fairy and brine shrimp). These shrimp species are both bisexual and parthenogenetic. Recent analysis of mtDNA sequences confirms that mtDNA of the Anostraca is almost identical to that found in *Drosophila* species. Unfortunately, the viruses of these simple crustaceans have not been well studied. As presented in chapter 6, other crustaceans, such as *Penaeus* species, are known to support diverse families of both DNA and RNA viruses, so prevalent virus infections of Anostraca might be expected. However, for Anostraca species, it has only been reported that white spot syndrome virus DNA can be found associated with cysts, and no virus replication could be confirmed in adults. Thus, the virus might be passively absorbed to the shrimp cyst without replicating. *White spot syndrome virus* shares very limited sequence homology with other DNA viruses, including insect DNA viruses; thus, it does not appear to have been a direct ancestor of modern insect DNA viruses. Branchiopod species now appear to be the closest relatives of the insect order, inferred from *cox1* and EF-1α gene sequence analysis. The Crustacea and Blattaria are both sister groups of the Branchiopoda. The adaptation of arthropods to terrestrial habitats occurred in association with early vascular plants (~425 million ybp). For this early adaptation, there is evidence of insect predation on the early plants. The insects most closely related to cockroaches are the mantids and termites. Termites appear to have evolved from wood-eating Blattaria after symbiotic acquisition of gut bacteria for cellulose digestion. Viruses of cockroaches have not been well studied, as evident from the very few reports in the literature on this topic. However, some positive-strand icosahedral ssRNA viruses have been observed in cockroaches. Yet it seems clear that acute infections of cockroaches by viruses do not constitute a widespread phenomenon. Persistent virus infections have not been observed in Blattaria, and it remains an open question whether this lack is common to this insect order.

From the perspective of plant evolution, it is most useful to consider the insects that both are predators of plants and are vectors for plant viruses. The largest groups of plant-eating insects are the Coleoptera and Orthoptera. The next order that most commonly feeds on plants are the Thysanoptera (thrips). In terms of virus transmission and predation, aphids (members of the order Hemiptera, the true bugs) are clearly the most successful plant-feeding insects, and it is estimated that they are responsible for the transmission of up to 50% of all insect-transmitted plant viruses. Over 600 aphid-transmitted virus species are known. For the most part, these plant viruses do not replicate in their aphid vectors. In a few species, however, plant virus can persist within aphid tissue and be transovarially transmitted (these aphids are called persistent propagative vectors). The aphid-plant virus relationship is examined in greater detail below.

In addition to plant predation, many insects have predatory and parasitoid relationships with other insects and feed on or parasitize those insect prey. As this relationship presents clear opportunities for virus transmission, this topic is also considered in greater detail below.

As was mentioned above, insect-plant coevolution is especially apparent due to the role of insects as pollinators of angiosperms. Among the insect orders, the Diptera (flies), Lepidoptera (moths), and Hymenoptera (bees, wasps, and ants) are the three major orders of plant pollinators. These orders, however, do not constitute the most efficient vectors for plant virus transmission, in comparison to the true bugs and the beetles. However, insect success as a pollinator clearly has a strong behavioral component, so insect behavior is very important for angiosperm reproductive success. For example, flower recognition by insects can be plant specific. The molecular basis for insect behavior is poorly understood. However, as noted below, parasitoid wasps (Hymenoptera) are known to greatly modify and control behavior of larval lepidopteran hosts in very specific ways. As will be made clear, there is a viral component to this behavior alteration.

Some insect viruses are highly insect and vector restricted, implying intimate molecular relationships between the insects and their viruses. Below I describe a few situations in which the nature of such linkages has been studied.

Insect DNA viruses

The major groups of viruses of insects are clearly distinct from those found in both plants and animals. The most apparent distinction is that insects support the replication of many large DNA viruses, none of which are found in plants. The great majority of these viruses have relatively large dsDNA genomes that encode their own replication proteins. Some of these DNA viruses are found in other animal species, but some (ascoviruses and PDVs) are unique to insects. Another distinction with plants is the relatively small number of disease-causing RNA viruses of insects. Yet insects are a main vector for such viruses. The evolutionary relationships between insect viruses and viruses of other orders are clear in most cases. Within the large DNA viruses of insects, the iridoviruses show a clear relationship to the viruses of aquatic vertebrates, such as *Lymphocystis disease virus* of fish. No related virus of land animals is known for this family, however. The ascoviruses are unique to insects and have little sequence similarity to other DNA viruses; thus, the immediate ascovirus predecessors are obscure. However, ascoviruses do show more distant sequence relationships to other large DNA viruses (e.g., phycodnaviruses of algae and T-even phages) when comparisons are based on DNA replication proteins. In addition, ascovirus biology is similar in several important respects to that of the PDVs, as is discussed below. It should be noted that there is one animal DNA virus that appears to occupy a unique position between insect DNA viruses and the DNA viruses of land vertebrates: *African swine fever virus*. This virus has the distinction of being the only DNA virus known to replicate in both insects and animals. It is also phylogenetically distinct in that it is often an outgroup for both vertebrate and insect DNA viruses and frequently shows sequence relationships to several other virus groups that otherwise fail to show strong links to any other types of viruses.

The baculoviruses are a very important and large group of insect viruses that show clear relationships to viruses of aquatic animals, so it seems likely that their lineage can be traced to those viruses. Since baculoviruses are so important, they are examined in greater detail below. However, as discussed below, the nonoccluded "baculoviruses" (such as *Heliothis zea* virus [Hz-1]) are sufficiently different from both true baculoviruses and other DNA viruses (both biologically and by sequence) to be considered a separate DNA virus group, with little homology to any aquatic or vertebrate animal virus. The entomopoxvirus family of insect viruses is clearly related to the poxviruses of vertebrate animals. Entomopoxvirus replication appears to be especially associated with metamorphosis of the host insect. These viruses are found in various flying insects. The PDVs are multisegmented circular DNA viruses of parasitoid Hymenoptera. These viruses are endogenous DNA viruses that replicate via circular dsDNA. PDVs present an especially interesting case from the perspective of virus-host coevolution and are examined in some detail below. Finally, insects also support the replication of densoviruses, which are ssDNA viruses that replicate via rolling circular mechanisms and show clear relationships to the parvoviruses of animals and the geminiviruses of plants. One curiosity to note about insect DNA viruses in general is that although insects support the replication of many types of DNA viruses, there are no herpesviruses that infect any insect. This is especially intriguing since herpesvirus replication proteins are clearly similar to the DNA replication proteins found in both algal viruses and bacterial viruses (e.g., phage T4). However, as discussed in chapter 6, aquatic herpesviruses often show a clear preference for nervous tissue, and insects might present very inhospitable cellular habitats for herpesviruses.

Insect RNA viruses

As mentioned above, positive-strand ssRNA viruses are highly common in angiosperm plants. Relative to this plant virus abundance, insect orders show an overall paucity of diseases caused by RNA viruses, especially in some specific insect orders. Yet, picorna-like viruses (nonsegmented, positive-strand ssRNA, polyprotein, and icosahedral) are known to infect several orders of both pollinator and plant-predator insects, including Hymenoptera (bees and wasps), Diptera (flies and mosquitoes), Hemiptera (aphids and leafhoppers), Lepidoptera (moths and butterflies), and Coleoptera (beetles). The best studied of the picorna-like insect viruses are *Cricket paralysis virus* (CrPV) and its relatives *Drosophila C virus* and *Gonometa virus*. CrPV has a 3′ poly(A) tail and a 5′-terminal protein. CrPV was discovered during mass breeding of Australian field crickets as causing a hind-limb paralysis disease.

The virus is also a major problem for silkworm growers in Japan. CrPV can be found on the surfaces of cricket eggs, and this feature seems to be important for natural transmission. Curiously, CrPV is hard to find in natural field populations of both *Drosophila* and crickets. This suggests that CrPV has recently adapted to an acute infection in these hosts. However, CrPV can be found in natural populations of *Drosophila immigrans*, but in this host the virus replicates without causing disease. These results strongly suggest that CrPV naturally has a persistent life strategy with specificity to the *D. immigrans* host but that it is able to jump species and cause acute disease in other hosts.

Tetraviruses are isolated exclusively from lepidopteran hosts. Curiously, these viruses do not grow in insect cell cultures. Some tetraviruses show a high host specificity in which closely related viruses (i.e., with 90% replicase sequence identity) have distinct host ranges. Vertical transmission of tetraviruses is known to occur in some hosts, and in these cases the virus cannot be eliminated from lab-bred hosts. Clearly tetraviruses have a persistent life strategy in wild host populations. However, even in this relationship, host population structure can affect virus-host dynamics and virulence. For example, in the case of *Nudaurelia capensis β virus* when high host densities are attained, horizontal transmission becomes efficient and results in enhanced transmission and high host mortality. It seems likely that similar (but uncharacterized) viruses and virus-host relationships apply to Orthoptera (grasshoppers and locusts), Isoptera (termites), and Acarina (mites), as unclassified viral agents have been reported to infect some of these hosts.

Bees and RNA viruses

There appears to be a curious general link between RNA viruses and bees. In the 1960s in Great Britain, there was an outbreak of sacbrood paralysis that was specific to the European honeybee. The causative virus was an ovoid or ellipsoid positive-strand ssRNA virus, chronic bee paralysis virus (CBPV). However, in spite of this initial isolation of an acute, disease-causative agent, most bee viruses are now known to establish persistent infections in their hosts with mild or inapparent disease. These viruses have received little attention, since their effects on bee populations are often minor. The ability of these viruses to induce disease in particular bee colonies is often dependent on the parasites of the colony, such as mites, which transmit CBPV to adult bees. CBPV is common throughout the world and can be found endemically in healthy bee colonies. In the endemic state, it appears that the queen bee is needed to provide protection to the hive. CBPV is also frequently found with an associated virus (CBPVA), which is also an isometric RNA virus. Clearly, virus-virus interactions are common in the persistent state. CBPVA multiplies at low levels and uses CBPV as a helper, but it also interferes with CBPV replication. Under permissive conditions, CBPV reproduction is most evident in the female reproductive caste (i.e., the queen bees) but can also be highly produced in other species, such as ants (especially in the head tissues). Some viruses infecting bee heads are also known to induce aggressive behavior. Other RNA sequences with clear sequence relationships to CBPV have also been reported. Some of these viruses can be highly productive, such as sacbrood virus, which can replicate in huge quantities in bee larvae. Acute paralysis virus (APV) is another common bee virus that persists in adult bees and causes no disease in nature. However, when persistently infected bees also become infested with mites, APV replication is induced by the mites and kills the bees. This apparent mite induction of APV multiplication occurs in association with mites feeding on bees. Another mite-related bee infection is seen with Kashmir bee virus (KBV). This virus is naturally found to persist in *Apis cerana* bees. In association with infestation by the *Varroa jacobsoni* mite, this virus becomes activated, resulting in overt infection, especially in European bees. Others viruses, such as blackqueen cell virus (BQCV), can specifically kill queen larvae but depend on a microsporidian vector for this killing. In addition, several other bee viruses are asymptomatic and distributed worldwide, such as deformed wing virus, slow bee paralysis virus, and bee viruses X and Y. Interestingly, the asymptomatic infections with some of these viruses are associated with shorter life spans for the infected hosts. A few additional virus types for bees have also been reported, including rod-shaped DNA viruses and one iridovirus, *Apis* iridescent virus, which causes clustering disease in bees and is isolated in Old World hives. Overall, persistent infection of bees by RNA viruses is common and can be accompanied by complicated biological interactions with the other agents and parasites that are associated with bee disease. It is worth noting, however, that bees are not native to the New World, and thus New World angiosperm evolution is linked to other insect pollinators, such as moths. In other words, the biology of virus-bee interactions is mainly a by-product of human activity associated for the most part with commercial hives.

Nodaviruses

Nodaviruses are a distinct family of insect positive-strand ssRNA viruses that differ from the picorna-like viruses in that they are bipartite (having two genome segments), with both genome segments encapsidated into one virion. Similar viruses are found in plants but not in invertebrates, so a relationship of nodaviruses to other plant viruses seems likely. One of the nodavirus RNA segments codes for an RdRP; the other segment codes for the capsid protein. These RNAs are distinct from the picorna-like virus genomes in that they have capped 5′ ends (as opposed to covalently linked peptides), although they retain the 3′ poly(A) tail. Some of these viruses have surprisingly broad host specificities. Flock house virus is one such example, but curiously the natural distribution of this virus is not known. This virus not only replicates in many insect hosts but also infects even mice. Flock house virus is known to kill mosquito larvae and also infects adult honeybees.

It has been proposed that some animal viruses may have evolved following recombination with a plant virus. A. Gibbs and G. F. Weiller have proposed that some plant viruses may have switched hosts from plants to vertebrates. One possible example of this is the ssRNA circoviruses of vertebrates, which may have acquired a genome segment from a plant nanovirus. The N-terminal regions of Rep proteins are similar in these two families. However, the C-terminal region is similar to that of the 2C RNA binding protein of picorna-like viruses (such as caliciviruses), which infect only vertebrates.

dsRNA insect virus

There are several examples of dsRNA insect viruses worth mentioning. Silkworm midgut virus is a dsRNA virus that is distinct from vertebrate dsRNA viruses. Cytoplasmic polyhedrosis virus (CPV) has a wide host range, with cross-species transmission. It can be highly infectious and is able to be transmitted from one host larva to another. This appears to indicate that the virus has an acute life strategy. Curiously, the virus is not lytic to cells in culture. As a final example, *Drosophila*-enhanced sensitivity to CO_2 can be due to infection with an alphavirus, which is a bisegmented dsRNA virus. The natural distribution of this virus is not known.

Insect vectors for plant viruses

Most plant viruses need an insect vector for transmission, but in almost all cases the virus does not replicate in that insect. The most common vectors for plant viruses are homopterans, with their piercing and sucking mouthparts. Arthropods, nematodes, and Fungi are also numerically important vectors for plant viruses. Various kinds of virus-vector interactions appear to occur. In most cases, the plant viruses must interact specifically with insect mouthparts. These mouthparts distinguish the numerous insect vectors, as they are the most variable part of insect morphology. Three major mechanisms of plant virus transmission by insects are known. These transmission mechanisms for the insect vectors reflect different levels of intimacy among the plant, virus, and insect and are termed circulative persistent, propagative, and noncirculative. The first (circulative persistent) involves virus binding to insect mouthparts, then virus movement through insect gut wall into insect salivary glands, and finally virus crossing through insect cellular membranes to infect a plant during insect feeding. The virus can also persist in the specific host. However, in general this persistence is really a form of virus storage within the insect cells, in that the virus generally does not replicate in the insect vector. Viruses that are transmitted in this way include luteoviruses, geminiviruses, and fungally transmitted furoviruses. In contrast to circulative persistent transmission, during propagative transmission the virus does replicate. The term *noncirculative* applies to virus that associates with the cuticular lining of insect mouthparts or foregut. The virus then deadsorbs to infect plants during insect feeding in an almost mechanical manner without moving through insect tissues. Aphid transmission of potyviruses can occur this way. Overall, the nature of the insect vector is an important and often specific determinant of the virus and can be used to identify a plant virus. Thus, it is often the case that the best protection against plant viral disease is via insecticide that destroys insect vectors (e.g., aphids).

Homopteran species can serve as vectors for various plant viruses. There are two major groups of vector insects, the tree- and leafhoppers and the aphids and white flies, which diverged around 180 million ybp. Aphids in particular seem to have their own patterns of association with plant viruses, and some aphid species transmit only one virus. Closteroviruses show some coassociation with aphid, mealybug, and white fly vectors. White flies are also often associated with geminiviruses. These flies can produce salivary gland toxins that are general growth retardants, but only female flies make the toxin. Some viruses of leafhoppers (such as *Nilaparvata lugens reovirus,* a fiji-like virus) may represent the earliest type of insect virus, as they are insect egg transmitted and show no multiplication in the host plant (maize). In this case, the roles may be reversed, and plants appear to be the vector leading to persistent virus infection of the insect. There are a few other similar examples, such as *Rhopalosiphum padi* virus transmission to rice and barley leaf via aphids.

Aphid biology

Aphids have an unusual biology in several respects. For one, some aphid species bear live young. Also, aphid species have a tendency to reproduce by parthenogenesis. This can result in predominantly clonal reproduction, with only occasional sexual reproduction for the species. The extensive diversity of aphids is striking. There are about 15,000 species of aphids in about 2,000 genera. Of these, 21 aphid genera have been established as vectors for plant virus transmission, especially for the luteoviruses. It is interesting that aphids can also transmit some rhabdoviruses (sowthistle yellow vein virus and lettuce necrotic yellows virus), suggesting that such rhabdoviruses may have originated in the insect predecessor of aphids. In fact, these viruses are most closely related to *Vesicular stomatitis virus*, an animal rhabdovirus. This has led some to suggest that plant and animal rhabdoviruses may have both originated in some early insects. Along these lines, the insect vector is normally highly specific to the plant virus (such as sowthistle yellow vein virus and potato yellow vein virus), involving long latency and also transovarial transmission via insect eggs. Such viruses can sometimes be adapted to insect-only transmission via insect passage and conversely can be adapted to plant-only replication following serial plant passage. Virus entry into plant or insect cells is clearly distinct, suggesting that insect or plant cell entry characteristics can be under independent selection.

Aphids tend to be circulative persistent insect vectors for plant transmission. Thus, they have a more intimate relationship with the plant virus than do many other vectors. Leafhoppers and plant hoppers tend towards plant virus persistence as well. In some aphids, specific plant virus transmission is restricted to one or a few aphid species. Because circulative persistent transmission requires the plant virus to reach the salivary gland, the virus must avoid degradation within insect cells. The specificity of this process is often the reason for virus vector restriction. The basis of this specificity is discussed below. Sometimes, a second virus can help a specific aphid species transmit an otherwise aphid-restricted virus to a host plant, such as potato aucuba mosaic virus needing potato virus Y for aphid transmission. Thus, virus-virus interactions are known to be important for some circulative persistent insect vectors.

Endosymbiont virus specificity

Another common and curious distinction concerning aphid biology is that most aphids support bacterium-like endosymbionts, such as *Buchnera* spp., which are related to *Escherichia coli*. With a few exceptions, all aphids, mealybugs, and tsetse flies have endosymbionts, although these endosymbionts all have different origins. Some aphids also have secondary symbionts whose identities vary with the specific host lineage. Overall, three types of bacterium-like symbionts are known, and these can be found in the sexually reproduced eggs of aphids.

The best studied of the *Buchnera* aphid symbionts is the pea aphid, *Acyrthosiphon pisum*. This aphid can be parthenogenetic and thus can be observed in clonal populations in field isolations. Analysis of mtDNA sequences shows little variation in these field isolates. These are important crop-feeding aphids, and it appears that they have undergone substantial species diversification in the last 100,000 years. *Buchnera* DNA is present within the pea aphid at about 120 copies per aphid cellular genome. The DNA is circular, contains a single origin of replication, and is transmitted maternally. This 200- to 250-Mbp genome corresponds to about one-seventh the size of the *E. coli* genome and thus is smaller than many large DNA viruses. However, unlike with bacteria, the 16S and 23S rRNA transcription units are separated in this genome. The maternally transmitted *Buchnera* endosymbionts are also often associated with two additional small DNA plasmids. It is curious that unlike mtDNA, these *Buchnera*-associated plasmids show considerable sequence variation. These plasmids contain the APSE-1 bacteriophage DNA sequence, as well as inverted repeats and *rep* genes that are consistent with belonging to the phage-like Rep A1 family. The presence of the *rep* genes and repeat sequences makes these endosymbionts similar to virulence plasmids of bacteria. It thus seems likely that the *rep* genes and inverted repeats can account for the high level of natural sequence variability observed in these plasmids.

Virus-like origin of endosymbiotic plasmids

Phylogenetic analysis suggests that the *Buchnera*-associated small plasmids may have a single evolutionary origin. Furthermore, these plasmid phylogenies are incongruent with the phylogeny of chromosome-borne genes. Plasmids, however, are not exchanged among aphid hosts, and their lineage is host restricted. These observations suggest that the plasmids have independently colonized specific host lineages. Two well-studied plasmids are the trpEG and leu plasmids, which carry genes for amino acid biosynthesis. These plasmids can undergo amplification and express elevated levels of biosynthetic enzymes for Trp and Leu, respectively. Other smaller noncoding plasmids that resemble defectives have also been identified, but these currently have no known functions. As noted in "Insect-plant-virus radiation" above, these plasmids also express GroEL, an hsp60 homologue. GroEL proteins are essential for binding plant viruses within the aphid cell, and they provide the molecular specificity for the virus-aphid interaction, such as the aphid specificity for luteovirus-like *Pea enation mosaic virus*. The plasmid-encoded GroEL proteins appear to allow the plant virus to persist in the aphid cell and avoid degradation. Interestingly, the GroEL gene has the most variable sequence of all aphid and endosymbiont ORFs, suggesting that it is under strong selective pressure to change. Although it seems that plant viruses might be involved in this GroEL diversity, it is not obvious how a plant virus could positively select for plasmid-derived gene diversity in an aphid vector. Some aphids have been reported to produce significantly more young when feeding on virus-infected plants (i.e., infected with beet mosaic virus or beet yellows virus), and this could presumably provide some selective advantage for virus-plasmid interaction. In other cases, however, aphids feeding on virus-infected plants may have shortened life spans. The pea aphid is also subjected to parasitization by *Aphidius ervi* Halliday (Hymenoptera, Braconidae) (a parasitoid wasp; see below). In a most amazing case of intimate but complex interaction between parasites, endosymbionts, a host, and possibly an endogenous virus, the parasitoid wasp larvae redirect aphid host reproduction and metabolism to favor the development of juvenile parasitoids. Presumably this occurs via the bracovirus that is coinjected along with wasp larvae, as this virus is known to be the main parasitoid-derived agent that alters aphid biology. In whatever way this as-yet-uncharacterized process works, it must ultimately cause the *Buchnera* endosymbiont to increase plasmid-mediated expression and amino acid biosynthesis (especially that of tyrosine) within the aphid. Thus, it may be that the fitness consequences to the host of these endosymbiotic, "phage-like" plasmids are by and large determined by interactions with other parasitic agents, such as parasitoids and plant viruses.

Wolbachia spp. are bacterium-like endosymbionts of some hybrid aphid species. It is interesting that *Wolbachia* also has virus-like phage particles, called WO particles, that appear to be carried as a prophage. *Wolbachia* has received much attention by evolutionary biologists because it can be responsible for sex distortion in these hybrid aphid species and can bring about sexual incompatibility among populations. Such distortions may represent a first step in the speciation of host.

Families of Insect DNA Viruses

Insect viruses were first noticed early in the 1800s. In 1808, there occurred an outbreak of silkworm jaundice that was shown to be due to a transmissible infectious agent. We now know that this outbreak was due to a polyhedrosis virus, a large DNA virus that generated refractive crystal-like bodies in hypertrophied silkworm nuclei. The virus infects gut epithelium, causing the silkworms to stop feeding and thereby leading to large losses of infected hosts. In this section, I describe the major families of viruses that infect insect hosts.

Iridoviruses

Iridoviruses are known to infect both insects and aquatic animals, but not higher animals. Thus, these viruses are clearly expected to have been in the oceans, infecting hosts such as shrimp, prior to the evolution of insects. Insect iridoviruses are known to favor Diptera (e.g., *Anopheles* spp., flies, and *Drosophila*). In 1953, the first report of an insect iridovirus described a virus infecting the larvae of crane flies and the North American *Aedes* mosquito. It was observed that this virus could be transovarially transmitted by some, but not all, hosts. In aquatic dipterans, such vertical transmission is common. Currently, Chilo iridescent virus and Tipula iridescent virus are the best-studied insect iridoviruses. Phylogenetic analysis suggests that older versions of this virus family appear to have smaller genomes, such as Coleoptera iridoviruses (e.g., *Costelytra zealandica* iridescent virus), which appears to be older than lepidopteran iridoviruses (e.g., *Wiseana* iridescent virus) and has extra DNA in its genome relative to *Chilo* iridescent virus. The viral DNA of iridoviruses is packaged into a nucleosome-like structure. The DNA is about 110 to 155 kbp long and is terminally redundant and circularly permuted. The level of redundant DNA corresponds to between 4 and 50% of the total genome. With respect to genome and replication topology, iridoviruses clearly resemble P22 phage. In natural host populations, high rates of infection (70 to 90%) can be observed. Thus, these viruses are highly prevalent in nature. However, in some specific populations, such as the blackfly, virally associated pathology is rarely observed. Yet these populations can test 37% positive for virus DNA by PCR-based DNA analysis. This suggests that host-specific persistent but inapparent infections are common. Iridoviruses have also been observed in coinfections with picorna-like viruses. However, the biological significance of these mixed virus infections has not been evaluated.

Entomopoxviruses

Orthoptera (crickets and locusts) appear to have a notable affinity for entomopoxviruses. There are currently no obvious reasons for this tight virus-host relationship. Although entomopoxviruses are clearly related to the orthopoxviruses of vertebrate animals, phylogenetic analysis suggests that the entomopoxviruses are basal to and more variable than the orthopoxviruses. This would suggest that the insect entomopoxviruses evolved first from large DNA virus predecessors present in the oceans. Predecessor candidates include iridoviruses or relatives of baculoviruses, although both of these virus groups are highly diverged from the entomopoxviruses. Entomopoxvirus sequences are most related to those of *African swine fever virus* and granulosis viruses (GVs). In Orthoptera, entomopoxviruses are known to infect more than 31 species of insects. Entomopoxviruses can also infect some species of Lepidoptera and Coleoptera. Interest in entomopoxviruses was initially focused on their possible uses as biocontrol agents, especially since they infect insect families that are major predators of crop species. However, the close similarity of entomopoxviruses to animal poxviruses has prevented their development into biocontrol agents.

Baculoviruses as models for DNA virus evolution

Baculoviruses as a group include possibly the largest number of known insect viruses. These viruses have dsDNA genomes of 90 to 180 kbp. Baculoviruses are mostly pathogenic in their arthropod hosts. The occlusion morphology that results from infection is used to classify baculoviruses into two major groups. Nucleopolyhedrosis virus (NPV) infection results in intracellular polyhedral bodies that contain large amounts of virus. GV infection results in ovoid occlusions with few virions. Because of their possible use as biocontrol agents, baculoviruses have been well studied. The entire phylogenetic tree of all family members has been evaluated and is reproduced in Fig. 7.4. Although in the past the tree shape depended somewhat on the specific genes used to generate the tree, it is now possible to include all of the genes in one comprehensive phylogenetic analysis. From such a comprehensive analysis, it was reported that 63 core genes were common to all baculovirus members. Within this comprehensive tree, 17 gene losses in specific branches could be seen, and 255 gene acquisitions were observed. Of the acquired virus genes, 80% were new to the GenBank database. The conclusions seem clear: for the most part, baculoviruses have evolved from a common core set of genes (mostly replication and structural genes) through the acquisition or creation of novel genes. This virus family is mostly composed of acute, lytic viruses. Although the viruses are not highly host restricted, some host specificity can be noted. For NPVs, all subgroup II-A viruses are from noctuid hosts (night moths); thus, there are some virus-host connections. Some viruses harbor TED "gypsy" elements in their DNA, but how this element affects virus or host biology is not clear. Nonetheless, this

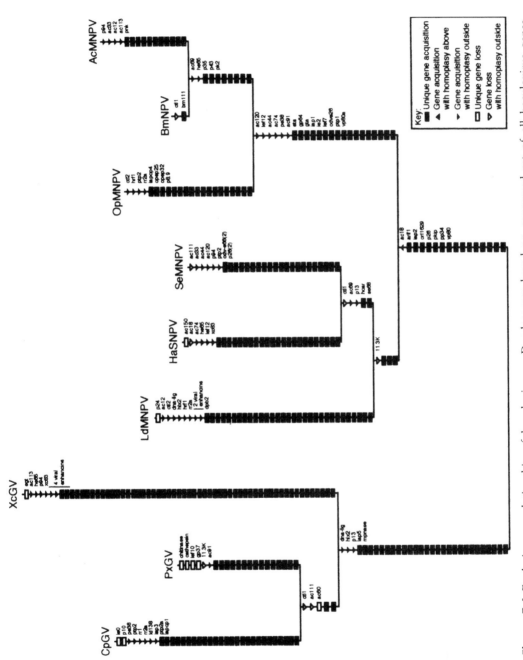

Figure 7.4 Evolutionary relationship of baculoviruses. Dendrogram based on conserved set of all baculovirus genes. Reprinted from E. A. Herniou, T. Luque, X. Chen, J. M. Vlak, D. Winstanley, J. S. Cory, and D. R. O'Reilly, *J. Virol.* **75:**8117–8126, 2001, with permission.

observation shows that baculoviruses themselves can be colonized by retrovirus genomes. Although the viruses are generally lytic, some can persist in the larvae of their hosts. These later kill the host during postlarval development. In these cases, it is likely that the lytic viral genetic program is linked to the differentiation of the host tissues.

NPVs

NPVs include some of the largest dsDNA viruses. As cytoplasmic viruses, they have no RNA splicing but do have 5′ leader sequences. The viral DNA polymerase is unusual in that it is not aphidicolin sensitive, which identifies these viruses as a lineage distinct from T4, phycodnaviruses, and herpesviruses. Most NPVs are not species specific, but their host range is still rather narrow. These viruses are highly lytic, and they appear to have a strictly acute life strategy. Infected larvae undergo virus-induced liquefaction of internal tissues along with lytic virus replication. This results in dissolved larvae that contaminate the soil and the environment and leads to the infection of other larvae via feeding. In field situations, about 5% of host populations tend to be infected, but in some situations of high host population density, infection rates can be as high as 90%. *Autographa californica* NPV (AcMNPV) is the best studied of the NPV group. For the most part, this is due to the existence of a cell culture system for growth of this virus, which has allowed experimentation to proceed. AcMNPV has been extensively used as a system for recombinant expression, which has the advantage (relative to recombinant protein expression in bacteria) of allowing posttranslational protein modifications (such as glycosylation) to occur. The polyhedrons are NPV group antigens.

GVs

There are no vertebrate DNA viruses that show a clear evolutionary relationship to GVs. Even the most conserved genes of most GV lineages, such as the DNA polymerase gene, are not highly related to those of other eukaryotic DNA viruses, probably owing to the fact that GVs have an unusual gamma-like DNA polymerase. That is to say, the baculoviruses have distinct DNA polymerase phylogenetic patterns. The DNA polymerase sequences most similar to that of the baculoviruses are from the entomopoxviruses (e.g., TC1). For the most part, the GVs show a narrow host range, with high rates of infectivity and mortality. Like other baculoviruses, GVs appear to adhere to a strictly acute life strategy. Natural infections with GVs are poorly studied. *Pieris rapae* granulosis virus is the best studied of this group, and natural infections of *Pieris rapae* have been reported. One particular epizootic event in southern California was observed in which >90% of the host larvae in a 1,000-km^2 range were infected. This is a highly impressive level of infection for an acute agent. In this case, the virus was transmitted by the parasitoid *Apanteles glomerates*, which was shown to account for up to 84% of all GV transmission. These observations highlight several important points. Clearly these acute viruses can have significant impacts on host populations in the field. Also, the role of a parasitoid vector for insect virus transmission had not previously been well appreciated. Such high efficiencies of vector-mediated transmission suggest that the reproductive success of this GV itself is very much dependent on the behavior of the parasitoid vector. Plants are also involved in communicating to the parasitoid. For example, it has been reported that following predation by insects, some plants can emit chemical signals that attract parasitoid wasps, thereby assisting the wasps in finding their hosts. Parasitoid species are surprisingly diverse, with 25,000 species. Such a large number of species raises the possibility that these wasp species could provide a major vector function for insect viruses, akin to the insect equivalent of aphids as vectors for plant viruses or mosquitoes as vectors for vertebrate viruses. If so, we can expect some very intimate molecular linkages between the parasitoids (i.e., insect virus vectors), their transmitted viruses, and their larval hosts. Finally, there is some evidence that mixed infections of GV and NPV have a distinct biology relative to those of pure infections. Mixed infections with GV and NPV (and some epizootics) in the field have been reported. It seems possible that these mixed infections involve some type of complementation between the two viruses, as some larvae are 50-fold more susceptible to mixed infections than to single infections. However, such double infections are usually not observed within the same cell; rather; they occur in the same larva. Thus, the double infections may complement at the organismic level. These mixed infections show high levels of field mortality, suggesting that the mixture affects the virus-host phenotype.

Nonoccluded baculoviruses

Until recently, nonoccluded baculoviruses had not been well studied, although reports of their observation have been scattered in the literature for some time. These viruses resemble baculoviruses morphologically, but they lack the cellular accumulation of occlusion bodies typical of baculovirus infections. Several nonoccluded baculoviruses have now been better characterized, including Hz-1, *Oryctes* virus, and gonad-specific virus. In addition, viruses resembling Hz-1 have been reported to persist in some hosts. In fact, it now appears that persistent infections may be the norm for this family of viruses. Gonad-specific virus is known to infect the reproductive tissues of the *Helicoverpa zea* moth and is associated with ovarian atrophy and infertility in some hosts. This virus is sexually transmitted and is generally latent. Although infertility can result, some moths are fertile when infected. The virus is a rod-shaped capsid and contains a circular, supercoiled dsDNA of 236 kbp. This virus was initially isolated from cultured cell lines that were persistently and inapparently infected. Hz-1 can infect and persist in many insect cell lines. Hz-1 has low sequence homology to GVs (only 3% similarity) and shows only 0.1 to 1% similarity to other DNA viruses. These viruses have unique genes not found in the GenBank database. Thus, although they are called nonoccluded baculoviruses, they are distinct and are currently an unclassified virus family. Of the 154 ORFs in Hz-1, only 29 showed some relationship to baculovirus and/or cellular genes. Of these 29, 16 encoded replication proteins, transcription factors, or structural proteins. Eighty-one percent of Hz-1 ORFs show no homology to any other genes. Nonoccluded-baculovirus DNA is also distinct from baculovirus DNA in that it has high AT content and repeat sequences (tandem, highly repeated 21- to 75-bp elements). Interestingly, the Hz-1 DNA-dependent RNA polymerase is most related to the pSKL plasmid DNA polymerase. In other words, it has a "yeast-like" RNA polymerase. The viral DNA-dependent DNA polymerase most resembles that of the insect hosts, whereas the DNA ligase is most similar to that of humans. Also distinct from most other DNA viruses, Hz-1 encodes several host-like nuclear metabolic enzymes.

Persistence by Hz-1

Because Hz-1 readily establishes a persistent state, it has been possible to study some of the molecular characteristics associated with this persistence. In several respects, Hz-1 persistence strongly resembles persistence of human herpesvirus 1. During Hz-1 persistent infection, only a noncoding, "persistence-associated" RNA is expressed. This is similar to the latency-associated transcripts of herpes simplex virus 1 persistence. The expressed Hz-1 sequence is called PAT-1 (persistence-associated transcript) RNA. However, there are some distinctions between Hz-1 and herpes simplex virus 1. Mainly, Hz-1 persistence appears to also involve defective interfering virus production: plaque-purified Hz-1 does not readily establish persistent infections, suggesting an important role for a defective virus in the natural life strategy (and fitness) of Hz-1. This poses a theoretical conundrum to evolutionary biology in that inhibition of virus reproduction (i.e., nonmaximal replication) is important for natural survival. This reinforces the concept that persistence, not replication, may be crucial for the fitness of some viruses. It is worth emphasizing that defective virus-mediated persistence could result in a virus-host relationship in which little if any virus is made, while all hosts are infected. This situation would suggest a way to evolve fully defective viruses that establish stable persistent infections, such as for the PDVs as discussed below. As an example of Hz-1 interaction with defective viral material, consider the following experiment with Hz-1 grown in cell cultures of adult ovarian tissue of *Heliothis zea*. If these infected cells are also transfected with NPV DNA (even UV-killed DNA), Hz-1 production will be induced, even though NPV will not produce inclusions. Thus, there seems to be good evidence for virus-virus interactions. Yet it should be noted that Hz-1 expression is mainly silent during persistence, and the discovery of

Hz-1 was essentially fortuitous. This raises the questions of how common similar persistent infections by other viruses are and how many more persistent viruses like Hz-1 have yet to be discovered.

Like Hz-1, OrL is a member of the nonoccluded-baculovirus family. This virus, which was initially isolated in the Philippines, is known to infect *Oryctes rhinoceros* beetles in a strictly acute, nonpersistent way that is highly lethal to the beetle hosts. Why do the highly similar viruses Hz-1 and OrL differ so greatly in their abilities to cause disease? The *O. rhinoceros* beetle is not a native species of the Philippines; rather, it is an accidental pest species introduced there by humans. OrL infections emerged after introduction of the beetles and resulted in considerable mortality to them, especially the larvae. The virus replicates widely in beetle tissues, especially in the midgut epithelium, and usually produces a fatal infection. Since it causes highly lethal infections, OrL is considered the best field model for the use of a virus as a biological control agent of a pest insect species. However, OrL did not originate from the *O. rhinoceros* hosts and can be found in native Philippine beetle populations, in which it causes little disease. Thus, the virus appears to maintain a persistent life strategy in its natural beetle host, but it can undergo a species jump to the *O. rhinoceros* beetle host and cause highly acute infections.

Ascoviruses

The ascovirus family is associated with parasitoid wasps of the Hymenoptera. These wasps function as vectors in transmission of the ascoviruses to lepidopteran hosts. The family *Ascoviridae* includes four species, three of which are opportunistic and one of which is obligate for the Hymenoptera. In some cases, ascovirus replication has been linked to iridovirus coinfection. Ascoviruses are sexually transmitted in a mechanical manner from a parasitoid wasp to the lepidopteran host via oviposition (i.e., egg laying). In terms of wasp biology, *Diadromus pulchellus* ascovirus (DpAV) is beneficial to *Diadromus pulchellus* wasps, in that it can enhance wasp development and has no adverse effects on the wasps otherwise. However, in *Itoplectis tunetana*, DpAV infection has a most adverse effect, causing an acute infection with rapid virus replication and nuclear lysis.

Ascoviruses have large bacilliform virions with circular dsDNA of about 116 to 190 kbp. The ascoviral DNA polymerase clearly clusters with those of iridoviruses by phylogenetic analysis and is also linked to those of phycodnaviruses and T-even phages. In terms of virion structure and vesicle formation, the ascoviruses resemble entomopoxviruses and iridoviruses. *Spodoptera frugiperda* ascovirus 1a and *Heliothis virescens* ascovirus 3a are found in noctuid hosts, whereas DpAV-4a is found in wasp hosts. In *D. pulchellus* wasps, DpAV is transmitted vertically only by the female, and it is not integrated into the wasp chromosome. Curiously, DpAV is very poorly transmitted by mechanical and other means, but it is highly infectious after natural transmission via oviposition. DpAV prevents the development of larval wasp hosts (i.e., the moth pupae). In addition, DpAV contains genes that suppress moth immunity, including a gene that encodes a protein highly related to scorpion short toxin. In many respects, ascoviruses resemble PDVs biologically, but ascoviruses are not genomic proviruses, as discussed below. Also unlike PDVs, DpAV-4a replicates well in moth larvae. The capacity for ascoviruses to replicate well in moth larvae but persist in wasps suggests that ascoviruses have a hybrid and host-dependent life strategy, which is persistent in the wasp vectors but acute in moth larvae.

Parasitoid wasps and PDVs

The relationship between parasitoid wasps and their hosts was noted by Darwin as an example of nature's cruelty. Various parasitoid wasp species (the endoparasitic Hymenoptera) inject their eggs via an ovipositor into a larval host (i.e., a lepidopteran larva), inside which the wasp larvae develop, thereby consuming the parasitoid host from within. These parasitoids are highly successful species. There are an estimated 200,000 parasitoid wasp species that have various biological strategies for parasitizing their hosts (e.g., endo- and ectoparasitoids). Parasitoid Hymenoptera, using caterpillar and lepidopteran hosts, are ubiquitous in terrestrial habitats. Wasps were an early insect group to evolve and are predecessors of ants, another highly successful insect group. Fossil evidence indicates that the oldest parasitoid wasp dates from approximately 140 million ybp. In some Hymenoptera, the wasps are known to harbor endogenous viruses that are involved in the parasitoid lifestyle. These viruses were first seen in 1967, when virus-like particles were observed in the hypertrophied nuclei of parasitoid wasp oviducts. In 1981, it was shown that the DNA of this "virus" was present in every female of the wasp species being studied. Since then, various observations have suggested that the ovaries of ichneumon wasps are susceptible to a wide variety of silent virus infections, most of which are not well characterized. These viruses are now known collectively as the PDVs, which exist in two groups (bracoviruses and ichnoviruses) (see Fig. 7.5). About 30,000 parasitoid wasp species appear to harbor PDVs. Given the evolutionary history of these wasp species, this virus-wasp relationship appears to have existed for a period of about 60 million years. At least three independent lines of PDVs also appear to have evolved, indicating that this virus-host relationship is polyphyletic. In addition, other viruses (ascoviruses, entomopoxviruses, and unclassified viruses) are known to exist in parasitoid reproductive tissues; however, these agents are poorly characterized, and the significance of these viruses is unknown.

The diverse and stable PDV-host relationship establishes that the combined fitness of a molecular genetic parasite and its host can result in highly stable lineages. In fact, the PDV-insect relationship is one of the most complex and highly evolved interspecific interactions known for insects. The

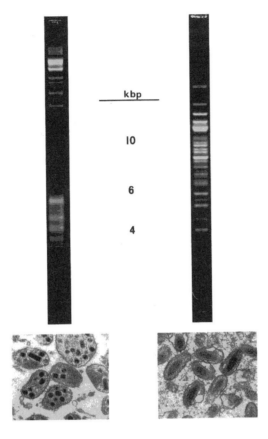

kbp

10

6

4

Figure 7.5 PDVs of parasitoid wasps. (Left) Bracovirus; (right) ichnovirus. Reprinted from D. B. Stoltz, p. 167–187, *in* N. E. Beckage et al. (ed.), *Parasites and Pathogens of Insects* (Academic Press, Orlando, Fla., 1993), with permission.

PDV *Campoletis sonorensis ichnovirus* (CsIV) is the best studied of these viruses. PDVs are composed of numerous circular dsDNA genomes, packaged into one type of virion coat protein. During transmission from a parasitoid wasp to a larval host, the virus is injected along with the wasp egg into the larval host. CsIV capsids are complex: they are composed of more than 15 peptides. The virions are rather distinct from other DNA viruses and so far only show some structural similarity to two unclassified viruses observed in 1977 in fire ants and whirligig beetles by R. J. Avery. The PDV genomes are very distinctive. They are composed of numerous small supercoiled dsDNAs that range from 6.2 to 18 kbp. Each virion is packaged with 10 to 30 segments, of which 22 are nonredundant. The virus DNA is episomal, and the genome segments are present in uneven molar ratios. In this feature, PDVs are not similar to any other insect DNA virus. Although their genomes are still being evaluated, it appears that these viruses do not code for their own polymerases. When comparing the two major groups of PDVs, it is clear that the bracoviruses have fewer and bigger DNA genome segments than ichnoviruses. About two-thirds of the PDV DNA is noncoding and resembles defective genomes. This noncoding sequence is distributed among all genome segments. PDV DNA also contains introns and includes genes from multigene families, including genes for innexins or "vinexins" that code for the gap junction structural proteins involved in the selective transfer of small molecules and ions across cell membranes. These proteins may be needed to affect histocytes (or the encapsidation response of plasmocytes); since vertebrate neutrophils and macrophages are known to use these molecules to communicate via gap junctions, it is assumed that similar mechanisms apply to insect cells. The normal insect larval host immune response occurs via fused melanized hemocytes that surround and suffocate the parasitic wasp egg. The PDVs are known to prevent this response, thereby allowing egg survival. Some PDVs (e.g., *Microplitis demolitor* bracovirus) are known to induce high-level expression of virus genes in larval hemocytes. PDV DNA is replicated from host genomic copies within the female reproductive tissues, where it is packaged into virions. The PDV DNA is thus amplified into episomal forms from proviral sequences that are clustered in the wasp genome and expressed as episomes in the larval host. The mechanism of DNA amplification is not known but could provide important clues as to the evolutionary origins of PDVs. A rolling circular DNA amplification has been suggested, which would imply rolling circular replicon virus ancestors since such a process is not a cellular replication mechanism. Excised and amplified PDV DNA appears to be able to reintegrate into the genome, but this does not appear to be an essential part of the viral life cycle. No sequence relationships to ascoviruses or iridoviruses have been identified, and a possible progenitor of PDV has yet to be established.

The wasp-PDV relationship is such a striking and successful virus-host relationship that it is worth further considering the possible origins of PDVs. A main conundrum is that PDVs express their genes in the larval hosts, not the wasp vectors. This suggests that the PDV genes were originally adapted for functions needed in lepidopteran larvae and further suggests that they most likely originated from a virus that infects and is expressed in these larval hosts. However, PDVs persist in the wasp; given the mechanistic difficulty of jumping persistent infections into a new host species, this wasp persistence suggests that the PDV progenitor virus was also adapted to wasp persistence. Both of these biological characteristics are known to apply to the ascoviruses, which persist in wasp reproductive organs but are amplified and expressed in lepidopteran host larvae. However, PDVs also appear to resemble defective viruses, and in this they resemble the nonoccluded baculovirus Hz-1, especially during Hz-1 persistence in reproductive tissue. Although the numerous small, supercoiled, circular PDV DNA genome segments and possible rolling circular replication resemble characteristics of no other insect viruses, these PDV characteristics do resemble the characteristics of other animal DNA viruses, such as the TT virus, in terms of replication mechanism and size. The TT viruses are a recently discovered group which shows no homology to any other virus group. Finally, the facts that PDVs have three independent evolutionary origins (one for ichnoviruses plus two for bracoviruses) and that these groups are paraphyletic to each other but bracoviruses are monophyletic suggest that PDV colonization was highly favored by selection at the origination of the wasp orders in order to allow multiple virus colonization events. Where PDVs have been identified in a host order, it is observed that all wasps within each of the host groups have PDVs. No wasp descendant within these groups appears to have lost PDV sequences during evolution. Thus, the acquisition of PDVs appears to be a clear example of virally mediated acquisition of a stable and most complex phenotype.

Teratocytes provide another example of a strikingly complex phenotype involved in the manipulation of the host larva physiology that is associated with endoparasitic wasps. Teratocytes are large wasp cells originating from the serosal membrane surrounding the embryo of the female wasp. These cells are injected along with the wasp egg and PDVs into the host larvae. Once in the larval host hemolymph, these cells undergo polyploidization, growing into giant cells and generating microvilli. These cells are thought to absorb nutrients and provide them to the growing wasp embryo. They also affect the host larval endocrine system, preventing host larval metamorphosis. Thus, the purpose of teratocytes appears to be to manipulate the host larval physiology in a way that creates an environment favorable for wasp embryo development. In this function, wasp teratocytes are trophic, immunosuppressive, and able to alter host endocrinology, all in support of the wasp embryo. It is worth noting that all these characteristics are also found in the placental tissues of viviparous mammals. As is discussed in chapter 8, these complex placental functions may have derived from colonization by molecular genetic parasites.

It is interesting that there is a behavioral component to the PDV-larval host relationship. PDVs and teratocytes express genes that directly affect the behavior of infected host larvae, inducing immediate lethargy and subsequent climbing activity. Induced alterations include release of ecdysteroids, juvenile hormone, and neuropeptides, all of which affect host development and alter host behavior. Interestingly, the plant being eaten by the larval host can also affect the wasp behavior. As previously mentioned, some plants, when eaten by larvae, can respond to the salivary gland secretions of those predatory larval hosts by producing volatile compounds that attract parasitoid wasps.

Typically, virus origins and fitness are evaluated from the perspective of virus or host reproductive success. However, as mentioned earlier, persistence and the sometimes associated virus-virus interactions might also be important for understanding how PDVs originated. Some virus-virus interactions are known to occur between PDVs and baculoviruses. For example, coinfection with the baculovirus AcMNPV and the PDV CsIV allows AcMNPV to replicate in a normally resistant host, *Helicoverpa zea*. This effect appears to be mediated by CsIV depletion of the hemocyte population, thereby facilitating AcMNPV infection. This example further indicates the importance of hemocytes for baculovirus infection control. Mixed infection has also been shown to be important for baculovirus coinfection with the PDV *Tranosema rostrale ichnovirus.*

Insect parvoviruses and densonucleosis viruses

Infections with small DNA viruses are also known for insects. Generally, insect parvovirus infections are acute and fatal. Some of these viruses have the ability to encapsidate both positive and negative template strands, such as the greater wax moth and silkworm densoviruses. However, there are some examples of viruses that may establish persistent infections, such as *Bombyx mori* densovirus 2, which chronically multiplies in the lepidopteran mulberry pyralid, *Glyphodes pyloalis,* with no corresponding disease.

Insect Genomes

Like the genomes of all other animals, insect genomes show considerable levels of colonization by DNA transposons and retroposons. Below, I evaluate the changes in insect retroposon composition associated with insect speciation and genome evolution. I also examine the possibility that insect-derived retroposons might have contributed to the large-scale retroposon colonization of plant genomes, as described above for the angiosperm plants. Not all retroposons have been conserved during the evolution of related lineages, and some retroposon elements that were present in aquatic animals appear to have been lost in insects. These include the Poseidon and Neptune (Penelope) elements common in fish, shrimp, sea urchins, and frogs but absent from *Drosophila* and human genomes. In addition, as discussed in chapter 5, some organisms, such as *Neurospora*, actively destroy or eliminate repeated retroposon sequences. Thus, the ability of retroposons to be maintained in genomes after initial colonization is not necessarily ensured. It would appear that some type of selective pressure must exist to maintain these elements in those lineages in which they are common.

The *Drosophila* genome, retroviruses, and speciation

Speciation and diversification are associated with germ line acquisition or colonization of DNA transposons and retroposons in both insects and plants. However, we currently lack a solid explanation for this observation. The concept of selfish DNA has often been applied to explain the accumulation of such genetic agents, as discussed in chapter 1. And yet the lineage-specific maintenance of the sometimes large numbers of nonreplicating, defective versions of transposons is difficult to explain by the selfish-DNA hypothesis. With the completion of the *D. melanogaster* genome, we can now evaluate an entire genome with respect to these transposon elements. *D. melanogaster* has 178 full-length LTR retroposons (also called ERVs). Most of these full elements belong to the Ty1/copia or gypsy element family. In addition, members of other retroposon families are present at low numbers, such as the

ROO elements and the BEL element (which is also seen in maize). These two elements appear to be a more recent acquisition in the *Drosophila* genome and show a high degree of similarity to the more abundant plant elements. In *Drosophila*, retroposons exist in 17 distinct families, with two of these families present in the greatest numbers: the copia/gypsy family and BEL/ROO family. In *D. melanogaster*, the Ty1/copia family is present at about 10 copies per genome. It is important to note that no flies lack the gypsy element, but as noted above, we currently lack an explanation for this conservation. Furthermore, all *D. melanogaster* strains have defective gypsy proviruses in pericentric heterochromatin, and a few wild strains of *Drosophila* have additional gypsy elements that are active in euchromatin. The latter are generally present at fewer than five copies per genome. In *Drosophila*, the telomeric region (e.g., the telomeric transposon TART) is clearly distinct from the telomeres of vertebrate animals and resembles retroviral RTs in sequence and in functions (i.e., to maintain the telomeres). The RT nature of these fly telomeres suggests that they may have a viral origin. In addition, TART-related LINE-like (Ty3/gypsy) elements have been reported to be expressed in oviduct tissues. Furthermore, this oviduct ERV expression can be induced by mating pheromone. Euchromatically located, LTR-containing retroposons are also present, but these appear to be much younger than *D. melanogaster* as a species, as they are not congruent with *Drosophila* genomes. In some *D. melanogaster* strains, there is a ZAM element, which is an *env*-containing full copy of the gypsy element. ZAM is, therefore, a full retrovirus. It is overexpressed in the follicular cells that surround the oocyte and can be sexually transmitted. In the Rev I strain of *D. melanogaster*, the mobilization of ZAM appears to have occurred following a spontaneous insertion by another transposon. These activated viruses can cause segregation distortion following mating between virus-producing and non-virus-producing *Drosophila* strains.

Plant and insect retroposons

As mentioned above, plants lack authentic retroviruses. And yet plant genomes are known to have acquired many copies of retroposons that resemble the gypsy-like and copia-like retroviruses found in insects. In fact, it is precisely these blocks of retroposon elements that differentiate recent events in plant genome evolution, such as the divergences between maize and teosinte, maize and sorghum, and barley and rice. Even wild rice DNA can be differentiated from crop rice DNA by such retroposon blocks. It should also be recalled that in gymnosperm lineages, the clades of gypsy elements were monophyletic, suggesting a congruence between plant orders and retroposon block acquisition. Cereals, in particular, appear to have evolved more recently than other angiosperms and show genomes that are especially high in copia-like and gypsy-like retroelements. Two events appear to have been required for these genomic retroposon block changes. First, genome colonization by a specific family or variant of retroviruses must have occurred at the origination of these plant species. Second, diversification (with deletions), amplification, and expansion of genomic retroposons must have been relatively rapid and perhaps remains an ongoing process. This amplification and diversification especially pertains to the defective derivatives (both LTR and non-LTR) of the colonizing retroposons. Diversification has occurred on a relatively recent timescale. This poses the questions of why there is an association between speciation and retroposon colonization and what roles the colonizing retroposon and its subsequent expansion play in the origination of plant species. As discussed below, the gypsy elements can be fully expressed as infectious retroviruses in insect reproductive tissues. Furthermore, these retroviruses can be passed as sexually transmitted agents. Given the coevolutionary relationship between insects and plants (as mentioned above), it thus seems plausible that insects might have been the source of the gypsy elements that colonized various plant genomes. This would require insect-derived retroviruses to somehow find their way into plant germ line DNA.

We can now speculate as to how insects might have contributed to the evolution of flowering plants. Much of our understanding of insect retroviruses and retroposons comes from studies of *Drosophila*. As described below, *Drosophila* speciation may directly relate to the expression of transposons and retroviruses by reproductive tissue. However, these dipterans were not yet present at the evolution of the early flowering plants (nor were many of their larval insect hosts). Flies (and ants) most likely evolved from solitary wasp predecessors (which included the very numerous parasitoid species). Parasitoid wasps are well established for being able to produce various types of endogenous viruses (i.e., PDVs) and exogenous viruses (e.g., ascoviruses) in association with their reproductive tissues. In some cases, these viruses are essential for wasp reproduction and egg survival. These viruses can even control wasp sexual behavior.

The earliest flowering plants are thought to be represented by orchids, which are highly diverse in species. In "The Coevolved Trinity of Plant, Insect, and Virus" above, the idea was proposed that insects may have been directly involved in the origination of flowers. If this speculation is correct, can we observe in extant wasp and orchid species some remnants of relationships consistent with this insect role? Parasitoid wasps must be able to implant their eggs in appropriate hosts. Although larval insects are currently their normal hosts, it would seem possible that at some distant time they had a similar relationship with plants as hosts for egg development. Some relationship like this seems essential if large numbers of genetic elements, such as retroposons, from insect lineages are to have entered the germ line of plants or led to the evolution of plant embryos/flowers. Some orchids, such as Australian *Chiloglottis* orchids, rely exclusively on specific wasps (e.g., *Neozeleboria cryptoides*) for pollination. These orchids emit volatile compounds that sexually attract specific wasp species and make flower structures that induce male wasp mating behavior, which results in flower pollination. This complex relationship has been called "sexual deception" and is considered a relatively unique relationship. Many other orchids also use wasp pollinators. In addition to orchids, we also have good evidence that wasp sexual behavior can be directly controlled by viral parasites. If insects were indeed involved in the origination of flowering plants, then relationships like the one described above may actually be much older than previously considered and may represent remnants of early wasp-plant relationships. Future phylogenetic analysis of these wasps, their viruses, and orchid genomes and retroposons may help clarify this possibility.

Insect speciation and ERVs

D. melanogaster and various other species of *Drosophila* have been used extensively by evolutionary biologists to study the issues related to speciation. *Drosophila simulans* and *D. melanogaster* are currently estimated to have diverged about 2.3 million ybp, and this has served as a useful time reference to estimate species diversification. *Drosophila* species diversification can be considered from an Old World-New World perspective, as these groups of species are distinct. The Old World *D. melanogaster* species has successfully colonized the New World. And with the completion of the *D. melanogaster* genomic sequence, we can now begin to evaluate the patterns of transposon and retroposon acquisition during the evolution and speciation of these flies. As many of these retroposon elements are found in heterochromatin regions of the genome (especially the X and Y chromosomes), it has been proposed that they simply reflect selfish transposons that lack phenotype and thus accumulate in inactive, less damaging regions of the fly chromosomes. However, more recent evaluations of this idea have not supported the selfish-DNA concept. Not all families of retroposons are concentrated in regions of heterochromatin; thus, there seems to be no link between the stability of the retroposon and its location in heterochromatin. In contrast to the plant retroposon sequences, flies conserve intact ERVs: 10 of the retroposon families of the *Drosophila* genome are complete with all ORFs, including the *env* ORF. Examples include the retroposons 17.6, 297, gypsy, Idelfix, opus, Quasimodo, ROO, springer, Tirant, ZAM, and Osvaldo. The conservation of the *env* gene strongly suggests that there is some functional selection for this gene. The fact that various other retroposon family members include all the normal retroviral genes except for *env* also suggests a functional explanation for *env* conservation. The presence of full ERV sequences is not unique to *Drosophila* genomes, as TED from lepidopteran species is also a complete retrovirus. As is always the case, larger numbers of defective copies of these retroposons are also found in insect genomes. However, all species of *Drosophila* have maintained some prototypical versions of the gypsy ERV related to that found in *D. melanogaster*. In *Drosophila obscura* strains, the *env* sequences are monophyletic, indicating a conserved and vertical mode of transmission. However, *Drosophila hydei* and *Drosophila virilis* have gypsy *env* sequences related to the *D. obscura* element that are not conserved in a monophyletic clade and appear to have been acquired by more recent horizontal transmission. In these species, an interrupted *env* ORF has been maintained. Thus, intact (i.e., *env*-containing) gypsy ERVs show both conservation and congruence with their hosts as well as recent colonization and incongruence, and these patterns are related to the specific *Drosophila* species. For example, *D. obscura* has an *env*-containing gypsy element related to the *D. melanogaster* element (GypsyDm). This element appears to be recently acquired and is inactive. It seems likely that sexual transmission from a stably colonized host genome to another recently infected and sexually isolated host genome may account for these transmission patterns. For example, the gypsy-like Osvaldo retroposon (with known similarity to the animal retroviruses human immunodeficiency virus 1 and simian immunodeficiency virus) is maintained in *Drosophila buzzatii* and is known to undergo transpositional activation in hybrids between *D. buzzatii* and its sibling *Drosophila koepfera*. In *D. melanogaster*, the gypsy element is normally persistent as an inactive and inapparent provirus under the control of the *flamenco* gene, which must be repressed to activate the gypsy virus.

Speciation and DNA transposon colonization

Another genetic element that appears to distinguish Old World and New World *Drosophila* genomes is the P element. P elements are 2.9-kbp transposable DNA elements with 30-bp repeats at both ends. They contain four ORFs, which include genes coding for a transposase and a repressor of immunity. This organization and these gene coding functions clearly resemble those of phage Mu, and P element transposons appear to be defective relatives of phage Mu lacking any capsid genes. In this lack of capsids, P elements also resemble the hypoviruses of Fungi. Despite the lack of capsids, there is evidence of P-element mobilization and colonization of new genomes. P-element activity is normally regulated via differential RNA splicing in which four introns are transcribed in non-germ line cells. Failure to remove these introns via RNA splicing results in a truncated transposon that fails to express gene products and is inhibitory to other transposons, rendering the transposons immobile and persistent in the genome. Differential splicing in germ line tissue would seem necessary for P-element germ line colonization. Like for the gypsy element, P-element phylogenetics can be either congruent or incongruent in various *Drosophila* species. In fly species of African origin, such as *D. melanogaster*, P-element phylogeny is incongruent with the host genome, strongly suggesting a recent colonization by the P elements. However, in American fly species, such as *Drosophila willistoni*, P-element phylogeny is congruent with its host genome. Thus, it appears that *D. melanogaster* lacked P elements prior to American colonization. In keeping with this hypothesis, P elements are also associated with hybrid dysgenesis between *Drosophila* species that contain and lack P elements. By evaluating codon usage differences, estimates of the time of transfer of P elements to the *D. melanogaster* species have been made. This transfer appears to have been a relatively recent event. However, it seems clear that hybrid dysgenesis is associated with both P-element DNA transposition and LINE-1 retroposon genome transposition. The implication of these observations is that genomic colonization by DNA transposons and retroposons can result in segregation distortion and initiate a process of sexual isolation among previously interbreeding insect populations. Segregation distortion could represent the initial genetic event needed to generate a distinct species, but it essentially results from a virus strategy for the competitive colonization of host genomes.

Here it has been noted that the major patterns of genomic change and the major distinctions between the genomes of closely related species both appear to involve changes in and acquisitions of retroposons and DNA transposons. Furthermore, there are striking contrasts in the distributions of endogenous retroposons among various lineages, such as *Drosophila*, human, and plant genomes. In the fly genome, most families of retroposons are found at a relatively low frequency (1 to 10 copies/genome) but at variable sites. In human genomes, most retroposon elements are numerous and fixed in their location, and polymorphisms in location are rare. Other insects can clearly have much larger genomes with a much greater content of repeated elements than *Drosophila*. One striking example of this is the mountain *Podisma* grasshoppers, which can have genomes 100-fold larger than the human genome. (Surprisingly, these superlarge genomes do not spontaneously delete DNA at high rates.) Clearly, the relationship between genome colonization and host speciation (or competition with other genetic parasites) is poorly understood, but such colonization is associated with and possibly responsible for the events leading to speciation.

Recommended Reading

Invertebrates and Viruses

Beckage, N. E. 1998. Parasitoids and polydnaviruses. *BioScience* **48**:305–311.

Beckage, N. E., and D. B. Gelman. 2004. Wasp parasitoid disruption of host development: implications for new biologically based strategies for insect control. *Annu. Rev. Entomol.* **49**:299–330.

Cheng, C. H., S. M. Liu, T. Y. Chow, Y. Y. Hsiao, D. P. Wang, J. J. Huang, and H. H. Chen. 2002. Analysis of the complete genome sequence of the Hz-1 virus suggests that it is related to members of the *Baculoviridae*. *J. Virol.* **76**:9024–9034.

Darai, G. 1990. *Molecular Biology of Iridoviruses*. Kluwer Academic Publishers, Boston, Mass.

Dowton, M., A. Austin, and the International Society of Hymenopterists. 2000. *Hymenoptera: Evolution, Biodiversity and Biological Control*. CSIRO Publishing, Melbourne, Australia.

Gaunt, M. W., and M. A. Miles. 2002. An insect molecular clock dates the origin of the insects and accords with paleontological and biogeographic landmarks. *Mol. Biol. Evol.* **19**:748–761.

Herniou, E. A., T. Luque, X. Chen, J. M. Vlak, D. Winstanley, J. S. Cory, and D. R. O'Reilly. 2001. Use of whole genome sequence data to infer baculovirus phylogeny. *J. Virol.* **75**:8117–8126.

Kurstak, E. 1991. *Viruses of Invertebrates.* Marcel Dekker, New York, N.Y.

Liu, H., and A. T. Beckenbach. 1992. Evolution of the mitochondrial cytochrome oxidase II gene among 10 orders of insects. *Mol. Phylogenet. Evol.* **1**:41–52.

Pfaff, D. W. 2002. *Hormones, Brain, and Behavior.* Academic Press, Amsterdam, The Netherlands.

Rahbe, Y., M. C. Digilio, G. Febvay, J. Guillaud, P. Fanti, and F. Pennacchio. 2002. Metabolic and symbiotic interactions in amino acid pools of the pea aphid, Acyrthosiphon pisum, parasitized by the braconid Aphidius ervi. *J. Insect Physiol.* **48**:507–516.

Stasiak, K., M. V. Demattei, B. A. Federici, and Y. Bigot. 2000. Phylogenetic position of the Diadromus pulchellus ascovirus DNA polymerase among viruses with large double-stranded DNA genomes. *J. Gen. Virol.* **81**:3059–3072.

Turnbull, M., and B. Webb. 2002. Perspectives on polydnavirus origins and evolution. *Adv. Virus. Res.* **58**:203–254.

Whitfield, J. B. 2002. Estimating the age of the polydnavirus/braconid wasp symbiosis. *Proc. Natl. Acad. Sci. USA* **99**:7508–7513.

Transposons and ERVs

Alberola Trinidad, M., and R. De Frutos. 1996. Molecular structure of a gypsy element of *Drosophila subobscura* (gypsyDs) constituting a degenerate form of insect retroviruses. *Nucleic Acids Res.* **24**:914–923.

Canizares, J., M. Grau, N. Paricio, and M. D. Molto. 2000. Tirant is a new member of the gypsy family of retrotransposons in *Drosophila melanogaster. Genome* **43**:9–14.

Dimitri, P., and N. Junakovic. 1999. Revising the selfish DNA hypothesis: new evidence on accumulation of transposable elements in heterochromatin. *Trends Genet.* **15**:123–124.

Leblanc, P., S. Desset, F. Giorgi, A. R. Taddei, A. M. Fausto, M. Mazzini, B. Dastugue, and C. Vaury. 2000. Life cycle of an endogenous retrovirus, *ZAM*, in *Drosophila melanogaster. J. Virol.* **74**:10658–10669.

Lerat, E., C. Rizzon, and C. Biemont. 2003. Sequence divergence within transposable element families in the *Drosophila melanogaster* genome. *Genome Res.* **13**:1889–1896.

Plants and Viruses

Bremer, K. 2000. Early cretaceous lineages of monocot flowering plants. *Proc. Natl. Acad. Sci. USA* **97**:4707–4711.

Gibbs, A. 1999. Evolution and origins of tobamoviruses. *Philos. Trans. R. Soc. Lond. B* **354**:593–602.

Gray Stewart, M. 1996. Plant virus proteins involved in natural vector transmission. *Trends Microbiol.* **4**:259–264.

Kenrick, P. 2000. The relationships of vascular plants. *Philos. Trans. R. Soc. Lond. B* **355**:847–855.

Lockhart, B. E., J. Menke, G. Dahal, and N. E. Olszewski. 2000. Characterization and genomic analysis of tobacco vein clearing virus, a plant pararetrovirus that is transmitted vertically and related to sequences integrated in the host genome. *J. Gen. Virol.* **81**:1579–1585.

Matthews, R. E. F., and R. Hull. 2002. Matthews' *Plant Virology*, 4th ed. Academic Press, San Diego, Calif.

Raubeson, L. A., and D. B. Stein. 1995. Insights into fern evolution from mapping chloroplast genomes. *Am. Fern J.* **85**:193–204.

Stensmyr, M. C., I. Urru, I. Collu, M. Celander, B. S. Hansson, and A. M. Angioy. 2002. Pollination: rotting smell of dead-horse arum florets. *Nature* **420**:625–626.

RNA Virus Evolution

Argos, P., G. Kamer, M. J. Nicklin, and E. Wimmer. 1984. Similarity in gene organization and homology between proteins of animal picornaviruses and a plant comovirus suggest common ancestry of these virus families. *Nucleic Acids Res.* **12**:7251–7267.

Gorbalenya, A. E., F. M. Pringle, J. L. Zeddam, B. T. Luke, C. E. Cameron, J. Kalmakoff, T. N. Hanzlik, K. H. Gordon, and V. K. Ward. 2002. The palm subdomain-based active site is internally permuted in viral RNA-dependent RNA polymerases of an ancient lineage. *J. Mol. Biol.* **324**:47–62.

Holmes, E. C. 2003. Molecular clocks and the puzzle of RNA virus origins. *J. Virol.* **77**:3893–3897.

Koonin, E. V., and A. E. Gorbalenya. 1992. An insect picornavirus may have genome organization similar to that of caliciviruses. *FEBS Lett.* **297**:81–86.

Zanotto, P. M., M. J. Gibbs, E. A. Gould, and E. C. Holmes. 1996. A reevaluation of the higher taxonomy of viruses based on RNA polymerases. *J. Virol.* **70**:6083–6096.

Evolution of Terrestrial Animals and Their Viruses

Introduction

Vertebrate emergence from the oceans

As vertebrates emerged from the oceans to become land-dwelling animals, numerous basic changes in their physiology and organ structures were necessary. The ability to breathe air required a physiological adaptation to respire directly from the air, a requirement fulfilled by the development of lungs. Living in a nonaqueous habitat also required the development of a skin able to retain moisture and prevent desiccation. Also, limb and skeletal structures needed to become sufficiently robust to support the full mass of the animal, since buoyancy is not significant in terrestrial environments. All of these adaptations were some of the basic evolutionary events leading to the origination of the terrestrial vertebrate animals. With these developments, the host animal habitats presented to virus species would also have changed. The creation of lungs and keratinized skin, for example, would have provided novel and distinct habitats for animal viruses. And as will be seen, numerous viruses have indeed adapted to both the terrestrial animal lungs and skin. For example, in humans respiratory infections are the most common type of modern virus infections. In chapter 7, I briefly considered some of these issues of host change during land adaptation with respect to the viruses that infect terrestrial insects and plants. I noted that the loss of the aqueous habitat would require viruses to evolve nonwaterborne transmission mechanisms as well as resistance to desiccation and solar damage. Many of the same issues apply to the viruses of terrestrial vertebrates. Curiously, however, the types of viruses that infect insects are often very distinct from those that infect vertebrates. Thus, it does not appear that the common terrestrial habitat led to common virus-host relationships. Another difference is that insects and early vertebrates were poikilothermic, whereas mammals and birds are homoeothermic. Successful virus infections are often associated with host temperatures, especially for fish and amphibian viruses. And in mammals, fevers are associated with the host defense response against virus infections. Thus, homeothermy can also be identified as a distinct thermal virus habitat that developed in birds and mammals.

Viruses of fish, host immunity, and virus evolution

As discussed in chapter 6, oceanic vertebrate animals represent the first animals to have developed the adaptive immune system. The acquisition of this complex and protective genetic system would certainly have affected the relationship between these hosts and their viruses. But how might the selective pressures brought about by infection with various types of viruses have affected the various host lineages that later developed? In the preceding chapters, I have described a strong tendency that suggests links between host diversity and virus diversity: the more diverse are the host species of a particular order, the more diverse tend to be the corresponding types of viruses that infect that host, e.g., bony fish (chapter 6) and flowering plants (chapter 7). Scientists cannot at this time offer a validated explanation for this observation. It cannot be determined whether viruses simply adapt to more diverse hosts or host diversification itself is related to viral loads. Yet the invention of the adaptive immune system would seem to pose a major problem for any virus and for any putative linkage of virus to host evolution. The adaptive immune system should severely limit these hosts as potential virus habitats. Is there now any evidence that supports this expectation? What patterns of virus evolution are seen in the vertebrate lineages, and do they suggest any linkage to the creation of the adaptive immune system? Sharks and skates are jawed, cartilaginous vertebrates that have no developed bony skeletal structures, and they represent the earliest animals to have an essentially complete adaptive immune system. It indeed appears that these vertebrate hosts, which are not highly diverse relative to the bony fish, are infected by relatively few types of viruses. However, until recently there were few data available to evaluate this issue because these organisms were not well studied. Can we now examine any molecular data that might illuminate the issue at hand?

One gene that defines the adaptive immune system is the V-H gene, which is essential for the DNA recombinational activity that generates the needed diversity of the adaptive immune response. Recent comparative studies of V-H genes have been used to generate a phylogenetic tree of all vertebrates based on the evolution of the adaptive immune system. This tree depicts five major groups of V-H genes: A, B, C, D, and E. Of these, the E group is monophyletic and includes the genes of cartilaginous fish (such as sharks and skates). Group D corresponds to the genes of bony fish. Group C is more mixed, containing fish, amphibian, reptile, bird, and mammal genes. Groups A and B are less mixed and contain genes of mammals and amphibians. It was especially interesting that the examples of the earliest representatives of the V-H genes (group E) were not found in the higher oceanic or terrestrial vertebrates. Furthermore, the V-H gene version found in bony fish (group D) was not found in terrestrial vertebrates. And some examples found in terrestrial vertebrates (groups A and B) were not found in oceanic vertebrates. These observations appear to identify a pattern in which early V-H genes of the adaptive immune system were initially replaced by subsequent versions of these genes found in bony fish and then replaced again by different versions found in the terrestrial animals. These transitions or gene replacements correlate with changes in host diversity. It is therefore of interest that the radiation of virus diversity was most apparent in the transition from cartilaginous fish to bony fish, and that this transition also corresponded to a replacement of the V-H genes as well as a radiation in host diversity (i.e., the resulting teleost fish are now much more diverse than their vertebrate ancestors).

Teleost fish radiation

It is now estimated that bony fish and aquatic lower chordates comprise about 18,000 species. The large majority of these species are teleost fish. This compares to about 6,300 species of reptiles, 9,000 species of birds, and 4,000 species of mammals in terrestrial habitats. Although bony fish represent a major radiation in vertebrate species relative to terrestrial vertebrate species, it should be remembered that this compares to 50,000 species of Mollusca and 6,100 species of echinoderms in the oceans. Also as a basis for comparison, there are an estimated 69,000 species of higher plants and 751,000 species of insects in terrestrial habitats. As a means for calibration of species generation, it is worth recalling that wasp hosts with specific polydnaviruses account for over 20,000 species of parasitoid wasp (chapter 7).

Diversity of fish viruses

It has previously been observed that bony fish harbor many virus types, which include most of the virus families that are also found in terrestrial animals. In terms of known virus families, fish seem to lack only the parvoviruses and the arboviruses. In teleosts, a large number of viruses are DNA-containing, LCD-like viruses or idiroviruses. Lymphocystis disease virus (LCDV) alone is known to infect over 140 fish species, resulting mainly in benign skin and proliferative connective tissue diseases. Fish also harbor a large number of RNA viruses. Curiously, two-thirds of fish RNA viruses are thought to be rhabdoviruses, which, outside of bats, are not common in other host species and represent a rather unique virus-host association. There are virus-host patterns in vertebrates that are strikingly different from those in insects or plants. For example, although DNA viruses are common in both mammal and insect orders, there are no ascoviruses or iridoviruses in mammals, yet these are common in both insects and cold-blooded vertebrates. Considerations of general virus-host patterns have allowed us to identify several clear virus-host themes in the vertebrates. One theme is that mammals have maintained the ability to host most of the diverse families of viruses that were found in bony fish, whereas amphibians and reptiles apparently have not (discussed below). One family of virus that is of special interest, due to its close coevolution with its hosts, is the herpesvirus family. Fish are susceptible to several lineages of herpesviruses. Some of these lineages are unique to fish, whereas other herpesvirus families link fish herpes to mammalian herpes. Fish herpesvirus infections, like fish retrovirus infections, are readily recognized in field surveys because they are often associated with benign skin growths, a feature not seen in herpesvirus infections of mammals.

Relationship of fish herpesviruses to the viruses of mammals

Channel catfish herpesvirus (CCHV) is clearly a herpesvirus as determined by its morphology and replication strategy, yet it shows almost no sequence similarity to other herpesviruses. For example, the CCHV thymidine kinase gene is not similar to the host version of this gene, nor is it similar to the thymidine kinase genes of other herpesviruses. This group of fish herpesviruses therefore does not appear to be an ancestor of the current mammalian viruses. In addition, there are other fish herpesviruses that appear to be members of the CCHV-like herpesvirus family, such as ictalurid melas herpesvirus. These two viruses, CCHV and ictalurid melas herpesvirus, are biologically interesting in that they both establish latency in fish leukocytes and result in persistent virus shedding in fish. Due to their distinct characteristics, the CCHV-like herpesviruses appear to represent a lineage of fish herpesviruses that has no apparent direct descendants in mammalian (or avian) herperviruses. Thus, this family appears to have a distinct evolutionary origin that has been maintained in fish. However, salmonid herpesvirus (SalHV) has a genome that clearly resembles those of mammalian alphaherpesviruses. SalHV-2 shows low virulence in both coho salmon and some trout. This virus also establishes an inapparent infection in kokanee salmon. Some fish appear to be able to shed the virus persistently (especially via sexual products), providing a source of virus for infection of other species. Interestingly, coinfection of salmon with the retrovirus lymphocystitis virus can result in SalHV-1 reactivation. In other species, such as rainbow trout, SalHV-1 is lethal. In contrast, steelhead herpesvirus (SHV) is biologically converse to SalHV-2, in that steelhead herpesvirus shows low virulence in rainbow trout but is virulent in salmon. Numerous other related herpesviruses of fish are also known (e.g., acipenserid herpesvirus, pleuronectid herpesvirus, and percid herpesvirus), but the biological relationships and virus characteristics of these are in general not well established. However, this set of herpesviruses is known to show discernible similarity to the herpesviruses of mammals; thus, these salmon herpesviruses may represent ancestors of the mammalian herpesviruses.

Are fish herpesviruses and human herpesviruses related?

Humans are especially well studied with respect to herpesviruses and are known to host seven (or possibly eight, depending on criteria) different human-specific species of herpesvirus. Most (but not all) of these herpesviruses are phylogenetically congruent with human evolution. These herpesviruses appear to represent an ancient lineage of viruses and have even been linked to bacterial DNA viruses (the T-even phages). It seems likely that this evolutionary lineage would also include the salmon (but not catfish) herpesviruses. As previously noted, similarities between the animal herpesviruses and the DNA viruses of green algae (phycodnaviruses) and filamentous brown algae (phaeoviruses) are also very clear. Thus, it seems clear that some forms of aquatic herpesviruses do show similarity to human herpesviruses. In addition, herpesviruses have been observed in mollusc, lamprey, and shark species. Although we know relatively little about these aquatic herpesviruses, we do know that some maintain common biological characteristics, such as ganglion infection, suggesting that these viruses likely share a common heritage with the mammalian viruses. Also, herpesviruses frequently persist in specific species but do not integrate into host chromosomes. Both of these characteristics appear to have been maintained by herpesviruses of animal hosts. It is curious that unlike the large DNA viruses of bacteria and filamentous brown algae, viruses of vertebrates do not normally integrate their DNA into host chromosomal sequences.

Another ancient virus lineage that links aquatic vertebrates to terrestrial vertebrates is the retroviruses. As noted in chapter 6, the retroviruses are prevalent, and autonomous retrovirus infections were first apparent in the bony fish. These fish viruses account for a significant fraction of the infections of commercially important species. Relative to these fish retroviruses, however, most retroviruses of terrestrial vertebrates are not as frequently associated with autonomous acute infections. Instead, they tend to occur more commonly as inapparent endogenous retrovirus (ERV) infections, sometimes associated with cellular proliferation, especially for cells of the immune system. Some of these ERVs are highly associated with the host genomes and tightly linked with the lineage of their hosts. Whereas some ERVs can be highly prevalent (at 100% levels), others are much more restricted in their distribution. In addition, there is significant variation in the quantity of ERVs observed for a species, especially in the quantity of defective ERV derivatives that are found in the specific lineages of terrestrial vertebrates. In general, amphibians, reptiles, and birds have significantly smaller numbers of ERVs and ERV-derived retroposons than do mammals. Of special note is that not only do ERV sequences tend to be much more numerous in all mammal lineages, but also many of these ERVs are specific to and maintained within a particular mammalian lineage. Thus, each mammal lineage is associated with its own peculiar versions of ERVs and ERV-associated long interspersed repetitive elements (LINEs) [e.g., poly(A) retroposon derivatives of the same ERV family]. Figure 8.1 is a schematic summary

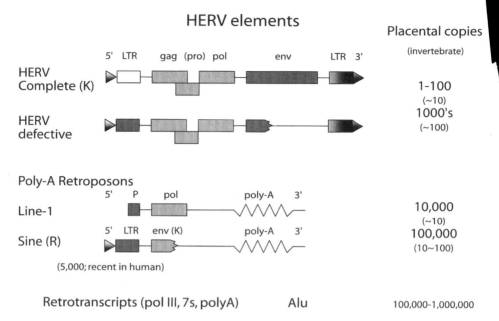

HERV elements

		Placental copies
		(invertebrate)

HERV Complete (K) — 5′ LTR   gag (pro) pol   env   LTR 3′

1-100
(~10)

HERV defective

1000's
(~100)

Poly-A Retroposons

Line-1 — 5′ P   pol   poly-A 3′

10,000
(~10)

Sine (R) — 5′ LTR   env (K)   poly-A 3′

100,000
(10~100)

(5,000; recent in human)

Retrotranscripts (pol III, 7s, polyA)   Alu   100,000-1,000,000

Retron/retroposons - self splicing integrating RT introns; present in phage and mitochondria, not prokaryotes

Figure 8.1 Schematic of HERVs and related elements (LINEs, SINEs, and *Alu*). The number of copies per genome present in invertebrates is indicated in parentheses.

of human ERV (HERV) elements and copy levels as well as HERV relationships to LINEs, short interspersed repetitive elements (SINEs), and retrotranscripts (*Alu* elements). Also indicated parenthetically are the relative ERV copy numbers found in invertebrates. It is worth noting that the mammalian sex chromosomes (X and especially Y) are especially highly colonized by ERVs. This issue and how it relates to host evolution are considered in much greater detail at the end of this chapter.

In summary, we can clearly see overall broad and well-maintained patterns between the terrestrial vertebrates and their viruses, and these patterns have undergone noticeable shifts with the evolution from oceanic to terrestrial hosts. There have been few prior attempts to explain these associations. Thus, the discussions below provide a unique perspective on these virus-host relationships.

Overall Characteristics and Evolution of Terrestrial Host Lineages

Although I have outlined some general differences between terrestrial animals and their predecessors, it is now worthwhile to consider the biological characteristics of the individual lineages of terrestrial animals so that I can subsequently evaluate how these lineages relate to the viruses that infect them. Terrestrial vertebrates are much less diverse in terms of species numbers than are insects. However, terrestrial vertebrates have a diversity similar to that of fish species. The biological adaptations needed for the transition from an aqueous habitat to a land habitat link the evolution of all the land animals, including insects and vertebrates. In particular, this transition is expected to have affected egg and larval development. To protect the embryos from the harsh and dry land environment, terrestrial eggs would have needed to be returned to a water habitat for development (as with amphibians), to acquire a nondesiccating covering (as with reptiles and birds), or to become internal (as with mammals). Similarly, aquatic larval forms would also have needed to either be returned to a water habitat for development (as with amphibians) or be lost from the animal's life cycle as a result of a more complete transition to the terrestrial habitat (as with reptiles, birds, and mammals). As aquatic viruses were often associated with both the larval forms and eggs of their hosts, the terrestrial viruses would have needed to adapt to these changes in larva (embryo) and egg biology. In addition, and of relevance to virus replication strategies, the transition to land would also have generally affected host population structures, since the common "school" population structures of fish are more difficult to attain on land (at least for the earlier terrestrial vertebrates, the amphibians and reptiles). Changes in population structure from high density to more dispersed would disfavor acute viral agents and favor persistent viral agents. An example of this issue applies to rodents and birds. Rodents, the most diverse and best-studied mammals, seldom exist in gregarious populations in natural settings. Birds, on the other had, do frequently exist in large flocks. Such variations in population structure correlate to virus life strategies in that large populations are much more able to support acute viral agents, whereas nongregarious host populations are better suited to sustain persistent viral agents. As noted previously, persistence is also associated with genomic colonization, sexual transmission, and old-to-young transmission, and I also consider these issues below with respect to the various vertebrate lineages.

The amphibians

Amphibians include about 4,800 species. Of the amphibian species, about 90% are frogs (4,200 species), indicating that frogs are by far the most successful amphibians. Amphibians appear to be monophyletic, so their descent from a common ancestor seems most probable. However, amphibians are distinct from and paraphyletic to the early four-legged vertebrates that were present in the Paleozoic era (such as the Labyrinthodontia, an extinct order that gave rise to the amniotes). Thus, it appears that amphibians diverged early from the lineage that went on to develop into mammals. Amphibians are poikilothermic: they have not developed thermal regulation. They generally have two life phases, an aquatic larval stage and a terrestrial adult stage. In addition, their eggs are laid and developed in an aquatic habitat, and egg fertilization is external. Although the large majority of amphibians have aqueous larval forms, some have evolved to lose this stage in their life cycle.

For example, 20 species of frogs are known to have lost the tadpole phase and also to have acquired a terrestrial egg. Unlike the evolution of amniotes, this change has occurred multiple times, not simply once. Another distinction of amphibians is that sex is environmentally determined and not the direct result of specific sex chromosomes. Amphibians have advanced visual and auditory senses relative to those of fish, and they also have distinct teeth. Their breathing occurs through a positive-pressure mechanism, not aspiration as in higher vertebrates. The skin of amphibians contributes significantly to respiration and is also often associated with the secretion of poisons and mucus. In connection to these characteristics of amphibian skin, it is interesting that for fish species, viral skin infections are highly common and are also often associated with mucus and toxin production. Among the amphibians, salamanders are the most basal species and are distinguished by internal egg fertilization. As discussed below, all amphibians are susceptible to iridoviruses, and most also host herpesviruses and adenoviruses. All of these viruses are associated with skin infections and persistence in tumor tissues. Curiously, no RNA viruses have yet been described for salamanders.

The amniotes

As mentioned above, amniotes have a monophyletic origin but are paraphyletic to amphibians. Amniotes include sauropsids and mammals and are distinguished by having eggs with reparatory (amniotic) membranes that are adapted to nonaqueous habitats. Adaptation to nonaqueous habitats also requires that the laid eggs have protective, nondesiccating coverings (i.e., hardened or leathery shells). This covering in reptiles and birds is distinct. I previously noted in the discussion of insect egg development (chapter 7) that an egg covering resembles the "walling-off" process of the innate-like immune response. In the case of internal fertilization, such as occurs with birds and marsupials, the egg shell can provide a barrier that seals off the allogeneic embryo from the mother's adaptive immune system. In the case of bird eggs, the amniotic membrane is frequently used as a tissue for the production of various types of viruses, and this forms the basis for the use of eggs for vaccine production. The shell coverings are made of various materials and are deposited by maternal cells following internal fertilization of the egg by sperm. The specific nature of the shell is distinct for the various lineages of amniotes, although in all cases, it must still allow respiration via the shell membranes. During amniote evolution, there was an early divergence of reptiles from synapsids, which subsequently diverged to yield pelycosaurs and later monotremes, marsupials, and mammals.

Sauropsids

Sauropsids, which are nonmammalian terrestrial vertebrates, include five groups: turtles, squamates, crocodilians, birds, and beaked reptiles. Of these, the turtles are the most basal. The squamates are a more recently evolved lineage that includes lizards and snakes. Crocodiles are also sauropsids, but they are paraphyletic to turtles and basal to sphenodons (dinosaurs) and birds. The characteristics that are common to all of these sauropsid lineages include production of terrestrial eggs. Although some of the genomes of these animals are poorly studied, it appears that overall their genomes are smaller than those of mammals and have many fewer repeated and ERV DNA sequences. Of the sauropsids, the birds are best studied. Furthermore, it is now clear that the bird and crocodile genomes are similar. In the following paragraphs, I consider reptiles and birds in greater detail.

Reptiles. Reptiles are a diverse class of terrestrial species, with 6,300 known members. Reptiles have internal fertilization of nonaqueous eggs. The eggs are shelled, have a large yolk to store nutrients for the embryo, and also provide a protective egg membrane. In most, but not all, species, sex is environmentally determined. This sex determination is similar to that of fish in that the temperature at which the fertilized reptilian egg is incubated can often determine the sex of the resulting offspring. For example, for alligators elevated temperatures result in male offspring, whereas for turtles elevated temperatures result in female offspring. Reptiles have no larval stage and are cold blooded. They are also characterized by the presence of a highly keratinized skin that contains distinct forms of keratin protein and resists desiccation. This keratinized skin is unique in undergoing periodic shedding with growth of the organism. The keratinized skin cells are produced by terminal differentiation of basal cells, which results in highly committed gene expression. In mammalian skin, this process of skin differentiation is associated with virus reproduction. Reptiles, like birds, have nucleated red blood cells (RBCs), which can support some types of viruses. Crocodiles are the basal clade of reptiles and appear to have diverged before the divergence of the lineage leading to dinosaurs and avians. The evolutionary origins of turtles are not clear, but it appears that they diverged early from the other lineages. The iguana-like lizards are basal to other lizards and snakes, which include about 1,000 species. Gekkotan lizards (1,500 total species) are also rather basal and include some species that are snakelike and lack substantial extremities. Snakes are the most recently evolved reptiles and also the most diverse (2,900 species). These rapidly diversifying species are all predators to other species, including lizards and other snakes. As mentioned below, snakes host many virus types and are especially well noted for hosting paramyxoviruses which cause lung infections.

Birds. Birds are the most diverse of the terrestrial vertebrates, with about 9,000 known species. Birds are homeothermic and evolved from the dinosaur lineage. Birds are distinguished by the presence of feathers, bills, external eggs, and complex reproductive behaviors. They have nucleated RBCs, which can host nuclear DNA viruses. As mentioned above, bird genomes are smaller and less variable across known species than those of other tetrapods and show lower rates of molecular evolution (although they do have high rates of point mutations). These genomes have many fewer endogenous retroviral elements (and other repeats) than do those of other animals or higher plants. Unlike reptiles, birds have genetically determined sex that is governed by particular sex chromosomes. However, in contrast to the sex of mammals, the sex of birds is not determined by the presence of a Y chromosome but is instead dependent on chromosome pairs. The female bird is heterozygous, with ZW sex chromosomes, and the male is homozygous, with ZZ chromosomes. As mentioned below, birds support many virus types.

Mammals

Current estimates indicate that there are approximately 4,692 extant mammalian species and an additional ~5,000 mammalian species that have become extinct. The defining biological characteristics of mammals are the mammary glands, hair or fur, and endothermy (i.e., production of heat through increased metabolism to maintain body temperature). Also, all mammals have four-chambered hearts. The mammary gland appears to have evolved from an adaptation of a mucus or sweat skin gland for secretion of nutrients to feed the embryo and early young by the mother. Many mammals show some link between mammary gland development and retrovirus production, such as the murine leukemia virus (MLV) association with mice (discussed below). Mammals are called therians and closely regulate their body temperature (endothermy). Accordingly, they do not determine sex of the offspring via environmental temperatures as do fish and reptiles. Rather, almost all mammals use an XY chromosome system for sex determination (with a few interesting exceptions, discussed below). Since all mammals are endotherms, they can generally respond to virus infections by increasing their core body temperature. This fever response does not occur in fish, amphibians, or reptiles. Mammals have been studied in the paleontological record by their dentition. All therians breathe by aspiration, using a diaphragm (i.e., a negative-pressure mechanism). Unlike bird RBCs, mammalian RBCs lack nuclei. Mammals evolved early, up to 148 million years before present (ybp) (before the dinosaurs), with all of the characteristics listed above. However, prior to 65 million years ago, there was only one mammalian lineage, which most likely resembled a rodent-like, egg-laying monotreme. Multituberculates are an early mammalian lineage that can be identified via dentition. Although they survived the dinosaur extinction and subsequent mammalian radiation of about 65 million years ago, the multituberculates all became extinct about 30 million years ago for unknown reasons. The lineages of mammals alive today are the monotremes, marsupials, and placentals, with the last being the most diverse.

Monotremes. Monotremes include few extant species. These egg-laying mammals are distinguished by having one orifice used for both sex and waste excretion. The two extant families of monotremes are the Ornithorhynchidae (platypus) and the Tachyglossidae (echidnas). In the monotremes, ovulation is not determined by hormonal cycling as in the placentals but is genetically programmed. As for most mammals, monotreme sex is determined via an XY chromosome system. Interestingly, monotreme Y chromosomes are very small, on the scale of 10,000 bp. There are few studies of monotremes and their viruses, and little is known about monotreme genomes. However, it is known that these genomes seem to lack most of the LINE-like elements found in other mammals, although some ERVs are clearly present.

Marsupials. Marsupials are metatherians and bear live young. One hundred forty genera are known. Although this group is much more diverse than the monotremes, the marsupials are still much less diverse than their sister group, the placentals. Mitochondrial DNA analysis confirms that monotremes, marsupials, and placental mammals are sister groups. Although they bear live young, the duration of direct embryo-mother attachment is short (7 to 14 days) relative to that for placentals and always corresponds to less than one estrous cycle for the mother. Thus, marsupial embryos do not confront the immunological dilemma of rejection by the mother's adaptive immune system to nearly the same degree as do placental embryos. The marsupial embryo has a shell membrane, which is made by the maternal oviducts, encloses the embryo, and can allow for long-term storage of unfertilized eggs in some species (e.g., wallabies). The major source of nourishment for the early embryo is the large egg yolk sac. Lactation is prolonged and occurs within a patch of specialized secretory skin within the pouch where newborn offspring are protected during their first weeks of postpartum life. Ovulation (estrus) is controlled by estrogen. However, in contrast to the case with placental species, egg fertilization and pregnancy do not hormonally interrupt the estrous cycle; rather subsequent lactation by suckling young interrupts it. Thus, aside from control of lactation, the strong hormonal regulation of placental pregnancy is largely missing from marsupials. The marsupial reproductive tracts are distinctly different from those of the placentals. Female marsupials have three vaginal openings, which open one at a time for birth of young, and the males have a bifurcated penis. Marsupial blastocysts have no true trophectoderm (the cell layer that surrounds and supports the placental embryo). And marsupial embryos lack placentas, which develop from the trophectoderm for mammal embryos. Thus, although marsupials do have a trophoblast-like membrane surrounding their early embryos, this marsupial "trophoblast" is not protective, nor is it involved in embryo implantation, as is the trophoblast of placentals. Marsupial embryo parturition is very rapid relative to that of placentals and is associated with the presence of inflammatory cells at the uterine wall. Unlike for placental embryos, little is known about the expression of ERVs or autonomous retroviruses in marsupial embryonic tissues. And the virus-like particles often reported for placental embryos have not been reported for marsupial embryos. The genomes of marsupials show some ERVs, but their frequency is much lower than that found in placental genomes. As with the mono-tremes, marsupial sex is determined by an XY chromosome system, and the Y chromosome is very small (about 10,000 bp). Since this chromosome must contain the male developmental genes, there can be little room for the accumulation of the large numbers of ERVs, LINEs, and SINEs seen in placental Y chromosomes.

Eutherians (placentals). Eutherians include a diverse set of animals from rodents to primates, altogether comprising about 4,400 extant species. Early placental mammals appear to have been shrewlike carnivores that diverged into the extant taxa. The basal placental taxon thus includes the shrewlike mammals. An early taxon to have diverged from this lineage is the Afrotheria, which includes elephant shrews, elephants, aardvarks, manatees, anteaters, and sloths. However, current rodent species, which physically resemble shrews, evolved later. Rodents (about 2,300 species) represent the most successful and diverse of all the placental species. Human (primate) and rodent lineages diverged from a basal lineage that includes bats, carnivores, and the modern horses. In placental species, ovulation and lactation are hormonally controlled via the estrous cycle. The process of lactation is controlled mostly by the prolactin hormonal system. It is interesting that this regulatory system is very old in evolutionary terms and in its earliest appearance can be found in the mucus-producing skin cells of fish. This observation supports the idea that the origin of lactation stems from such mucus-producing skin cells. The most variable biological aspect of placental mammal biology is found in the anatomy of their reproductive organs (the uterus and placenta). Many characteristics (such as the number and placement of embryos) differentiate the various placental lineages. These reproductive characteristics are specific to and maintained by each placental lineage. In contrast, nonreproductive organs (kidneys, hearts, etc.) are much better conserved among the placental lineages. Placentals are characterized by long gestation times, which involve an extended and intimate contact between the allogeneic embryo and the mother. This situation poses the unresolved issue of how the mother's adaptive immune system fails to reject the allogeneic embryo. The trophectoderm is known to be directly involved in this protective process. The placental trophectoderm has a variety of distinctive characteristics. It is the first tissue to differentiate in the fertilized embryo before implantation and is involved in facilitating uterine implantation, uterine wall penetration, and protection from the mother's immune response. The trophectoderm differentiates into the placenta, thereby establishing blood flow and nourishment from the mother to the embryo (via syncytiotrophoblasts). As described in greater detail below, placental and reproductive tissues (especially the trophectoderm) of the placental mammals are also associated with high-level expression of various ERVs, producing RNA, gene products, and retroviral particles. These ERVs are generally host lineage specific and are highly related to the lineage-specific

LINE and SINE elements found in all placental species. Overall, retroelements are significantly more abundant in placentals than are similar elements in marsupials, and they are also much more abundant than distantly related elements found in avian species. Furthermore, placental mammals host a wide variety of both RNA and DNA viruses. Many of these viruses (especially the persistent ones) are highly species specific and phylogenetically congruent with specific placental lineages, showing long evolutionary relationships. Figure 8.2 presents a schematic summary of the evolution of the mammalian orders.

Summary

In summary, the terrestrial animal lineages have acquired an array of distinguishing biological characteristics that are maintained in a lineage-specific manner. Particularly variable are the biological characteristics associated with reproduction, sex determination, and egg development. As described below, there are also many lineage-specific associations with viruses. In addition, there are important distinctions among the ERVs of these lineages, which are discussed below.

Figure 8.2 Schematic of the time line for the evolution of mammalian species.

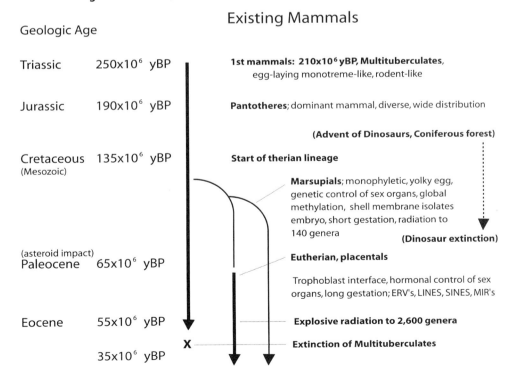

Viruses of Nonmammalian Terrestrial Vertebrates

Overall virus patterns

As noted above, with the evolution and diversification of the teleost fish species, there was a related diversification of the types of viruses that replicated in the fish species. As described below, mammals and birds have maintained most, but not all, of these fish virus families. However, some of the remaining terrestrial vertebrates show much more restricted patterns of virus replication. For example, the overall range of virus families found to infect reptiles and amphibians appears very reduced relative to that found in oceanic animals. Nonetheless, some relationships are maintained in amphibians. For example, the iridoviruses were the most common fish DNA viruses and are also commonly found in amphibians and insects (but not birds or mammals). However, the most common RNA viruses of fish, the rhabdoviruses, are not common in amphibians, reptiles, or birds (although rhabdoviruses are commonly found in some mammals, especially bats). As mentioned above, herpesviruses can be found in many fish and clam species. They are also common to essentially all mammals and birds (but are absent from insects). Poxviruses, in contrast, are not found in fish species but are found in insect, mammal, and bird species. Unfortunately, the literature available to clarify virus-host relationships in the terrestrial vertebrates is very uneven, and thus it seems possible that the virus sampling from these hosts is distorted. Virus studies are largely "mammal-centric" and also "acute-centric." That is to say, scientists have mostly studied the acute-disease-causing viruses of the mammals and birds that are either domesticated or commercially significant to humans (plus, of course, the medically relevant human viruses). Yet, as argued in chapter 1, it is the persistent virus-host relationships that most often provide long-term evolutionary stability. Since such inapparent relationships are generally not well studied, we have a very restricted body of literature that limits our understanding of such virus lineages. Below I examine the best-studied examples of viruses of the various terrestrial vertebrate species. As in previous chapters, I pay most attention to these well-studied systems. Also in this chapter, I examine the roles that persistent viruses and especially ERVs have had in the evolution of terrestrial vertebrates. This is especially true for the placental mammals. This theme relates to overall host genome evolution, including the evolution of the "lumpy genes" and LINEs in animal genomes.

Viruses of amphibians

Amphibians and their viruses

As noted above, salamanders are considered to be the most basal of amphibian species. Salamanders maintain a life cycle that is more aquatic than those of the other amphibian lineages. With respect to viruses, however, salamanders appear to be more like sharks, in that there are few reports of

viruses that infect them. Although salamanders are not commercially produced or otherwise grown in large numbers, and hence not well studied,
there is a limited set of studies to evaluate. Currently, these studies show a
dearth of viruses associated with salamanders. No acute or persistent RNA
viruses have yet been reported to infect any salamanders. Only infection
with one iridovirus has been described. In contrast, in the case of frogs,
iridoviruses are well established as important viral parasites. As mentioned
above, frogs constitute 90% of all amphibians. Frogs are known to support
the replication of iridoviruses, retroviruses, caliciviruses, poxviruses, herpesviruses, adenoviruses, and polyomaviruses. Thus, frogs are an established
host for a broad array of viruses. However, on closer inspection this appears to be a most curious list of viruses and seems to be especially lacking
in RNA viruses. With respect to DNA viruses, this list essentially resembles
the list of DNA viruses able to infect teleosts, with the notable exception
that poxviruses are now present. The most prominent DNA iridoviruses of
bony fish (LCDVs) are also prominent amphibian viruses. In fish, LCDV is
known to infect many species. Similarly, 30 types of iridoviruses of frogs
are known. A well-studied frog iridovirus is *Frog virus 3* (FV3), which was
originally discovered due to mass mortality of United Kingdom frog species.
FV3 is of special interest in that this virus, unlike fish iridoviruses, undergoes DNA synthesis in two stages. The initial FV3 viral DNA synthesis is
like that of fish virus—restricted to the nucleus. However, subsequent virus
DNA replication and amplification are cytoplasmic. Furthermore, the cytoplasmic replication process resembles that of phage, in that DNA is synthe-
sized as concatenated molecules that undergo headful processing into mature virions. Epizootic haematopoietic necrosis virus, another frog virus, is
even less dependent than FV3 on host nuclear systems. Unlike FV3, epizootic haematopoietic necrosis virus does not require host nuclear enzymes
for its replication. Thus, these DNA viruses of frogs, although both clearly
iridoviruses, also closely resemble the poxviruses in that they can support
extranuclear, cytoplasmic DNA replication. Consistent with this observation, the iridovirus-encoded eukaryotic initiation factor 2α (eIF-2α) is most
related to eIF-2α found in swinepox virus. The amphibian iridoviruses appear to have become less dependent on host nuclear replication systems
than are the fish and insect iridoviruses, and they show more similarities to
poxviruses.

Iridoviruses are known for their ability to cause high mortality rates in
tadpoles of various toads and frogs. Thus, they are acute agents in this situation. However, it appears that some frog species are persistently infected
with iridoviruses and that these species are the source of virus for acute infections of other species. For example, Venezuelan toads (*Bufo marinus*) are
often persistently infected with Gutapo virus, which is highly pathogenic to
tadpoles of other toads and other amphibians. It is worth recalling that the ascoviruses are the viruses most closely related to Gutapo virus, and no ascoviruses are known to infect any vertebrates. The only DNA viruses that have

yet to be reported to occur in frogs are parvoviruses, which curiously have also not been reported for fish. Although entomopoxviruses were observed to infect insect orders (especially locusts), the amphibian poxviruses are the first true example of vertebrate poxviruses. Since the entomopoxviruses appear more basal, it might be suspected that mammalian poxviruses may have evolved from these insect poxviruses. This possibility is developed in detail below in "The broad patterns of poxvirus evolution: insect to bird to mammal." In frogs, these viruses are associated with skin lesions and hemorrhages (pathologies also associated with mammalian orthopoxviruses).

Amphibian herpesviruses

Amphibians are also known to support herpesviruses. Ranid herpesvirus 1 and Lucké tumor herpesvirus have both been studied in some detail. These viruses are distantly related to fish herpesviruses. Infection with these herpesviruses, like infection of fish, is associated with tumors, which are not metastatic and are normally benign. Infection also frequently results in surface skin lesions, which are associated with mucus production and abnormal growth. These tumors are able to later produce virus, and this appears to provide a mechanism for virus persistence. That is, to establish persistence, the virus appears to induce a tumor. The viral DNA has been found to persist in tumors in the absence of infectious virus production when the frog is maintained at low water temperatures. At high water temperatures (a seasonal occurrence), virus production is again induced. Persistently infected cells are immortalized by the virus, in that cellular apoptosis is prevented by viral gene products. In general, frog herpesviruses appear to persist as tumors.

Amphibian adenoviruses

Frog adenovirus 1 (FrAdV-1) was initially isolated from a naturally occurring frog renal tumor, but it can also be isolated from healthy wild frogs. This virus can also infect fish species. In codfish, this virus can cause epidermal hyperplasia. In amphibians, FrAdV-1 appears to generally be a persistent virus. FrAdV-1 (26,163 bp) is the smallest of the sequenced vertebrate adenoviruses, and it lacks various 5' and 3' regulatory genes found in other vertebrate adenoviruses. By phylogenetic analysis, FrAdV-1 appears to represent an ancestor of both avian and mammalian adenoviruses (see Fig. 6.5). The latter viruses have apparently acquired additional, accessory genes during their evolution. This pattern of gene acquisition during evolution clearly resembles that of baculoviruses (discussed earlier) but is the opposite of the pattern of genomic evolution (associated with gene loss) presented for rodent and other mammalian poxviruses below. Recently, adenoviruses have been classified into three large groups: mastadenoviruses of mammals, avian adenoviruses, and atadenoviruses, which include amphibian and reptilian adenoviruses. In all of these groups, only 16 core virus genes are conserved, and these encode mostly replication and structural proteins. With the recent sequencing of FrAdV-1, it is now apparent that FrAdV-1 is most related to turkey adenovirus 3. Previously, turkey adenovirus 3 was not known to be closely related to any other adenovirus. As noted in chapter 6, adenoviruses also infect fish species, when they are associated with benign skin metaplasia. The fish adenoviruses show very limited similarity to other adenoviruses and appear to be most basal of all adenoviruses. However, the morphological similarities, the similarities of replication strategy, and the similar genome organizations shared by all adenoviruses and the PDR1 phage of *Bacteria* have been used to argue that bacterial phages were the direct progenitors of all vertebrate adenoviruses, even though the genome of PDR1 lacks observable sequence similarity to adenoviruses.

Other amphibian viral agents

Frogs also support replication of polyomaviruses (small nuclear circular DNA viruses). A leopard frog polyomavirus, which induces mainly benign skin growths but can also induce kidney tumors, has been described. These viruses have not been well studied. Although frog genomes have not yet been well characterized, retroviral elements are known to exist in frog genomes. Essentially all vertebrates appear to harbor some endogenous viruses related to various degrees to spumaviruses and MLV. The ERVs DevI, DevII, and DevIII have been reported to occur in dart poison frogs. However, these ERVs so far appear to be defective. Nonetheless, they do represent a distinct family of retroviruses unrelated to the seven currently recognized retroviral genera. The amphibian retroviral fragments are equally distant between the MLV and *Walleye dermal sarcoma virus* (WDSV) genomes. As WDSV is a fish genetic parasite, it would appear to most likely represent the ancestral retrovirus of the frog ERVs. These frog ERVs are only distantly related to avian retroviruses. It is interesting that transcripts of Dev sequences are present in high copy numbers in the ova of frogs. Thus, as will be seen with the reproductive tissues of other terrestrial animals (both vertebrates and insects), high-level ERV transcription is associated with amphibian reproductive tissue.

Viruses of reptiles

Reptilian RNA viruses

As noted above for the salamanders, other amphibians and the most basal representatives of the reptiles (the crocodiles and alligators) all appear to show a paucity of disease-causing RNA viruses. Since some farming of alligators occurs, the opportunity for observing virally induced disease in these species is considerably better than for salamanders or frogs. So far, there are few RNA viruses reported for alligators or crocodiles. However, very recently alligator farms in the southern United States have reported some alligator infections and deaths due to West Nile virus, which was only recently introduced into the United States. Yet it was not completely clear how these animals became infected, as they did not appear to have acquired the infection via mosquito transmission from environmental sources (the common route of transmission). Instead, it appears that the farmed alligators were infected by consumption of West Nile virus-infected horse meat. Thus, although it seems clear that alligators are seldom observed to be infected by natural means, alligators can be infected by at least some RNA viruses (that are also known to infect birds) as a result of farming practices.

In addition to a paucity of RNA viruses, no retroviruses have yet been observed in these reptilian species. However, as retrovirus infections can be rather inapparent, the failure to observe these viruses may be due to insufficient examination. Crocodiles are similar to various amphibians with regard to virally induced skin pathology that results from infection with both poxviruses and adenoviruses. For example, Nile crocodiles are known to support infections with an adenovirus, and such infections produce skin lesions.

In contrast, other reptiles, such as some lizards (iguanas) and many turtles and snakes, do appear to harbor both acute and inapparent infections with RNA viruses of various types, including caliciviruses and paramyxoviruses. In reptiles, paramyxovirus infection occurs primarily as an infection of the lungs. The lungs of snake species in particular appear to have been efficiently exploited by these viruses. It is not clear whether there is some feature of snake lung biology that might contribute to their propensity to support virus infections. In other reptilian species, paramyxoviruses can also establish persistent infections. One possible example of a persistent snake paramyxovirus has been reported for iguanas, which can harbor the virus as inapparent infections. Virus-infected iguanas also appear to be the source of paramyxovirus for acute infections of turtles. Interestingly, turtles generally appear to show a high incidence of acute infection, with paramyxoviruses residing in lungs and other pathologies. Thus, both turtles and snakes seem prone to paramyxovirus lung infections. Although snakes can be persistently infected with some paramyxoviruses, they are also susceptible to acute lung infections by various other paramyxoviruses. Numerous virus-mediated die-offs on snake farms have been reported, but the relationship between persistent and acute-disease-causing snake paramyxoviruses has not yet been evaluated. The diversity of snake paramyxoviruses is impressive. In fact, 16 reptilian paramyxovirus types have been described. Furthermore, phylogenetic analysis suggests that these snake viruses are related to and basal to *Sendai virus*. This result is most interesting because *Sendai virus* is phylogenetically basal to all the vertebrate paramyxoviruses. The implication is that reptiles harbor paramyxoviruses (both persistent and acute) that may represent ancestral paramyxoviruses of mammals and avians. Placental mammals are also known to be hosts for many paramyxoviruses. The paramyxoviruses are important pathogenic viruses of their mammalian hosts and establish mainly acute infections. Very few, if any, natural field studies or studies of domestic mammals show paramyxoviruses to establish stable persistent infections. This observation suggests that these mammalian paramyxovirus infections could be unstable on a long evolutionary timescale and dependent on prevailing host population structures. If snakes and other reptiles, however, harbor related viruses in a stable persistent way, they might provide an evolutionary and ecologically stable source of virus for adaptation to and infection of other mammalian species. That snakes support 16 known types of paramyxovirus upholds the idea that this virus-snake relationship may be basal and ancestral to the mammalian relationship with this family of viruses.

Snakes also support the replication of additional RNA viruses. These include caliciviruses, togaviruses, flaviviruses, reoviruses, and retroviruses. For example, rattlesnakes are known to support 16 types of caliciviruses. At least one of these (Cro-1 virus) is nonpathogenic in reptiles and frogs. Also in reptiles, reoviruses often show no disease and can be highly prevalent in natural populations (47% positive in some healthy iguana populations), indicating a persistent lifestyle in these hosts. These viruses are very similar to the strictly acute reoviruses of avians and mammals, but their relationships to those viruses have not been evaluated.

Turtles are also known to support bunyaviruses, togaviruses, flaviviruses, and retroviruses. Noteworthy is the occurrence of the "arboviruses," togaviruses, and flaviviruses in turtles. These are virus types that have not yet been reported to infect any bony fish populations. In addition, various field studies have suggested that these reptile-derived viruses may be the source of virus infections of other vertebrate species. Consistent with this idea, turtle, snake, and lizard blood samples from field isolates are often positive for Japanese encephalitis virus, St. Louis encephalitis virus, Powassan virus, and Venezuelan equine encephalitis virus.

In summary, although basal reptiles such as crocodiles appear to have a paucity of RNA viruses, RNA viruses of other reptiles (snakes, lizards, and even turtles) are numerous and diverse. These RNA viruses include both paramyxoviruses and caliciviruses. These viruses establish both persistent and acute infections in their specific reptilian hosts. These reptile viruses also appear to be basal to the similar RNA viruses that cause acute disease in mammalian hosts. It is likely that these reptilian viruses represent ancestral versions of these mammalian viruses.

Reptilian DNA viruses

With respect to DNA viruses, snakes have been reported to support infection with iridoviruses, herpesviruses, poxviruses, adenoviruses, and parvoviruses. Mostly, these viruses have not been observed in crocodiles. However, the Nile crocodile has been reported to be infected with an adenovirus. Turtles also support infection by various DNA viruses, including iridoviruses, herpesviruses, and polyomaviruses, the last two being associated with benign tumors. Snake parvoviruses are also known. In some reptiles (corn snakes and iguanas), parvovirus infection occurs concomitantly with other DNA viruses, suggesting that parvoviruses may depend on prevalent infections by other DNA viruses. Many vertebrate parvoviruses are helped by adenovirus superinfection, and this may be a general relationship. The snake parvoviruses exist in two major and distinct clades, and these clades show some geographical restriction in distribution, as well as some species specificity. In addition, the snake adenoviruses are interesting in a phylogenetic sense because they are basal to the adenoviruses from ducks and possums, which are themselves basal to the adenoviruses of cattle. However, the snake adenoviruses may be most closely related to frog adenoviruses (described above). Recent sequence evidence suggests that the snake lineage of adenovirus may have jumped species, resulting in a virus adapted to cattle; this would explain the existence of a distinct clade of atadenovirus that infects snakes and ducks as well as cattle.

Reptilian herpesviruses

Herpesviruses are also known for snakes and lizards. Boid herpesvirus 1, elapid herpesvirus 1 (a cobra virus), and iguanid herpesvirus 1 have all been studied. These viruses are often inapparent in their corresponding reptilian hosts, although the level and sites of persistence are not well studied. The iguana herpesvirus, for example, can be isolated from healthy green iguanas in the field. Lacertid herpesvirus 1 is another snake herpesvirus that is nonpathogenic in the ring snake. Elapid herpesvirus 1 in cobras is not highly pathogenic, although it is associated with inefficient venom production. In contrast to these nonpathogenic snake herpesviruses, herpesvirus infections in terrestrial tortoises and sea turtles are associated with high mortality and die-offs. For example, in terrestrial tortoises, *Chelonia mydas* herpesvirus induces mortality in up to 100% of cases. In sea turtles, green turtle fibropapilloma-associated turtle herpesvirus is highly prevalent and associated with tumors. It is interesting that this herpesvirus has a DNA polymerase gene with homology to the alpha herpesviruses of mammals and birds. Thus, as seen for various other orders, the pathogenesis associated with herpesvirus infection of reptiles is highly species specific.

Reptilian skin and virus growth

There appears to be a noticeable shift in the relationship of reptile viruses to their hosts relative to the relationships discussed for previous host-virus pairs. In reptiles, viruses do not tend to cause growth abnormalities in infected skin, as was described for fish in chapter 6. In vertebrate fish and many amphibian species, infections with the large DNA viruses (e.g., herpesviruses and adenoviruses of cod and white sturgeon) are mostly associated with epidermal and other tissue hyperplasia (i.e., excessive cell growth). This relationship also extends to include the smaller DNA viruses of fish, such as polyomaviruses (e.g., swordfish melanomas and winter flounder epidermal hyperplasia). This tendency to cause benign tumors was also a characteristic of polyomavirus infections of amphibians (e.g., leopard frog skin growths and kidney tumors). That DNA virus infections can affect skin growth in fish and amphibians suggests that the induction of cellular growth and differentiation programs and the inhibition of apoptotic pathways are inherent characteristics of the replication strategies of these viruses. It is interesting that for not all fish does the skin have basal cells that undergo differentiation. Shark skin, for example, lacks the basal epithelial cells present in bony fish and amphibians. Curiously, the only DNA virus known for sharks (dogfish herpesvirus) is associated with skin necrosis, not hyperplasia. Although reptilian skin has basal cells, the highly differentiated scale-producing (keratinizing) cells are terminally differentiated and could be impervious to virus infection or virally mediated growth control. In crocodiles, skin lesions, not hyperplasia, were seen with adenovirus infection. Similar skin lesions and oral lesions (but not growths) are seen with poxvirus infections of crocodiles and captive caimans. In snake species, herpesviruses and other DNA viruses tend to be nonpathogenic and do not induce tumor growth. In green lizards and Bolivian turtles, polyomavirus-like and herpesvirus-like virions can be observed in various tissues, but again, these infected reptile tissues are not hyperplastic. However, in sea turtles, internal tissues, not skin, were induced to form tumors by DNA viruses. Although this issue has not been systematically studied, the observations are consistent. What, then, accounts for this shift in the outcome of the biological relationship between DNA viruses and their hosts' skin from hyperplasia to skin lesions? Reptiles have highly keratinized, terminally differentiated skin that is periodically shed. As noted, such skin might be physically impervious to virus release, providing a biological barrier that would prevent use of reptilian skin for the purpose of virus transmission. In contrast, it seems clear that the corresponding skin lesions of fish and amphibians are able to transmit subsequent rounds of virus progeny. This interference, or keratin-inhibited virus transmission, might have provided a virus-host basis for the selective pressure that contributed to the evolution of highly keratinized host skin.

Reptilian retroviruses

Both ERVs and autonomous retroviruses are found in fish, amphibians, turtles, and snakes. However, autonomous retroviruses have yet to be reported for salamanders, crocodiles, and lizards. Many, but not all, of these autonomous reptilian retroviruses are related to ERVs found within the reptilian genomes. These reptilian retroviruses, however, are a distinct group and have been called epsilonretroviruses. This group includes WDSV of fish (an autonomous virus causing skin hyperplasia). According to phylogenetic analysis, this virus is the basal member of the epsilonretrovirus group, which appears to be monophyletic. It therefore seems most likely that an aquatic WDSV-like ancestor was the progenitor of all of these reptilian retroviruses. It is most interesting, however, that this family of autonomous retroviruses also shows significant homology to the HERVs (discussed below), suggesting some evolutionary linkage to mammals as well.

There are other types of reptilian retroviruses. The ERVs of pythons (PyERVs) include two closely related types. However, both types are not classifiable with other retrovirus families and are not related to known type B, D, or C retroviruses. Boid snake inclusion disease is due to infection with a retrovirus that is closely related to the PyERVs and may likely have been derived from such an ERV. In most pythons, the PyERVs are not well expressed: although strong expression is seen in *Python curtus,* expression is low in five other distinct boid species. This situation is reminiscent of the specificity of the herpesviruses noted above. Low virus expression and inapparent infection are characteristic of the virus when it infects its persistence-associated host, but high expression and disease are seen when the same virus infects a related but distinct host species. All reptilian species have genomic copies of ERVs, which are mostly species specific and usually defective. As has been argued previously, these defective copies could provide a mechanism to achieve stable persistent infections by suppressing the autonomous virus. If defective ERVs are crucial for the purpose of homologous retrovirus suppression, then this could provide a selective pressure for their maintenance. Without them, or with a sufficiently different set of ERVs, the host could be susceptible to high-level autonomous retrovirus expression. Such a scenario could explain the relationship between the genomic ERV-derived viruses and the species-specific snake infectious retrovirus disease mentioned above.

In summary, both genomic ERVs and infectious retroviruses are known for snakes. In addition, the infectious retroviruses are similar to and likely derived from these ERVs. It thus seems likely that there has been a long-term interaction among the ERVs that colonize specific reptilian hosts and that these ERVs retain the capacity to infect and induce disease in hosts not colonized by the same ERV. What is curious to consider is how a numeric balance of ERV colonization might be achieved. What keeps the numbers of ERVs and their defectives at a relatively low level in reptilian genomes compared to the high numbers seen in the genomes of all placental mammals? What type of events or selective pressures might lead to large-scale shifts in the level of ERV genome colonization?

Viruses of birds

Bird species have descended from ancestors that are shared by the reptiles and the dinosaurs. Birds are distinguished from those ancestors by the acquisition of homeothermy, feathers, the avian egg shell, and beaks. In addition, birds acquired genetically determined sex via ZW chromosomes, in which the female is heterozygous. Thus, like mammals, the entire lineage has both become homeothermic and acquired chromosomal sex determination. As mentioned previously, birds are the most diverse of terrestrial vertebrates (9,000 species), and the galliform birds are the best studied as well as representing a relatively basal taxon. Since bird genomes maintain high sequence similarity to crocodile genomes, it might be expected that bird viruses would resemble the viruses of reptiles. However, unlike crocodiles, birds are known to support a broad array of virus types. Unfortunately, our understanding of the avian viruses is biased towards viruses that cause diseases in commercial flocks. The situation is similar to that for flowering plants, in that our scientific literature is highly biased towards disease-causing viruses of domesticated species. And thus like for the agriculturally important plants, we know very little about the natural prevalence of persistent, species-specific virus infections of birds. Rather, avian viruses have been studied mostly in the context of chicken and turkey commercial flocks, with a focus on the acute pathogenic diseases that affect large commercial populations of genetically homogeneous birds. Yet even with this focus on these specific disease-causing viruses, very few field studies have evaluated the virus-host relationships in terms of the natural prevalence of these virus infections in natural avian populations. As DNA viruses were significant contributors to reptilian virus-host biology, I will consider DNA viruses of birds first.

Avian poxviruses

Although the iridoviruses are common large DNA viruses of fish, amphibians, and insects (as are ascoviruses) and therefore might be expected to also infect birds, there are no iridoviruses or ascoviruses known for birds (or mammals). With reptiles and birds, we instead see the emergence of poxviruses as a prevalent type of large DNA viruses. Poxviruses were also present in insect hosts, and it appears by phylogenetic sequence analysis that these insect poxviruses may be more diverse and basal to those of birds and mammals. The poxviruses of birds display essentially all the biological characteristics of the mammalian poxviruses. For example, avipoxviruses, like mammalian poxviruses, are associated with skin lesions, such as growths on exposed (unfeathered, less keratinized) skin, found on the claws and around the beak. In chicken flocks, these viruses can be highly pathogenic due to obstructive growths in the airway epithelia. Overall, avipoxviruses compose one of the largest and most complex of the poxvirus genera. Like many other poxviruses, they also show a broad species specificity, though their natural distribution is not well evaluated. In some field studies, 44% of the wild birds (such as Swainson's francolin of southern Africa) were observed to be infected with avipoxviruses and showed the associated but benign skin growths. Avipoxvirus DNA has been isolated from dermal squamous cell carcinomas in natural settings. However, in some situations, avipoxvirus DNA can also be isolated in a latent state from normal skin. Thus, these viruses may maintain a persistent life strategy in some specific hosts. It is also clear that not all bird species or populations are exposed to avipoxviruses, as these viruses were absent in many other field studies. The avipoxviruses that have been evaluated show some variation and species specificity in terms of pathogenesis. For example, the avipoxvirus isolated from Hawaiian crows was significantly less pathogenic in chickens than were other isolates. The avipoxviruses in commercial flocks appear to represent species jumps from poorly characterized natural sources. It seems likely that these viruses are maintained by some reservoir species in benign persistent states, from which they function as the source of viruses that adapt to other (commercial) bird species. If this view is correct, it can be proposed that the evolved gene functions of many avipoxvirus open reading frames (ORFs) will be to maintain a persistent benign state of virus production, rather than to promote the high-level virus replication and associated disease that are observed in commercial flocks.

Avian herpesviruses

The best studied of the avian herpesviruses are Marek's disease virus and infectious laryngotracheitis virus, which have important effects on commercial chicken flocks. Another bird herpesvirus is duck plague virus, which is associated with disease in commercial duck flocks. All of these viruses cause proliferative diseases of the lymphatic and other tissues. Resistance to Marek's disease is associated with the B-F region of major histocompatibility complex proteins. For the most part, these viruses seem to be typical herpesviruses in terms of genetic organization and virion structure. According to sequence analysis, all of these avian herpesviruses appear to be most similar to the alphaherpesvirus family. However, biologically, they do not resemble the mammalian alphaherpesviruses in that they do not establish latent or persistent infections in nervous tissues (e.g., ganglia). Instead, the avian herpesviruses have a distinct biology that more closely resembles that of the mammalian gammaherpesviruses (such as *Human herpesvirus 6* [HHV-6] and HHV-7), which are T-lymphotropic herpesviruses. It is interesting that HHV-6 and -7 have repeated terminal DNA, or telomeric, repeat sequence motifs, that resemble the repeated sequences of human telomeres. Among the herpesviruses, only Marek's disease virus has a striking resemblance to this sequence element. The avian herpesviruses are frequently associated with growth abnormalities, especially malignant lymphomas and atherosclerosis. In this biological aspect, they are clearly more like HHV-6 and -7 or the herpesviruses of fish species and not like other alphaherpesviruses of mammals. Although avian herpesviruses are also known to persist, the cellular sites and mechanisms of persistence are not known. Nor is it clear if persistence in the natural ecological setting is restricted to particular avian species.

Other avian DNA viruses

Adenoviruses are also known for avian species. Of interest is egg drop syndrome virus, which has pathogenic effects on chickens and egg production. However, there have been few field studies that have evaluated the natural biology of these adenoviruses, so little can be said about host specificity, species jumping, or virus persistence. Those few studies that have been conducted have not shown a significant prevalence of avian adenovirus in wild birds. Thus, we cannot now account for the natural source of this virus. As noted before, however, these avian viruses are clearly more related to the viruses found in frogs than they are to the adenoviruses of mammals. Thus, avian adenoviruses may represent a distinct lineage of adenoviruses that adapted to birds from amphibian or reptilian predecessors.

Avians also support papillomaviruses and polyomaviruses (small circular dsDNA viruses). Papillomatous lesions in male green finches have been reported, and these did not affect other birds, suggesting a tight species specificity. The interest in avian polyomaviruses is due to their effects on commercial aviaries, where they are especially associated with diseases of hatchlings, such as budgerigars. There are several interesting distinctions between the polyomaviruses of avians and those found in mammals, including the human polyomaviruses. Avian polyomaviruses are simpler than their mammalian counterparts and typically have smaller and simpler early genes (T antigens). Within the polyomavirus family, the two most conserved regions in all members are found within the region of the T antigen (associated with ATPase activity) and within the capsid-encoding region. These are both maintained in the avian polyomaviruses. Thus, the avian polyomaviruses appear to be related to the mammalian viruses, although their simpler genetic structure has led some to consider them as more representative of the progenitor polyomavirus. However, there are several reasons to think that they are not progenitors. For one thing, the biologically distinct characteristics that these avian viruses display suggest that they may include mostly acute replicators. Although both avian and mammalian viruses show preference for respiratory and excretory (kidney) tissues, avian polyomaviruses show a much lower host species specificity than do mammalian polyomaviruses. That is, mammalian polyomaviruses are highly species specific, whereas avian polyomaviruses are able to infect a relatively broad array of host bird species (although infection appears to require hatchlings). Also, all mammalian polyomaviruses appear to establish lifelong persistent infections, whereas the avian polyomaviruses appear to cause acute infections that do not result in persistence. Nonetheless, there may be some exceptions to this situation, in that persistence may be much more species specific in avians. One report indicated that sulfur-crested cockatoos in New South Wales, Australia, had a high prevalence of polyomaviruses. Thus, it seems possible that the avian polyomaviruses have adapted a simpler, acute life strategy for many avian host species but may also establish species-specific persistent infections in other, less well-studied hosts. Consistent with this idea, one study of wild sulfur-crested cockatoos in Australia reported that 64% of the birds were positive for avian polyomavirus infection, while other abundant wild species (e.g., galahs) and nearby domestic flocks were negative for this virus. Curiously, in neither avian or mammalian polyomaviruses do we see the common association of virus replication with benign tumor growth, which was seen for amphibian and fish polyomaviruses. This could be related to the highly keratinized (feathered) avian skin.

In summary, birds support the full complement of DNA viruses that are found in mammals. However, there seem to be clear biological distinctions between the virus-host relationships of avian and mammalian DNA viruses. The avian herpesviruses and poxviruses are generally associated with growth abnormalities in infected avian cells. Examples of neuronal persistence (such as occurs with type I and type II alphaherpesviruses and various fish herpesviruses) have not been observed in birds. However, we generally know little about the natural distributions of the avian viruses or their natural hosts, so little can be said about such important biological issues.

RNA viruses of birds: the influenza virus story

One of the most studied of all avian RNA viruses is influenza A virus. This virus is of intense interest not only because of its ability to cause widespread disease in commercial bird flocks but also because it can adapt to human hosts and cause major human epidemics. Avian species are now clearly known to be the original source of the influenza viruses that adapted to infect various other animals. Influenza virus is a segmented negative-strand single-stranded RNA (ssRNA) virus in which the segments exist in various alleles, allowing segment mixing or reassortment during mixed virus infection. This provides influenza viruses with the ability to form recombinant viruses between distinct parental virus types. That is, influenza viruses can reassort their eight subgenomic segments during mixed infection, thus allowing for a recombinational process to apply to these negative-strand ssRNA viruses. This reassortment provides a greater degree of virus genetic adaptation and is directly involved in facilitating influenza virus adaptation to other new host species, including both other avian species and mammalian species.

In replication strategy, the negative-strand influenza viruses resemble the rhabdoviruses, as discussed in chapter 6, and some similarity in the RNA replicases can be observed. Rhabdoviruses are also negative-strand ssRNA viruses and are found in plants, insects, and especially fish. However, rhabdoviruses cannot undergo recombination. Curiously, there are very few rhabdoviruses of birds. Yet rhabdoviruses do not seem to be the direct ancestors of influenza viruses. More likely direct ancestors would be the paramyxoviruses, which are also negative-strand ssRNA viruses but are structurally much more like influenza viruses than rhabdoviruses. As mentioned above, the paramyxoviruses cause prevalent lung infections in various reptilian species, especially snakes. The influenza virus appears to represent a segmented version of a paramyxovirus that has acquired the ability to undergo reassortment or recombination. Consistent with this idea, influenza viruses are not found in any lower organisms (including amphibians or crocodiles) but are restricted to avians and mammals.

Persistence of waterfowl influenza virus

Although much of our attention has focused on the ability of influenza virus to cause acute diseases in humans and commercial bird flocks, it now appears quite clear that waterfowl, such as various species of shorebirds and ducks, are the major sources of influenza viruses for other species. These water birds have a clearly different and persistent relationship with the influenza viruses. Influenza virus infections of waterfowl are long-term and nonpathogenic. Also, by far the largest numbers of alleles for all eight of the influenza virus genomic segments are found in waterfowl species, and some of these alleles are specific to waterfowl. In some species, such as Peking ducks, influenza virus infection is limited to intestinal tissues and establishes a persistent nonpathogenic infection in which virus is excreted. While in this host, influenza A virus shows a very low rate of genetic mutation and is phylogenetically congruent with the evolution of its host. This is a major distinction between the influenza A viruses isolated from waterfowl and those isolated from other species (including other birds). In fact, there has been almost no change in the genetic sequence of the virus that infects ducks for the last 85 years. Dendrograms of the isolates for this persistent virus display "frozen" evolution with little diversity and appear as dendrogram "sticks" rather than the usual "trees" characteristic of the influenza virus quasispecies associated with human infections. Thus, it appears that the selective pressures that apply to a persistent virus-host relationship exert a clonal or purifying selection on the virus, in which the coding sequence is maintained essentially unchanged. As it is clear that the influenza virus RNA-dependent RNA polymerase, which lacks proofreading function, has a high error rate, it must be that during persistence, the initial colonizing virus genome sequence is somehow maintained. However, as we know almost nothing about the mechanisms by which influenza viruses establish persistent infections in duck intestines, we are unable to understand how viral gene products contribute to this genetic maintenance or how influenza viruses avoid the host adaptive immune response.

Human-adapted influenza viruses from avians: influenza A virus segment evolution and human adaptation

The two influenza virus RNA segments that have received the greatest attention encode those surface proteins associated with human immune protection, the H (hemagglutinin) and N (neuraminidase) segments. It is reassortment of these segments (called antigenic shift) that is associated with loss of immunity and major human pandemics of influenza A virus infection. For example, the Spanish influenza pandemic of 1918, which killed more than 20 million people worldwide, was associated with the appearance of the H1 segment, whereas the 1957 Hong Kong pandemic was associated with the H5 and N1 segments. Phylogenetic analysis now clearly argues that these segments both originated from viruses present in avian species, which somehow adapted, possibly through an intermediate host, to infect humans. Generally, human-adapted influenza viruses have lost their ability to infect waterfowl. Also, acute and highly pathogenic infections by influenza A viruses are not restricted to mammals. Nonwaterfowl avian species are also susceptible to pathogenic infections by waterfowl-derived influenza A viruses, and such infections appear to be common in natural bird flocks. This has been most apparent in commercial flocks of chickens and turkeys, which can be decimated by avian influenza A virus infections. For example, the major outbreak of influenza A virus of 1997 in commercial chicken flocks was due to an H5N1 virus that was later determined to have most likely originated from geese. Similar to what occurs in human pandemics, these geese appear to be a reservoir for viruses that cause chicken and turkey epidemics. The water-fowl species harboring persistent influenza viruses vary geographically. In other parts of the world, such as Germany, wild Peking ducks are frequently infected with H6N1 viruses, whereas in Brazil many wild waterfowl support H1N1 or H3N2 viruses. Different flyways appear to be associated with different types of influenza viruses. Virus persistence in these waterfowl species appears to be very important for the long-term ecological stability of influenza viruses. However, as mentioned previously, the viral genetic functions involved in host persistence, although crucial for the long-term maintenance of the viruses, are not at all understood.

Species jumps and adaptation to acute infection

It has also been established that the genetic requirements for influenza virus replication in avian cells are distinct from those for mammalian cells. For example, all avian isolates appear to be able to agglutinate chicken, but not mammalian, RBCs. Thus, the viral gene function needed for infection or persistence in waterfowl does not directly result in genes able to function well in other species. However, domestic birds provide a concentrated, homogeneous host population that can allow influenza viruses to adapt to efficient acute replication strategies. As mentioned above, the H5N1 virus of 1997 adapted to be a highly lethal killer of chickens. In addition to losses from virus infection, millions of additional chickens were culled to prevent the possible epidemic spread of virus to humans. Although it is difficult to judge the success of this preemptive strategy, a human epidemic from this influenza virus did not result. Curiously, and unlike what is seen during most human influenza epidemics, antigenic drift is seldom involved during commercial flock epidemic outbreaks. Rather, most avian epidemics result from reassortment and acquisition of new segments. Lethality and disease in avian species are not restricted to domesticated birds in large flocks (such as turkeys and chickens). Wild peregrine falcons, which can be predators of waterfowl, have been reported to die of H3N2 influenza virus infection, as have owls and buzzards. The prevalence of influenza virus-mediated killing of birds of prey, however, has not been evaluated. Thus, the overall picture we are left with is that influenza viruses, although able to infect a wide variety of mammalian species, are strictly acute replicators in mammalian and nonwaterfowl avian species. As such, they require large populations for stable virus transmission. Once adapted to humans, however, these viruses may lose their adaptation to birds and become unable to replicate in duck intestines. The human-specific influenza B and C viruses are most likely examples of viruses that have adapted so thoroughly to acute replication in their human hosts that they have lost all ability to replicate in avian species. Yet, phylogenetically, these human-specific viruses still appear to have been derived from avian viruses in the recent past (i.e., the past 200 years). The various strains of human-specific influenza A viruses are, however, unstable and tend to be either lost or displaced from the population with time.

In summary, the avian-human influenza virus situation may be our best-studied example of the relationship between persistence and emerging viral disease. It seems clear that influenza virus has been maintained on an evolutionary timescale as a persistent infection of specific water birds. However, our focus on human disease has not led to an understanding of this asymptomatic virus-host relationship or why it results in such a remarkable genetic stability for the virus. It also appears that the ability of new versions of these avian viruses to emerge from persistence and adapt to human and other species is an unending phenomenon, resulting in endless waves of acute influenza epidemics associated with high rates of genetic change.

Avian paramyxoviruses

Birds also support infections with nonsegmented, negative-strand ssRNA viruses of the paramyxovirus group. As with influenza viruses, much of the attention the avian paramyxoviruses have received is due to the impact that their infections have on commercial bird flocks. In this regard, Newcastle disease virus (NDV) has posed the biggest problem and caused serious economic losses. Overall, it appears that NDV is introduced into commercial flocks from exotic feral birds, such as parrots, pheasants, and doves. The natural source of NDV, however, is not completely clear. Some field studies (e.g., in Germany and South Africa) report very low prevalence of NDV in wild native birds (less than 1%). However, one study from Australia reported that NDV was highly prevalent in wild anhingas and that these infections were not pathogenic, but infected birds did secrete virus via intestinal shedding. Also, one field study examined recent Australian isolates of NDV in sentry chickens and found that unlike the NDV that was responsible for acute respiratory disease in flocks, these recent isolates were much less pathogenic and initially were shed via the intestinal tract, not the respiratory tract. The implication of this observation is that NDV can exist in stable, nonpathogenic infections in specific bird species, but it is likely to adapt to other bird species and cause acute disease in these new hosts. In this characteristic, the avian paramyxoviruses may resemble the influenza A viruses. Other paramyxoviruses of birds include Sendai virus and avian paramyxovirus, which, like NDV, are also associated with lung infections. These viruses can sometimes be found in wild bird populations, but they are also frequently absent in other field studies. The host source of these viral infections remains unknown, as it does not seem possible to maintain acute epidemics in some of these wild bird populations. In terms of phylogenetics, it should be emphasized that *Sendai virus* is phylogenetically basal to all of the mammalian paramyxoviruses. As noted above, most of the mammalian paramyxoviruses do not establish persistent infections in their hosts. However, due to the clear similarity of *Sendai virus* to the more basal paramyxoviruses found in snakes, it seems likely that the evolutionary trail can now be proposed for the origins of the mammalian paramyxoviruses and the influenza viruses. The paramyxoviruses of reptiles (as currently found in snakes) were the likely progenitors of the avian paramyxoviruses (e.g., *Sendai virus*) and the avian influenza A viruses, and both of these virus families were able to establish evolutionarily stable persistent infections in specific avian species. However, both of these avian virus families have also been able to jump species and adapt to new hosts (e.g., various mammalian lineages) as acute viral agents. Thus, the mammalian paramyxovirus lineage may trace its evolutionary origins from reptiles, through avians, to mammals, and during the course of this adaptation the reptilian and avian viruses appear to have gained the ability to establish species-specific persistent infections.

There are some other distinctions between the RNA viruses of birds and those of other species. As mentioned in chapter 6, rhabdoviruses of fish are very common. Yet avian rhabdoviruses are very rare. However, two novel and unclassified rhabdoviruses from birds have recently been reported during surveys for encephalitic arboviruses.

Avian arboviruses: positive-strand ssRNA viruses

In evolutionary terms, arboviruses were first observed in fish hosts, but they also clearly produce significant infections in birds. For bird hosts, the arboviruses are mainly mosquito transmitted. Arbovirus infections have frequently been associated with significant die-offs of various wild bird populations. Most recently, a crow die-off due to the establishment of West Nile encephalitis virus was seen in the eastern United States. Birds (and reptiles, as noted above) may be the main hosts for many of these viruses. Generally, arbovirus infections are not persistent in infected birds, so how the viruses are maintained in the ecosystem is not clear. However, large bird flocks might be of sufficient size to support a chain of transmission of acute arbovirus infections. In this case, persistence in bird hosts might not be necessary for virus stability, as the virus will migrate with its host. It has been suggested that in some cases, virus might persist in the insect vector as well, but this does not appear to be a general situation. Although there are also reports that some birds might harbor persistent infections, these cases are so far poorly documented. It is clear that arboviruses pose a significant biological parameter that affects the size and structure of bird populations. It has, however, not been well studied as to how such a relationship might also have affected bird evolution. There are some examples of other bird RNA viruses that appear to be able to persist in their hosts. For example, duck hepatitis virus (a picornavirus) appears to be highly prevalent and persistent in ducks. This virus may represent the evolutionary progenitor of other hepatitis viruses. Also, some bird coronaviruses, such as avian infectious bronchitis virus, can persist in infected birds. However, the ecologies of these viruses and their hosts are poorly studied, so I am unable to comment much on the evolutionary issues of these relationships.

Autonomous retroviruses and ERVs of birds

The long-term interest in the retroviruses of birds relates to the observations by Peyton Rous in the early 20th century that a transmissible virus caused infectious sarcomas in chickens. Years later, it was determined that the virus involved was a retrovirus, an avian sarcoma virus. Early on, it was also apparent that there was a genetic component to the occurrence of these bird tumors, which led to an intensive period of research into the genetic basis of avian tumor viruses that eventually included the study of ERVs of chickens. The literature has long developed under the notion that avian retroviruses cause prevalent infections and are genomic agents in all birds. However, with time it has become clear that, compared to the mammalian retroviruses, avian retroviruses are much less prevalent in nongalliform bird species than was originally suspected. With the contribution of genomics and sequence analysis, several evolutionary patterns have become clear. In most natural bird populations, retrovirus-associated tumors are rare. Unlike the situation with fish retroviruses (e.g., WDSV), few, if any, field studies of natural bird populations have reported a significant prevalence of retrovirus-mediated tumors. However, it has also become clear that the tumors seen in chickens are often associated with the presence of ERVs related to the specific breed of chicken. ERVs (and tumor production) can actually be bred out of most chicken lines. In evolutionary terms, the Moloney class of C-type retroviruses (MLV) are the most well-conserved lineage. As mentioned in earlier chapters, the Ty3/gypsy class of reverse transcriptase (RT) elements is part of the MLV class and has been conserved in most invertebrate and vertebrate animal lineages. The avian ERVs (e.g., avian sarcoma/leukosis virus [ASLV]) are also related to MLV. However, phylogenetic analysis suggests that the avian retroviruses have been acquired in 19 galliform birds (based on ASLV *gag* genes) but that these viruses appear to have originated from horizontal transmission from a mammalian source, followed by a rapid adaptive radiation into related avian lineages. *Reticuloendotheliosis virus* (REV)-related sequences (C-type, not *Avian leukosis virus* [ALV]-like) have been found in some wild birds, but most of these birds were healthy and lacked tumors. One report on Attwater's prairie chickens did document that REV was present in tumors, although captive flocks remained healthy and did not develop antibody to the virus. This pattern of ERV distribution in birds is distinctly different from that which is described below for the mammals.

Galliforms and retroviruses

Galliforms (pheasants, turkeys, chickens, etc.) are known to host three groups of retroviruses: ASLV, REV, and lymphoproliferative disease retrovirus (a lymphatic turkey virus). Avian retroviruses tend to cause lymphoid and hemopoietic proliferative diseases in domestic birds. Several classes of ERVs have been identified. One class, CH-1, is present at about 10 copies/cell in some chicken lines but can be bred out. Rous-associated virus-0 (RAV-0) is another complete avian ERV that contains an *env* gene and is related to *Avian myeloblastosis virus*. The pattern of ERVs in the red jungle fowl, the nondomestic ancestor of chickens, has not been well evaluated. It is known, however, that Art-CH elements (which are noncoding) are found in pheasants. *Tetraonine endogenous retrovirus* is currently the only known complete nonchicken ERV and is found in ruffed grouse. This ERV appears to have been acquired in this lineage early during phasianid evolution. Twenty-five distinct species of galliforms are known. For the most part, ERVs are not phylogenetically congruent with these species and appear to represent more recent colonizations of the germ line genomes. Avians appear to have a type of control over the activity of ERVs and autonomous retroviruses distinctly different from that of mammals. Another distinction between mammals and avians is that avians do not globally suppress the activity of retroviruses in early embryos, as do mammals. In chickens, this applies to both ERVs and autonomous retroviruses. For example, both RAV-0 expression in chicks and ALV infection of eggs result in high-level retrovirus expression in most of the organs of the resulting adult birds.

Avian genomes and retroviruses

Bird genomes are significantly smaller and less variable across bird species than are the genomes of tetrapods. For example, the mass of a bird genome is about 2.8 pg/cell, compared to 8.0 pg/cell for the mammalian genome. As the total numbers of genes between mammals and birds do not differ by this amount, much of this difference is due to greater amounts of noncoding DNA (including ERVs) in the genomes of mammals. One suggested explanation for this difference is that birds are under selective pressure to maintain light cells, which could limit acquisition of noncoding DNA. However, similar patterns are seen in crocodile genomes and the genomes of flightless birds, so it is not clear that the "light cell" hypothesis applies. Some have suggested that the evolutionary rate in bird genomes is lower than that of mammals and that this has limited the acquisition of noncoding DNA sequences. However, recent measurements indicate that bird genomes have three- to six-fold-higher levels of single nucleotide polymorphisms than does the human genome. One possible explanation for this observation is that bird genomes are older and hence have had more time to accumulate single nucleotide polymorphisms. Regardless of this possibility, it is clear that bird genomes can change and have changed significantly during evolution, but that the pattern of this change is distinct from that for mammalian genomes. The basic question is, Why have bird genomes been colonized by ERVs so much less than have mammalian genomes? As outlined above, there is direct evidence that avian retroviruses do exist and that avian genomes can be colonized by ERVs, so there seems to be no structural barrier to avian ERV acquisition. And yet there appears to be a paucity of ERV elements in avian genomes relative to all mammalian genomes, which are colonized by large numbers of lineage-specific ERVs and even larger numbers of ERV-derived degenerate retroposons (discussed below). There also appears to be some link between ERV colonization and the sex chromosomes, in that sex chromosomes appear to be especially prone to ERV colonization. Unlike crocodiles and amphibians, in which sex is determined by ambient environmental temperatures, birds are warm-blooded and have genetically determined sex, involving Z and W chromosomes. The female is heterozygous (ZW) and the male is homozygous (ZZ). In chickens, ERV-Z chromosome associations are known. In the broiler chickens (i.e., White Leghorn chickens), the endogenous virus EV21 is directly associated with the sex-linked large broiler body mass of the bird. In addition to the intact EV21, other complete ERVs are also known to be sex chromosome associated, such as EV3 as well as some defective retroelements. EV21 resides in both the Z chromosome and other autosomes. So it is apparent that the general tendency of ERVs to colonize sex chromosomes also applies to chickens. Furthermore, EV21 loss from the Z chromosome has a phenotype and is associated with early feathering. EV21 can also produce infectious virus that can be sexually transmitted to the eggs and hatchlings of females lacking EV21. Thus, EV21 can be transmitted both via the germ line and via virus progeny. What, then, might be the role of EV21? Can it protect against the known and related autonomous retroviruses? It is interesting that exposure of EV21-harboring birds to ALV did not affect the response of these chickens to this virus or to ALV-mediated tumor induction. In fact, the endogenous EV21 may actually have increased the incidence of ALV-induced tumors in hatchling infections. Thus, EV21 did not protect chickens against ALV or ALV tumors. Avian EV21-like ERVs are not uniformly maintained in all bird lineages. For example, some breeds of domestic geese (Chinese, Synthetic, and Embden) lack any ERV-related sequences as measured even by low-stringency hybridization to endogenous avian retrovirus polymerase (ASLV). Clearly, ERV colonization is not always favored in avian lineages.

In summary, it can be seen that infectious avian retroviruses are well established, especially in domesticated birds and the galliform species. However, these viruses are not common and are seldom, if ever, observed in natural avian populations. In addition, ERVs residing in the genomes of avians are well established, and some of these reside in sex chromosomes and can be sexually transmitted as infectious virus. However, bird genomes are much less colonized by ERVs and ERV defective derivatives (LINEs) than are the genomes of mammals. Also, avian ERV colonization is variable, not seen in all lineages, and not uniformly associated with the origination of various bird species.

Viruses of Mammals

Mammalian ERVs

It was shown above that amphibians, reptiles, and birds all carried at least some types of ERVs within their genomes. But it is in the mammals, especially the placentals, that an explosive and lineage-dependent increase in ERV colonization can be seen. This high-level colonization is especially apparent in both the X and Y placental sex chromosomes. We also see that in general, mammals support a broad array of other virus types as well. The early mammals were egg-laying, monotreme-like, and shrewlike organisms (i.e., multituberculates) that evolved even before the dinosaurs (Fig. 8.1). These early mammals, whose history can be traced to about 210 million ybp, survived the dinosaur extinction, only to become extinct themselves about 30 million ybp with the radiation of the placental species. The biggest distinctions between these early mammals and current mammals are the reproductive organs and the reproductive strategies. Although existing monotremes are still egg layers, the marsupials and placentals do not lay eggs and have developed distinct reproductive strategies for embryo development. All mammals have mammary tissues, which developed from modified skin glands and respond to prolactin by proliferation. Although this has not been exhaustively examined, most mammals appear to be able to produce infectious virus for various ERVs in association with the development of mammary tissue. The mouse virus that was first recognized in this association was mouse mammary tumor virus (MMTV), which produces milk-borne virus particles during lactation. We now know that MMTV represents a very old and conserved class of ERVs that are defective in all mammals.

Monotreme ERVs

Monotremes appear to be able to support infection by autonomous retroviruses as well. However, neither the acute viruses, the ERVs, nor the genomes of monotremes are well studied, so we are currently unable to fully classify the viruses and ERVs present in these species. Thus, we can conclude little about monotreme retrovirus-host relationships. Monotremes have a very small sex-determining Y chromosome, but it is so small (about 10,000 bp) that it would seem incapable of coding for more than one or a few intact ERVs. Clearly, it cannot be highly ERV colonized.

Marsupial viruses

Marsupials are better studied than monotremes with respect to their viruses and genomes, although the literature on these subjects is still rather thin. Marsupials are known to support the replication of various types of viruses, from herpesviruses to retroviruses to mosquito-borne arboviruses. For example, Wallaby herpesvirus appears to be a member of the type I alphaherpesviruses (which includes the avian Marek's disease virus of turkeys). There is also a Parma wallaby herpesvirus (PWHV). This virus is prevalent in field populations and has been isolated in 23% of wild wallabies. These herpesviruses tend to establish inapparent infections in their native hosts, although the tissue sites of persistence are not known. It is also not clear whether these viruses are able to cause acute diseases in other related hosts, as do the mammalian herpesviruses. PWHV replicates in all the marsupial cell lines examined to date, but it does not replicate in most eutherian cell lines. Thus, these marsupial viruses appear to recognize inherent differences in marsupial and placental host cells. Consequently, it is curious that PWHV has a broad marsupial species specificity. The possible mechanisms involved in cell type restriction have not been evaluated.

Retroviruses of marsupials are also known. A fat-tailed dunnart cell line is known to produce D-type and A-type *env*-containing retrovirus particles. However, so far no disease has been associated with these viruses. In addition, an ERV of koalas has been described; is ubiquitous to that species and resembles the gibbon retrovirus. In fact, it has been suggested that this koala virus may be the original source of *Gibbon ape leukemia virus,* having been transmitted during mixed captivity. Koala ERV has not yet been associated with any diseases of koalas. Although the examples are relatively few, it is nevertheless clear that marsupials can and do support retroviruses and also have ERVs. However, it is also clear that as a whole, marsupial genomes are significantly less colonized with retroviral elements (including LINEs and SINEs) than are the mammalian genomes, although these elements are more abundant in some marsupial lineages than in avian genomes. One survey study by J. Jurka et al. in 1995 clearly showed this result. This presents a curious situation because there does not appear to be a global system in marsupials that prevents ERV colonization. In other words, it might be expected that marsupial genomes should resemble those of mammals in having high levels of ERV colonization. Along these lines, it has been reported by R. J. O'Neill that in contrast to placental mammals, hybrid offspring between two reproductively isolated species of marsupials (swamp wallaby × tammar wallaby) undergo global activation and expansion of kangaroo ERV retroposon elements, resulting in new chromosomes not contained in either parent, and that this genome-wide kangaroo ERV retroposon reactivation might limit the success of such species hybrids. Regardless of this hypothesis, however, there appear to be no obvious barriers to ERV colonization in marsupials, and expansion has been observed. Like the monotremes, the marsupials also have a very tiny Y chromosome, which similarly appears to be unable to support substantial levels of ERV colonization.

In summary, although monotreme-like mammals are evolutionarily old (predating the dinosaurs), we know little about the viruses or genomes of these early mammalian predecessors. Existing monotremes and marsupials are known to support DNA viruses (herpesviruses) and retroviruses. However, little disease is associated with either virus type. Both of these groups of mammals also harbor lower levels of ERVs and LINEs and have much smaller Y chromosomes than do placental species.

Eutherian placentas and endogenous virus expression

ERVs are known for all vertebrates. Mostly, they are lineage specific and conserved, with only a few examples of what appear to be species jumps (as seen in snake species). With the evolution of mammals, we begin to see much more evidence for relatively recent species jumping of various ERV sequences. However, it is still the case that most of the older ERVs are conserved within mammalian lineages, which suggests that there has been a general increase in ERV colonization over time, and not ERV loss. While it appears that in all the mammalian lineages ERV transposition and genome colonization have become more active, we lack any explanation for this observation. However, this ERV activity seems to have occurred mostly early in placental evolution. For example, the integration of HERVs occurred after the split with the great ape lineage. And yet, these ERVs do not generally show polymorphisms. This suggests that most of these ERV integration events are not recent but are instead associated with the origins of specific mammalian lineages and that the ERVs have been stably maintained since that original integration event. Placental species represent the most diverse lineage of the mammals. Among the placentals, the rodents are the most diverse family. The relationship of the placental organisms to their viruses and the organization of their genomes are in general very well studied. Placental species support a very rich array of viruses.

The most variable feature of placental biology is the reproductive biology, along with differences in reproductive system-associated tissues. In particular, it is the biology of the uterus that varies among placental species. During pregnancy, the placental estrous cycle is not genetically controlled as in marsupials but is hormonally interrupted and regulated via embryo implantation. Despite their differences, all placentals share some biological features in regard to embryo development. The most distinguishing feature of placentals relative to marsupials is the presence of the placenta itself, which surrounds, nourishes, and protects the embryo in the uterus. Unlike the maternally derived shell and shell membrane that surround avian eggs, the placenta develops from embryonic tissues to make the trophectoderm, an endoreduplicated layer of cells that surrounds the fertilized embryo. This is the first layer of cells to differentiate in placental embryos and provides protection via immune suppression, facilitates penetration of the uterine wall, and mediates the establishment of the fetal-maternal blood exchange. These complex interdependent features were apparently all acquired at the origination of the placental lineage. It is worth noting that these features are reminiscent of those needed by parasites to colonize a host. In this case, the embryo resembles the parasite and the mother resembles the host. It is thus highly interesting that the placental organism's ova, trophectoderm, and uterus are all tightly associated with high-level production of ERV particles, retroposon (LINE and SINE) transcripts, and gene products (*env*). Following early mouse-based observations by R. Hubner and colleagues in a series of studies published in the late 1970s and early 1980s, Jay Levy and colleagues, including Jay Nelson, reported that human, baboon, and mouse embryonic tissues (syncytiotrophoblasts) all expressed large quantities of ERV particles (classified as xenotropic viruses in mice) and suggested that the expression of these particles might be part of the normal developmental program and evolutionary process of the placentals.

As discussed above, all placental genomes are highly colonized by lineage-specific ERVs as well as their corresponding LINE derivatives. Placentals also have a much larger Y chromosome (about 60 Mbp) than the other mammals, and both the X and Y chromosomes are highly colonized by lineage-specific ERVs and retroposons. Placental ERVs are expressed, and the trophectoderm and placenta in particular have been examined with regard to ERV expression. These studies have demonstrated that the trophectoderm and the early embryo are, in fact, globally derepressed for ERV expression. This is a paradoxical situation, since it has been argued by many that ERV repression (via DNA methylation) must be a genome defense mechanism that would be needed in the early embryo for protection against ERV colonization prior to the differentiation of the germ line. Ironically, the opposite is true. The early placental embryo is open to ERV expression, and it is not until after the germ line differentiates that the embryonic DNA undergoes global DNA methylation and ERV suppression. It is for this reason that early studies of retrovirus infections (e.g., endogenous Moloney MLV clone MOV3) of mouse embryos demonstrated that mice carried the retrovirus in the germ line but were suppressed for virus expression in adult male tissues. Curiously, MOV3 proviral sequences were subsequently observed to amplify in the progeny of females, not males. More recently, this issue of ERV methylation in mouse embryos has been circumvented by using lentivirus-based vectors (which have features that prevent DNA methylation) for infection. This derepression of ERV expression indicates that trophectoderm and early embryo tissues have undergone DNA demethylation, thereby allowing a global activation of ERV (and LINE) expression. The trophectoderm is also unique in that it is one of the few tissues that express paternally specific genes via an imprinting process. One such paternally expressed gene is IGF-II, a major modulator of placental growth. Also, in the trophectoderm of the female (XX) embryo, chromatin undergoes a methylation-mediated process of X gene inactivation (curiously mediated by an RNA sequence that resembles latency-associated transcripts of human herpesvirus 1). However, this derepressed state of ERV embryonic expression is transient. As soon as the totipotent embryonic blasts undergo commitment, their DNA becomes methylated and ERV expression is globally suppressed.

In summary, all placental genomes are highly colonized by lineage-specific ERVs and even more highly colonized by the ERV-derived LINEs and SINEs. Placental organisms also have much larger Y chromosomes than do other mammals, and these are also highly ERV colonized. The trophectoderm, which is unique to the placental embryo and develops into the placenta, expresses high levels of ERVs as a result of DNA demethylation in the early embryo. A schematic summary of ERV colonization of the genomes of the terrestrial vertebrate lineages is shown in Fig. 8.3.

The problem of ERV nomenclature

As mentioned above, observations that normal human placental tissues express high levels of ERV particles date back to the 1970s. At that time, it was observed that human trophectoderms (syncytiotrophoblasts) produced

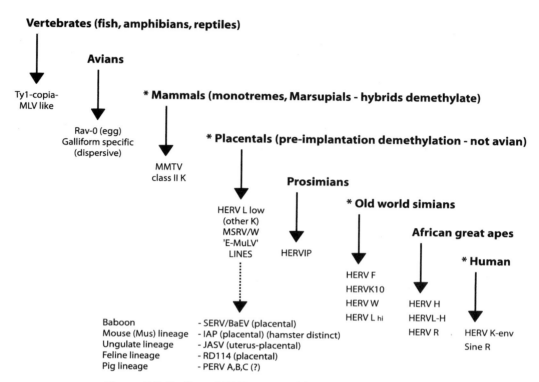

Figure 8.3 Outline of ERVs acquired by terrestrial vertebrates. Asterisks indicate significant ERV expansion. Rav-0, Rous-associated virus-0; JASV, jaagsiekte sheep retrovirus.

large numbers of retrovirus-like particles that could be purified directly from the tissue. However, efforts to show that these particles were active as infectious retroviruses were uniformly unsuccessful; thus, they appeared to be replication-incompetent or defective ERVs. Similar ERV particles are also made in mouse trophectoderms (called intracisternal A-type particles [IAPs]) and cat placentas (called RD114 particles). The initial nomenclature that was applied to these virus particles referred to ERVs as either ectropic viruses or xenotropic viruses. This historical nomenclature is confusing and unfortunate because although these two types are very similar ERVs, they differ in an important way. Ectropic viruses replicate in the cells of the native host, whereas xenotropic viruses do not replicate in the native cells; they replicate only in cells of another species. Thus, the particles being made in trophectoderms were xenotropic, since they were not made in the native cells. The ectropic and xenotropic virus types are essentially the same ERV (i.e., have the same internal genes), but they have acquired different Env proteins. All cells of all *Mus* species harbor xenotropic viruses, which do not replicate in these cells. These viruses maintain a lambda-like lysogenic state, in that the host persistently infected with a xenotropic virus is immune to that virus's replication. However, these viruses can often replicate in other host species. Ectropic viruses, on the other hand, are not uniform and are not found in all hosts, not even all lab mouse strains. They are products of selection and genome rearrangements. Basically, an ectropic virus is a xenotropic virus that has acquired an *env* gene that allows it to replicate in native cells. The historical emphasis on tropism apparent in this nomenclature has obscured the relationship between exogenous viruses and ERVs. Unfortunately, an awkward nomenclature also applies to other areas of ERV biology. The various forms of HERVs that were later discovered were given a bewildering series of names (e.g., RTVLH, HDTV, MSRV, ERV3, K-T47), and sometimes the same ERV would have multiple names. Later, a better naming scheme became common in which the letters that designate the corresponding amino acid codon for the tRNA primers used to synthesize the ERV RNA were used for nomenclature. According to this scheme, HERVs are given letters to designate the primer: K for lysine, W for tryptophan, R for arginine, L for leucine, and H for histidine. More recent analysis of HERVs using this classification scheme has identified 22 HERV families in the human genome. This scheme, however, is sometimes confounded by the existence of apparent chimeras of retroviruses which appear to have recombined two lineages of ERVs. Table 8.1 summarizes ERV classification with respect to the use of tRNA primer sites.

HERV-Ks and human evolution

Humans originated in Africa about 1 million ybp, and they can be differentiated from the other great apes by ERV acquisition. The human genome has 22 independent ERV families, and six members of the K family of HERVs

Table 8.1 Classification of ERVs[a]

Parameter	Feature(s)
Sequence similarity	RT—most conserved (RT captured in LINEs)
	Env—highly divergent types and receptors
	LTR—divergent types
Tropism	Ecotropic/xenotropic—mainly *env* determined
Morphology	Class I—C type, MLV-like (HERV-9, HERV-E)
	Class II—B or D type, MMTV-like (IAP, HERV-K via polymerase)
ERV mosaics	HERV-K family (6 HML groups, 18 groups of LTR clusters, K(10) parent is B-C chimera, K-T47D is B type)
	ERV-9 (156 elements, 14 subfamilies, 4 active)
	HERV-W (C-D chimera)
	JSRV (B-D chimera)
tRNA primers (nondefective)	E (Glu), F (Phe), H (His), K (Lys), L (Leu), R (Arg), W (Trp)

[a]The human genome has 22 independent ERV families, 6 new to humans.

(HERV-Ks) are new to the human lineage. HERV-W, discussed below, is one member of a large family of HERV-Ks. The human genome has 25,000 HERV-K-related long terminal repeats (LTRs). HERV-K-related sequences are also found in the human LINE-1 element. This is a non-LTR poly(A) retroposon that retains HERV-K RT polymerase sequences. SINE-R is also a HERV-K relative, in that it is a poly(A) retroposon with a 5′ LTR and retains some *env* sequence from HERV-K. The L1 LINE elements are present at about 100,000 copies/genome. See Fig. 8.1 for a schematic summary of HERV structure. These elements are highly transcribed in embryonic tissues. Curiously, as noted above, HERVs do not show polymorphisms in DNA integration sites among human populations, which suggests that HERVs were acquired early in human evolution and are stable with respect to their integration sites. The most abundant class of HERVs are the HERV-Ks. The HERV-K family shows sequence similarity to MMTV and also jaagsiekte sheep retrovirus (JSRV). HERV-Ks have been grouped into 18 LTR clusters, of which cluster 9 is found only in humans, whereas cluster 1 is much older and is found in Old World and New World monkeys. In the human genome, 10 HERV-Ks appear to have recently been acquired in evolution, and 9 of these are unique to the human lineage. The human genome has 20 to 50 copies of full HERV-K elements, yet none appear to be replication competent. As previously noted, there are also 25,000 HERV-K-related LTRs in the genome. Yet these HERV-K sequences have maintained a virus-specific nucleotide bias, which suggests that they are under positive selection.

The HERV-K(10) superfamily is composed of six HML groups and has about 50 members found in Old World primates (except for chimpanzees). The HML designation stands for human MMTV-like virus group (a hormone-responsive type of ERV) and is based on RT and *env* similarities. HERV-K(10) was originally observed to be produced in teratocarcinoma cells with very high levels of viral particle production and was associated with the differentiation of these cells. Teratocarcinomas are embryonic tumors that are able to differentiate into various normal tissues, including trophectoderm. It is interesting that there is no exogenous virus that resembles HERV-K(10). It is also worth noting that SINE-Rs are abundant human-specific retroposons that have HERV-K *env*-like sequences and are present at about 5,000 copies per cell. One of the highly conserved genes of HERV-K(10) is the dUTPase gene. HERV-K(10) is the phylogenetic parent of all HERV-Ks. Interestingly, these ERVs also encode a protease that properly cleaves a human immunodeficiency virus (HIV)-encoded protein.

HERV-K(OLD) is an HERV-K(10) superfamily member and appears to be the ancestor of the HML-2 group in humans. It has an intact ORF that codes for a viral Env protein. It also has conserved the central motif of the dUTPase gene in an active form. HML-2.HOM has a central *gag* sequence with a deletion that corresponds to 96 amino acids relative to the sequence found in Old World primates. This deleted sequence has undergone amplification in HERV-Rs in the human lineage.

HERV-W is another member of the large HERV-K superfamily that contains complete *gag* and *env* genes. And as for most HERVs, no replication-competent HERV-W has been seen. HERV-W is specific to catarrhines (Old World great apes). However, the HERV-W *env* is now known to contain the human *syncytin* gene, which encodes an essential and functional membrane protein involved in the fusion of trophoblasts and is highly expressed in the placenta. HERV-W resembles a type C-D chimera virus (which is MMTV-like) and is present at 20 copies per cell. Thus, HERV-W is one clear example of an acquired ERV that provides an essential function for the placenta.

Various other HERVs have also been studied. The HERV-H family includes about 1,000 elements and is thus one of the largest HERV families. However, within this family, it appears that only three members express their *env* genes. HERV-K T47D (D-type) has been reported to be expressed in placental and mammary carcinomas. This virus is interesting in that antibodies against its *env* sequence have been observed during pregnancy, and these antibodies were able to cross-react with HIV *env* sequences. ERV3 is also known as an HERV-R, which contains a complete *env* gene that is expressed in the placenta (syncytiotrophoblasts) and in differentiated embryos. It also appears that syncytiotrophoblasts are able to produce various other types of HERVs. For example, patients with leukemias and lymphomas have been reported to produce antibodies that were reactive to syncytiotrophoblasts. In the case of patients with trophoblastic tumors, one of these antibodies was shown to react to HERV-K Gag and Env proteins.

In summary, all placental species have acquired sets of intact ERVs in significant numbers, and expression of these ERVs in embryonic and reproductive tissue is common. Although classification has been confounded by a historically confusing nomenclature, these ERVs can now best be classified according to the intact versions that are conserved within each lineage. Table 8.1 presents a summary of this classification. Table 8.2 summarizes known intact human ERVs according to this nomenclature. LINEs and SINEs are related to these ERVs and are also lineage specific. In humans,

Table 8.2 Known intact HERVs

Family type	Other name(s)	Expression site(s)
HERV-E		Placental trophoblast
		Seminomas, monocytes
HERV-F	RTVLH-RGH	Placental trophoblast
HERV-H	p15E	Placental trophoblast
HERV-K	HTDV, K-T47D	Teratomas, seminomas, placenta
	K(10) family, IDDMK	Malignant trophoblast, Y
HERV-L	MHC-I[a] associated, Fv1-like	Placental trophoblast
HERV-R	ERV3, p15E	Placental trophoblast, adrenal gland, Y
HERV-W	MSRV	Placental trophoblast, Y (syncytin)

[a]MHC-I, major histocompatibility class I.

the HERV-K family is conserved and is the most human-specific ERV, including family members that are unique to the human genome. Some of these HERVs (e.g., HERV-W) code for proteins that have essential functions in the placenta. A summary of reports in the literature that describe expression of HERV genes is presented in Recommended Reading. The above ERV analysis has shown a bewildering array of associations between the origins of placental species, the acquisition of larger X and Y chromosomes, and the colonization of genomes by ERVs, which are often expressed in and can be functional in placental tissues. These associations suggest that there may or must be a more causal association between ERV acquisition and the origination of placental mammals, such as that seen with HERV-W. Therefore, it appears plausible that the massive ERV colonization seen in all placental genomes contributed directly to the complex life strategy of these placental organisms.

ERVs of other placental species

As mentioned previously, in addition to the human ERV associations outlined above, the placentas of most other mammalian species appear to show associations with ERV production. Baboon placental tissues produce an ERV that cross-reacts with antibodies against HIV-1 RT and gp41, as well as cross-reacting with simian immunodeficiency virus (SIV) p27.B. Thus, although little is known about this ERV, baboon placenta resembles the human placenta at least in ERV production. Rhesus monkeys also express a D-type ERV, which is 20% similar to *Mason-Pfizer monkey virus,* in their placentas. Simian ERV (SERV) is similar to the ERV found in baboons (baboon endogenous virus [BaEV]). Both of these primate ERVs are also related to HERV-W, which encodes syncytin, as described above. HERV-W sequences can be found in the genomes of great apes. However, intact BaEV is found only in baboons. A general pattern thus starts to emerge with recently acquired ERV sequences generally differentiating the various primate lineages from one another.

Mouse ERVs

Mice (*Mus musculus*) provide a particularly well-studied system for the evaluation of ERVs. All known *Mus* species have within their genomes IAPs. These ERVs have sometimes been classified as xenotropic viruses. Typically, these ERVs are classified according to their LTR sequence similarities, which are used to cluster the types of IAPs. The mouse genome has about 900 copies of IAP. IAPs were initially discovered in embryonic carcinoma cells, which are embryonic tumor cells that are able to differentiate. The IAPs are produced in high numbers in association with embryonic carcinoma trophoblast differentiation, and they accumulate in the cytoplasm but do not function as infectious viruses. However, some IAP sequences can also include intact *env* sequences. One such sequence is IAP-E, which is expressed in mouse trophectoderm cells. In addition, mouse oocytes are known to express the *env* sequence of an MMTV-like ERV at fertilization, but this expression subsequently declines during embryo development. Defectives of IAPs are also highly expressed from the mouse genome. The best known of these defective elements is VL30. Each mouse strain appears to have a unique and characteristic set of VL30 elements. VL30 replication-incompetent retroposons are derived from defective IAPs and are expressed in late embryos and also in many established mouse cell lines. Another mouse-specific ERV is MuRVY. This is a Y chromosome-specific ERV that is found in all *Mus* (but not other rodent) species and also contains an intact *env* sequence.

Hypothesis about mouse ERVs and live birth

The possible relationship between mouse ERV activity and embryo development has not been well evaluated. Other rodents (e.g., rat and hamster) also have their own types of IAPs, and these are highly conserved within the corresponding lineages. For the most part, these other rodent IAPs are distinct from those of *Mus*. For example, probes specific for the hamster IAPs do not cross-react with the mouse IAPs. The experimental evaluation of possible IAP function is made highly complicated by the numerous copies and versions of IAP sequences present in the mouse genome. This constellation of IAPs would interfere with most experimental genetic approaches that could be employed to inactivate them. However, it might be possible to use genetic methods to focus on the much smaller number of *env* genes sequenced, but this has yet to be done. An interesting issue to evaluate is whether IAPs have any direct role in normal mouse pregnancy, especially with respect to the nonrejection of the embryo or other "parasite-like" biological activities of the embryo. Given the established capacity of retroviral *env* genes to suppress immunological reactions, this possibility has been proposed in various forms (see Recommended Reading) and is here called the "retroviral hypothesis for the origin of live birth." The concept is that ERVs directly contributed gene functions that led to the origination of the placenta and the ability of the embryo to escape immunosurveillance. If this idea can be tested experimentally, it would seem that mice would provide the best system for this analysis. However, even for mice this hypothesis is difficult to evaluate experimentally due to the heterogeneity and complexity of ERVs. In humans, the fact that HERV-W codes for syncytin strongly supports a role for these ERVs in embryo development or placental function. Various experimental approaches might work for further evaluation of the live-birth hypothesis. For example, it seems possible that stimulators or inhibitors of ERV function could help elucidate such roles. However, there have been few, if any, systematic evaluations of this possibility. It is known that drugs that stimulate ovulation can also increase C-type particle formation, such as in mouse oocytes expressing an MLV-like *env* gene. RT inhibitors are not generally felt to affect pregnancy, but some reports have suggested that early events (implantation) in embryo development can be inhibited.

One study did attempt to globally repress all mouse IAP production in in vitro-produced blastocysts and observed that these IAP-repressed blastocysts failed to implant. However, no other direct evaluation has been attempted. There is, however, some indirect evidence on this issue. A correlation between embryo nonrejection and tumor growth has been noted. It has been observed that some tumors are able to grow in pregnant females but not in nonpregnant females or male mice. There is also some relationship between mating, the birth of offspring, and the induction of ERVs in some mice. The mating of female BALB/c mice with C57BL/6 males followed by immunization of offspring with paternal lymphoid cells results in the activation of the EmV1 ERV and the development of a subsequent AIDS-like or acute leukemia-like disease. However, this is seen only in the mixed offspring of multiparous female mice, not in virgin females. Nor is this ERV reactivation seen in the F_1 offspring of a BALB/c male and BALB/c female mating. Although it is difficult to interpret these observations, they do suggest a link between the nonhomologous male Y chromosome, ERV reactivation, immune recognition, and pregnancy.

Generation of infectious viruses from ERVs

It is worth noting that mating of BALB/c females with C57BL/6 males also resulted in the development of an ERV into an autonomous retrovirus able to cause disease. Similar observations had previously been made. For example, it has long been known that mouse ERVs (such as those in AKR mice) can become active in embryonic tissue, leading to the selection of a replication-competent retrovirus able to induce disease. In the AKR mouse, initial expression of a replication-incompetent AKR ERV occurs in the early embryo. This initial AKR ERV lacks a functional *env* gene. However, this defective virus expression does select for the genetic acquisition of altered *env* sequences that results in an infectious virus that initiates additional rounds of virus infection that will eventually select for the leukemia-causing virus. In both of these examples, the initial, embryo-expressed ERVs were noninfectious but allowed for the subsequent selection of infectious variants. This process is probably also involved in the generation of xenotropic viruses from ectropic viruses, described above.

Hypothesis about mouse ERVs and protection against exogenous retroviruses

Various researchers have suggested that ERVs are maintained in genomes mainly to inhibit infection by exogenous retroviruses, which can cause disease. This suggests that in order to maintain a positive selection for ERV function, an autonomous version of the virus should be prevalent in the host population. Some experimental support for this idea has been presented. However, other results fail to support this view. Surveys of wild mice have generally failed to find evidence of ongoing disease by exogenous retroviruses. MLV and MMTV are the viruses best studied in this regard. In feral mouse (*Mus*) populations, MMTV has been observed to be produced in the milk of about 50% of mice examined. Yet these MMTV-positive mice did not show evidence of disease or breast cancer from these infections. In contrast to this high incidence of MMTV (and other prevalent mouse viruses, such as mouse hepatitis virus and mouse parvoviruses), MLV is not normally isolated from most feral mouse populations. Yet some specific wild mouse populations have demonstrated pathogenic MLV infections. This occurred in an abandoned squab (young pigeon) farm known as the Lake Casitas farm in Southern California. Like most house mice in the United States, the feral mice of this region are mainly *M. musculus domesticus*, originally derived from northwestern Europe. It now appears that *M. musculus castaneus*, originally from East Asia, was also introduced to this region. *M. musculus castaneus* harbors the Fv-4 MLV-like ERV along with the Fv-4R locus, which confers resistance to Fv-4. The Fv-4R resistance locus is itself another defective MLV-like provirus that expresses a gp70 *env* gene that appears to be able to block infection by Fv-4 but not MLV. The Fv-4–Fv-4R combination can be considered a persistence or addiction module according to arguments made earlier in this book. When *M. musculus castaneus* mates with *M. musculus domesticus*, the F_1 offspring will sometimes acquire the Fv-4 endogenous virus without the corresponding protective Fv-4R locus, thus losing the stable persistence module for control of Fv-4. These F_1 mice will start producing Fv-4 virus, which results in premature death caused by lymphomas and paralysis. Through this process of addiction module breakup, a natural outbreak of pathogenic MLV-like infection was observed in this specific feral mouse population. As neither the Asian or European strain of these mice was native to or evolved in California, this event represented the biological meeting of two reproductively isolated lines of the same genus of mouse mediated by human activity. Both of these mouse lines had previously acquired distinct ERV persistence modules. And when these two mouse lines were introduced into a new interbreeding habitat, some of the resulting F_1 offspring were now incompatible with respect to ERV control. The resulting progeny were in a sense no longer fully viable or compatible with respect to controlling their respective ERVs. Some researchers have pointed to this result to argue that it demonstrates the emergence of acute retroviral disease from the genome. However, this result does not support the prevailing hypothesis that ERVs are conserved in order to protect against autonomous or exogenous retroviruses. Instead, it supports the hypothesis that the ERVs present in genomes are persistence modules, preventing the reactivation of the stable persistent ERVs, not protecting against prevalent exogenous retroviruses present in the ecosystem. The example of mating-dependent ERV reactivation recounted here resembles the EmV1 situation described above and the reactivation of plant genomic viruses presented in chapter 7. Such results can also be used to argue for the idea that ERVs also have a role in reproductively isolating lines of the same host species, contributing to the formation of species.

Summary of mouse ERVs

The summary of the combined ERV-mouse studies is complex. Mice, like all other rodents, have lineage-specific sets of ERVs (IAPs) that are both intact and defective (such as VL30). Many of these ERVs can be observed to be highly expressed in embryonic and placental tissues. Such ERVs are typically noninfectious (IAPs). However, in some lineages (AKR), these ERVs can acquire a functional *env* gene, resulting in autonomous and disease-causing retroviruses. In some mouse lines, these viruses can be induced by mating. Although a hypothesis has been presented that ERVs are maintained to inhibit exogenous retroviruses, this hypothesis often fails to explain observed results. Instead, it appears that specific sets of ERVs constitute persistence modules that suppress the exogenous replication of the very same ERVs. In other words, ERVs more often behave as regulators of themselves than of others. Another hypothesis has proposed that mouse ERVs were involved in the origination of live birth, providing the embryo with various characteristics needed for viviparity (e.g., immune evasion, cell fusion, and embryo invasion). Some experimental support for this hypothesis does exist.

ERVs of other mammals

The proposed role of ERVs in placental biology would be expected to be a general one applicable to all placental species. Although other placental species are not as well studied as mice or humans, some relevant results have been obtained. In domestic cats, the placental trophectoderm is observed to express high levels of the feline leukemia virus (FeLV) RD114 ERV. No RD114-related autonomous viruses are known; thus, there is no reason to suspect that RD114 provides protection against any existent exogenous retrovirus. Interestingly, the receptor for RD114 has been determined to be the neutral amino acid receptor, which is also expressed in placental tissue. The presence of this receptor in placental tissue is interesting from an immunological perspective. The receptor is involved in tryptophan transport. And low levels of tryptophan prevent T-cell recognition via indoleamine 2,3-dioxygenase production. Low levels of tryptophan also result in a Th1-type cytokine bias, and such a bias is characteristic of pregnancy. In contrast, a Th2-biased cytokine response is associated with abortion. In fact, interleukin 4 (IL-4), IL-5, and IL-10 are all inflammatory cytokines that can terminate pregnancy and are associated with high rates of embryo loss (up to 30%). These circumstantial observations suggest a link between the FeLV RD114 receptor and the immunological status of the placenta.

All species of ungulates have an ERV related to JSRV called enJSRV. Most ungulates have about 20 copies of this JSRV-related ERV per cell, but bovid species have only a few ERV copies. enJSRV is highly expressed as virus or viral gene products in various reproductive tissues, including the placental syncytium, the cytotrophoblast, and especially the sheep uterus. In fact, the Env and capsid proteins of JSRV are some of the most highly expressed proteins of the sheep uterus. However, their role in uterine biology has yet to be elucidated. It is known that enJSRV can adapt to become an exogenous virus causing lung and nasal adenocarcinomas in sheep and goats.

Pigs have three known ERVs (PERVs) that have correspondingly distinct *env* sequences. There are at least two complete copies of PERV in the pig genome. However, the expression of these PERVs in reproductive tissues has not yet been evaluated. The main concern in the study of these PERVs has been that they appear to be able to replicate in some human cells. This implies that pig tissue used for xenotransplantation might pose a risk of possible infection of the human recipient with PERVs found in the pig tissue.

Overall pattern of placental ERV acquisition

All placental lineages have acquired lineage-specific versions of ERVs. However, we know that some ERV families were acquired in host genomes well before the evolution of viviparity or the placental orders. For example, all vertebrate lineages have some version of MLV-like, Ty1/copia elements (see Fig. 6.4). Within the mammals, all appear to also have MMTV (class II K) elements. In terms of placentals, some ERVs are common to all species, such as low levels of HERV-L, MSRV-W, and related LINE elements. However, all placental lineages have also undergone an expanded colonization by their own lineage-specific ERVs and associated LINEs. This has been well studied in the context of human evolution. All prosimians have HERV-I and -P, and all Old World simians have HERV-F, -K(10), and -W. Old World simians also have amplified levels of HERV-L. African great apes, like humans, also have amplified levels of HERV-L and have additionally acquired HERV-W. It is interesting that ERV-L was also amplified in the extinct woolly mammoth. This ERV-L amplification did not occur in modern elephant species. Humans can be distinguished from chimpanzees by the acquisition of HERV-K(10) HMLs and the corresponding SINE-R element. As these examples show, ongoing ERV acquisition is characteristic of placental evolution. Overall, we see evidence of lineage-specific ERV colonization in all vertebrates, well before the evolution of placental species. These early ERV colonizers are still maintained, typically in small numbers. Additionally, much more numerous and specific ERV colonization events have occurred and are highly correlated with the vertebrate placental species. Consequently, species lineages can be distinguished based on these later ERV colonizations. For example, humans can be distinguished from their primate relatives by these specific ERV colonization patterns.

Y chromosomes, heterochromatin, and ERVs

In placental species, the Y chromosome is of particular interest due to its large-scale (6,000-fold) DNA expansion relative to the Y chromosomes of marsupial and monotreme mammals. Much of this expansion has resulted from colonization by lineage-specific ERVs. In the human Y chromosome (the only Y chromosome sequenced to date), we can find intact and defective HERV-K, -L, and -W and defective derivatives (SINE-R). As the Y chromosome does not undergo homologous recombination with other chromosomes, these elements are the most maintained and distinguishing genetic elements in placental species. This makes the Y chromosome an outstanding genetic marker for tracing ancestral lineages. Many of the Y chromosome ERV elements are found within regions of condensed heterochromatin. HERVs have typically been considered selfish DNA elements; hence, they are expected to be selectively neutral in their hosts, assuming that they do not interrupt active genes. As inactive genes are generally maintained within heterochromatin, selfish HERVs might be expected to favor colonization of heterochromatin. In autosomes, ERVs are also present in large numbers. Chromosome 21, for example, has 225 genes, 3,000 DNA elements, 50,000 repeat elements, and 2,000 retroviral elements. HERVs do not show polymorphisms at their sites of integration, which would be expected if they retained transpositional activity. The human genome has notable "gene deserts," or areas lacking coding sequences, that give it a more uneven distribution of genes than the genomes of lower animals. We might consider the entire Y chromosome to be such a gene desert, given its low ratio of coding to noncoding sequences. Curiously, the distribution pattern of these gene deserts is maintained in the placental mouse genome.

How might we understand the forces that led to the accumulation of human transposons? Most human transposable elements (TEs) do not appear to be mobile. The DNA-based transposons in the human genome all appear to have been inactivated, as have most LINEs. Yet humans have twice the amount of LINEs as do chimpanzees, and chimpanzees have more LINEs than gorillas and orangutans. Thus, although they are now mostly inactive, these ERVs have been acquired recently during the evolution of the human genome. Alu sequences are retroposon-processed, 300-bp pseudogenes which correspond to the nuclear estrogen receptor superfamily. Alu sequences are very highly repeated and found only in higher primates. This lineage-specific retroposon accumulation is difficult to explain if the bulk of genomic transposons are inactive. According to the selfish-DNA hypothesis, the equilibria of TEs within the genome should depend on their genetic stabilities, since TEs have no phenotypes. Also, to prevent interference with active genes, these elements should be overrepresented in heterochromatin. Thus, TEs in heterochromatin are predicted to be more stable. However, the results of genomic analysis indicate that no retroposon family is more unstable than other families. Therefore, the accumulation of retroposons into heterochromatin is not associated with their instability in euchromatin. However, since most retroposons are persistent silent elements, accumulation in heterochromatin might be a way to attain genetic silence and persistence. The Y chromosome is especially notable for containing high levels of heterochromatin and retroposons, and these retroposons are the most distinguishing elements among closely related lineages (such as human and chimpanzee).

Susceptibility of placental X and Y chromosomes to ERV colonization

Although the X and Y chromosomes share some limited regions of homology, recombination to repair or maintain parts of the Y chromosome does not occur. The regions of X-Y homology appear to have been acquired recently and after the divergence of the hominids and primates. It is now known that roughly one-half of all human Y genes are ampliconic (retroposon-like). These ERV acquisitions must have been frequent, recent, and specific to the human lineage. The HERV-K LTR occurs in 12 copies on the X chromosome, 10 copies on the Y chromosome, and several copies on the autosomes (e.g., 12q24). SINE-Rs are human-specific LTRs derived from HERV-K(10) and occur on the Y chromosome in high numbers. Other intact ERVs also accumulate on the Y chromosome. The Y chromosome has one of the few *env* ORFs of HERV-Ks, and this element appears to be a complete HERV that is also present in gorillas and chimps. In addition, an intact ERV3 and an HERV-H are also found on the human Y chromosome. Clearly, there is a strong bias towards ERV and retroposon colonization and accumulation in the human sex chromosomes. And it is this ERV accumulation by the Y chromosome that most distinguishes the human genome from the chimpanzee genome. Thus, we should consider the possible roles for HERV involvement in the evolution of human-specific attributes. These attributes include human cognitive function, human speech, and associative learning, which involves the formation of the social attachments that are so essential for human society. However, there are only a few studies that address the potential involvement of sex chromosomes or ERV-related sequences in such human attributes. For example, the Xq21.3 and Yp blocks have been linked to handedness and psychosis, both of which are distinctly human characteristics related to human language capacity. In addition, SINE-R.C2-like transcripts are homologous to cDNAs, isolated from brains of persons with schizophrenia, that have unknown functions. These issues are discussed further at the end of this chapter.

The Y chromosome shows curious variability in other placental mammals as well. In mouse species, MuRVY ERV is repeated on the Y chromosome at 500 copies per cell, but only in *Mus* species. This mouse sequence differs significantly from that of the Syrian hamster, in which the entire Y chromosome is heterochromatin and fully one-half of the chromosome is occupied by ERVs (IAPs) that have little sequence similarity to the mouse IAPs. This suggests that in most Syrian hamster tissues, genes are not expressed from the condensed Y chromosome. Thus, this Y chromosome is expected to have little, if any, coding potential. However, since all silent chromatin is derepressed following DNA demethylation in an early embryo, even this otherwise silent Syrian hamster Y chromosome is most likely actively expressed during early embryo development. What, then, is the importance of so many ERV sequences on the Y chromosome? There seems to be little need to express Y genes in most tissues. Do all placental species need to maintain sex chromosome ERVs for embryo-specific expression? If so, what function might be fulfilled by such expression? And also, how is it possible for the more recently evolved voles to have entirely lost their Y chromosome if these chromosomes maintain some function in the embryo? Do they make embryonic ERVs from other genetic locations? Clearly, there remain some major mysteries about placental ERVs and the Y chromosome in embryos.

My emphasis on the Y chromosome should not detract from the fact the ERVs and their related LINEs also colonize the other autosomes. In the case of these chromosomes, however, recombination would be expected to eliminate or correct the accumulation of such sequences. Nonetheless, there is a broad autosome ERV colonization, which curiously varies among species. For example, human and mouse chromosomes are both colonized by LINEs and SINEs. However, there is an overall difference between the genomes in that these retroposons correspond to nearly 20% of the human genome but only 8% of the mouse genome. In addition, although these human and mouse elements are distinct, they are curiously often at the same chromosome positions (especially the SINEs), including their positions on the X and Y chromosomes. Such differences might suggest that the human genome has more active retroposons than does the mouse chromosome. Yet various measurements of DNA mobility suggest the opposite, that the retroposons of the mouse chromosome are much more active than those of the human chromosome. It thus seems clear that the lineage-specific ERV colonization events of all placental species occurred in association with the originations of each of these species and were not due to subsequent transpositional events.

Summary of ERVs and sex chromosomes

What can be summarized with respect to X and Y chromosomes and ERVs? One can start by posing several questions to obtain some perspective. Why are the sex chromosomes of placentals so prone to colonization by these agents? What is so different about the Y chromosomes of nonplacental mammals that would lead to this big difference? The recent sequencing of the human Y chromosome provides a picture that resembles a dizzying house of mirrors. The redundant, inverted, and complex nature of the repeated sequences made this chromosome technically challenging to sequence, and this also made it very difficult to understand the significance of the repeated sequences. The conundrum is why the Y chromosome contains mainly retroposons, with so few protein-coding sequences (perhaps as few as 20 to 30), and is in some cases entirely condensed yet still conserves the retroposons within host lineages. The patterns of ERV colonization in the sex chromosomes represent the biggest genetic difference among otherwise closely related species. Thus, the sex chromosomes represent the most dynamic of all chromosomes and reflect the large changes that occur during speciation. Yet, the Y chromosome is, due to the absence of recombination, the most stable, and ironically, it can be used to directly trace male lineages for many generations. Can these ERVs and their defectives have resulted from colonization by persistence modules that force stability? If so, how might such persistence modules affect host evolution? Many of the retroposons of the Y chromosome are highly expressed in embryonic, reproductive, and placental tissues, and perhaps therein lie some answers to the questions posed above.

HIV and retrovirus evolution

HIV-1 is a newly emerged human retrovirus that has caused a worldwide pandemic. Understanding the origins of this major human disease has taken much effort and time, but it now appears that a most plausible scenario for the origination of this virus can be presented. The HIV disease, AIDS, has an extended character in that it can take years to kill its human host. However, this disease basically represents an extended acute viral disease, which requires a long time to accumulate lethal levels of host damage. Thus, it is important to first understand the "extended-acute" nature of this human disease. During the time HIV takes to kill, disease is caused mainly through the depletion of stimulated human CD8-positive cytotoxic T lymphocytes (CD8$^+$ CTLs), which ultimately kill the human host due to immunological incompetence. However, this extended duration does not accurately resemble a typical persistent infection, in that it is an ongoing productive and lytic infection. This acute infection is highly efficient relative to most virus infections. HIV-1 replication in CD8$^+$ CTLs is so highly productive of progeny virus that each infected cell yields about 10,000 virions. This staggering degree of virus production always results in the lysis of the infected CTLs. Typically, there is no true biologically silent or latent state during HIV infection. Antiretroviral drug therapy keeps the numbers of infected cells to low levels, but the lytic replication in CTLs is still maintained at low rates. Although there have been some reports that inactive or latent HIV-1 genomes are present in resting CTLs, these silent genomes do not contribute to the biology of subsequent HIV replication. These latent, or "stored," HIV genes do not appear to function later in typical HIV infections. HIV-1 infections also display the high degree of genetic variability that is associated with many acute infections. This high-level genetic variation is mostly limited to the *env* sequence, and the resulting HIV clades vary in sequence by about 10% per year. The variation follows a punctuated pattern, which seems to result from the deletion of T-cell sets due to virus replication and subsequently leads to evolution of new receptor and Env specificity. Thus, HIV establishes an extended-acute lifestyle in its human hosts.

A persistent origin for HIV?

Unlike the situation with AKR ERV induction in mice (discussed above), no human genomic HERVs appear to have been the predecessor of HIV-1. In other words, HIV is not a recombined or reactivated form of a HERV, suggesting that it originated in another species. The closest relative and likely progenitor of HIV-1 is the simian virus SIVcpz, found in certain populations of chimpanzees. Both the human and chimpanzee lentiviruses have *vpu* genes, which distinguish these viruses from all other retrovirus families. Some troops of wild chimpanzees (*Pan troglodytes*) in certain regions of central Africa are observed to support SIVcpz infections but do not develop an AIDS-like disease. In chimpanzees, SIVcpz is nonpathogenic and persistent. However, the distribution of SIVcpz in wild chimpanzees is highly restricted (being absent from much of central Africa), and it does not appear that SIVcpz establishes a long-term prevalent infection in most wild chimpanzee populations. This observation leads to the question of where SIVcpz itself may have originated. SIVcpz is also present in various African monkey species. In all, about two dozen different African monkey species have been shown to harbor SIV in wild populations. When African monkeys are infected by SIV or HIV, their cells respond sluggishly to the infection. Thus, these monkeys do not seem to be able to support the highly productive lytic infections of HIV seen in human $CD8^+$ CTLs. However, if African SIV is used to infect Asian monkeys, these animals do develop an AIDS-like disease, suggesting that virus persistence is specific to particular monkey species. However, the versions of SIV found in African monkeys do not appear to have been a direct predecessor of HIV-1.

Other simian retroviruses have also been described. SIVgsn can be isolated from spot-nosed monkeys in Cameroon and is highly divergent from SIVsyk. And it has a *vpu* homologue that was previously thought to be unique to SIVcpz and human lentiviruses. In addition, SIVgsn has a complex genome structure, similar to that of HIV, and it has an *env* gene that is related to SIVcpz *env*. Other types of SIVs that can be found in other monkey species (such as SIVmonNGI in *Cercopithecus nictitans*) also have a *vpu* gene. Recent results from sequence analysis of the additional SIVs now support the idea that chimpanzees were indeed the source of the SIV that evolved to be HIV-1 in humans. However, chimpanzee infection with SIV is neither prevalent nor old, so it appears that these viruses were introduced only recently to specific chimpanzee populations. This sequence evidence suggests that SIVcpz is itself a recombinant between two SIVs specific for other species of monkey. Both of these monkey species are preyed upon and eaten by chimpanzees. In general, SIVs are ubiquitous and persistent in monkeys and appear to represent evolutionarily old infections. In contrast, HIV causes acute disease in humans. Like all human acute viral infections, HIV-1 appears to be traceable to a persistent virus in a nonhuman but highly specific host. HIV-2 also appears to trace its origins to viruses that persistently infect monkeys, in this case to the SIVsmm that infects sooty mangabey monkeys as its natural host. As is typical of most persistent infections, SIVsmm shows little sequence variation in its native host but is highly variable in humans. It is interesting that SERVs are members of the baboon virus complex, which resemble lentiviruses and are fixed in the genomes of all Old World monkeys. This complex may define a lentivirus-based persistence module in the genomes of these monkey lineages. Monkeys thus appear to be the natural and original source of lentivirus systems that can undergo adaptation to cause acute human diseases. SERVs are absent from the genomes of apes and humans, and they appear to be ancestral to BaEV and SRV.

In summary, HIV-1 is a newly emerged human virus that appears to have originated in another primate species but then adapted to humans. The human disease, although of extended duration, nevertheless resembles an acute lytic infection. Although a virus that can be found as a persistent infection in some chimpanzees appears to be the direct progenitor of HIV-1, this chimpanzee virus itself appears to have been recently acquired from a mixture of two stable and persistent monkey viruses, each from a distinct species of monkey. Chimpanzees are known to eat these other two types of monkey.

Herpesviruses of placental species

The herpesviruses appear to cause common infections in animals from clams to fish to most aquatic and terrestrial vertebrates. These hosts span all the way from oysters to wallabies to humans. As far as it can be determined, all placental species appear to harbor species-specific versions of herpesviruses. Herpesviruses are classified into three major groupings according to biological and sequence similarities: the alpha-, beta-, and gammaherpesviruses. The alpha- and betaherpesviruses are, for the most part, phylogenetically congruent with their hosts, suggesting long-term virus-host coevolution. The gammaherpesviruses are similar in several respects to the betaherpesviruses, but gammaherpesvirus genomes are more variable, suggesting that more recent species jumps have occurred for this virus lineage from betaherpesvirus ancestors. Thus, herpesviruses might be more broadly classified into the alphaherpesviruses and the beta- and gammaherpesviruses. All herpesviruses appear to have descended from one common ancestor. It is assumed that this ancestor is best represented by the herpesviruses of either clams or vertebrate fish. Within herpesvirus genomes, the most conserved regions are five blocks of genes that code for both structural and enzymatic functions. The biological properties of the virus groups are also well conserved (lymphotropism, latency in neurons, etc.). From the perspective of human biology, it is interesting to consider why humans host so many types of herpesviruses, each of which appears to have an old relationship with the human host. Eight types of human-specific herpesvirus are known: herpes simplex virus 1 (HSV-1), HSV-2, cytomegalovirus (CMV), varicella-zoster virus (VZV), Epstein-Barr virus (EBV), HHV-6, HHV-7, and HHV-8. All of these viruses are ubiquitous and result in lifelong infections. VZV and HSV persist in different types of ganglia (neurons). Alphaherpesviruses (HSV-1, HSV-2, and VZV) infect mucosal epithelial cells and are latent in neurons or ganglia. Betaherpesviruses (HHV-6 and HHV-7) persist in and are tropic to T lymphocytes. Betaherpesviruses show some genetic similarity (at telomeric repeat sequences) to Marek's disease virus and also lymphotropic bird virus, suggesting some bridge-type connections in evolution of these viruses. Gammaherpesviruses are biologically more similar to betaherpesviruses and are represented by EBV, HHV-8, mouse herpesvirus strain 4 (MHV-4), and wildebeest virus.

HHV prevalence in most primates

Human herpes-like viruses have been observed to infect most other species of primates. For example, ancestors of EBV are present in cercopithecine species, and the EBV nuclear antigen is conserved in simian lymphocryptoviruses. However, there are some differences between human and primate herpesvirus evolution. Monkey B viruses (e.g., *Cercopithecine herpesvirus 1*) are alphaherpesviruses. These are endemic in Asian macaques (Old World only) and most similar to human HSV. The virus found in marmosets appears to be basal to these primate herpesviruses. Herpes B viruses are acquired at sexual maturity; thus, they may be transmitted by close contact. In their native monkey hosts, these viruses establish persistent infections that normally cause no disease but do undergo inapparent mucosal reactivation. However, when transmitted to humans, infections with these viruses are always symptomatic, causing acute central nervous system (CNS) disease and resulting in death rates as high as 70%. The herpes B virus from rhesus monkeys may be the most lethal to humans. Various macaque species have strain-specific herpes B viruses that can be found at a prevalence of 80 to 100%. Oral lesions have been reported to occur in monkeys, but no genital lesions are seen. This is in contrast to the segregated biology of oral HSV-1 and genital HSV-2 in humans. This separation of HSV oral and genital habitats seems to have been a human-specific development. The nearest relative to monkey herpes B virus is HSV-1, and its second nearest relative is VZV. It seems possible that behavioral differences in humans, such as frontal sex, may have led to isolation of oral and mucosal habitats and divergence of the two HSV types. The mammalian alphaherpesviruses are coevolving with their hosts. However, the evolutionary patterns of alphaherpesviruses and their avian hosts are different and generally not coevolving. Curiously, no alphaherpesviruses have yet been described for any rodent species.

Beta- and gammaherpesvirus evolution

Although more recently discovered, Kaposi's sarcoma-associated herpesvirus (KSHV, or HHV-8 in the new nomenclature) may present a good model to understand gammaherpesvirus evolution. KSHV is a gamma 2 virus (rhadinovirus) and is the most recently isolated human gammaherpesvirus. It was recovered from an AIDS patient, and it appears that HIV immunosuppression led to HHV-8 reactivation. However, the virus is otherwise very inapparent, and within a host, the virus genome is stable. Four stable HHV-8 genotypes are known worldwide. This family of related viruses is found in New World and Old World monkeys, in humans, and in chimpanzees. Recently, additional chimpanzee- and gorilla-specific versions of this virus family have also been described. KSHV is endemic in central Africa and clusters according to its human host populations. Thus, it has been used to trace human migration out of east Africa into the rest of the world (similar to the use of human papillomavirus [HPV] and JC virus [JCV]). KSHV transmission occurs primarily via sexual intercourse. The virus shows very low rates of recombination. According to phylogenetic analysis (via DNA polymerase similarities), the gammaherpesviruses appear to have descended from the betaherpesviruses. It appears that EBV evolved in Africa from ancestral viruses present in ceropithecines, as noted above. Both the beta- and gammaherpesviruses most likely descended from alphaherpesvirus predecessors. Broad phylogenetic analysis indicates that as a group, the gammaherpesviruses are the most complex of all herpesviruses and also the most recent to have undergone large-scale genetic changes. Currently, it appears that the gamma 2 viruses are clearly coevolving (with New World and Old World hosts), but at a much higher rate of change than are the alpha- and betaherpesviruses. HHV-8 is similar to murine gammaherpesvirus (MHV-68), which is well studied and known to persist in the spleens and peritoneal cells of mice. Thus, gammaherpesviruses can also be found in rodents, unlike the alphaherpesviruses, which have no known rodent hosts.

Herpesviruses of other placental species are also known. Tree shrew herpesvirus is associated with malignant tumors in some situations, although most shrew infections with tree shrew herpesvirus 2 are inapparent. In cattle, bovine herpesvirus (BoHV) has been identified. As a herpesvirus, BoHV is unusual in that it crosses the placenta from the mother to infect the fetus. BoHVs exist in two types, BoHV-1 and BoHV-4. BoHV-4 appears to have descended from an ancestral virus of the African buffalo and is coevolving with its cattle hosts. Phylogenetic analysis suggests that domestic cattle were initially infected by herpesvirus after a species jump of an African buffalo virus-like virus about 700,000 ybp. As the original wild species that led to domestic cattle is unknown or extinct, it is not clear if this herpesvirus jump is associated with the predecessor of domestic cattle. BoHV-4 is also lethal in sheep. Farmed deer are also susceptible to herpesvirus infections, which result in malignant catarrhal fever. This deer herpesvirus is the closest relative to the herpesvirus of bison. Elephants also harbor their own species-specific, persistent, and inapparent versions of herpesviruses, and these viruses can jump elephant species to cause acute and fatal diseases (generally in young elephants). This can be observed when Asian and African elephants are housed next to each other in zoos. Also, the herpesviruses of both the Asian and African elephants can cause fatal diseases in the other elephant species. A feline herpesvirus has also been identified. This virus has been shown to persist in small cat populations, thereby establishing that its maintenance is not host density dependent.

In summary, the alphaherpesvirus are evolutionarily the oldest herpesvirus lineage and appear to be the ancestors of the beta- and gammaherpesviruses. This virus family can be traced back through algal viruses and bacterial phages. Overall, these herpesviruses conserve biological characteristics and core replication genes. Mostly, they have persistent life strategies and establish lifelong persistent infections, although they often jump species to cause acute diseases in related hosts. Their evolution is mostly congruent with that of their hosts, in some cases being highly host congruent. Curiously, the gammaherpesviruses, but not alphaherpesviruses, are found in rodents, yet the gammaherpesviruses are the most recently evolved. Humans support an unusually large number of herpesviruses.

Poxviruses: a major example of DNA virus emergence

Poxviruses of moles, rodents, and domesticated animals

The poxviruses represent the largest and most complex of all vertebrate viruses. Interest in this family of viruses stems from the ability of the human-specific smallpox virus (*Variola virus*) to cause major human epidemics. These epidemics have had devastating effects on human populations, especially in the New World following the introduction of smallpox virus in the 1500s by Spanish conquistadors, which resulted in a virgin soil epidemic. In historical terms, smallpox virus is responsible for more human disease and death than any other virus. Smallpox virus initiates human infection through the respiratory route, but it results in a systemic disease characterized by the infection of sebaceous glands of the skin, producing the characteristic skin pox lesions. Transmission is not as highly contagious as that of measles virus and occurs during the pox phase but not during the prodromal phase. Thus, transmission is not inapparent, nor does the virus persist in humans or any other host. Smallpox virus is a strictly acute disease agent, and it is thus dependent on host population structure and density to maintain a chain of transmission. Because of this, smallpox virus was susceptible to eradication by vaccination. This eradication was accomplished by the World Health Organization in 1978, using vaccinia virus as a live vaccine. Vaccinia virus infection normally results in one small pustule at the site of infection and does not lead to systemic infection. *Vaccinia virus* is sufficiently similar to smallpox virus to provide immunological cross-protection, but it is also similar to cowpox virus in that it is able to replicate in other species, such as cows, horses, rodents, and chickens (in eggs).

Until recently, the question of the origin and evolution of human small-pox virus was difficult to address. It is clear that smallpox virus could not have been maintained by small preagricultural human populations (about 10,000 ybp), suggesting that it adapted to the human host after this time. The earliest reference to the disease can be found in Sanskrit texts from ancient India; thus, smallpox appears to have originated in the Indian subcontinent several thousand years ago. Radiation into Europe appears to have occurred via North Africa during the Arabian expansion. However, it was clearly absent from the New World and the Pacific Islands. It seems highly likely that smallpox virus initially evolved in another host and was able to jump species and establish itself as a successful acute viral agent specific to humans.

Other vertebrate poxviruses

Poxviruses are known to occur in numerous other vertebrate hosts, including monkeys, cows, horses, and camels. In all of these hosts, the corresponding poxvirus infections are also acute and do not persist. Although these viruses can clearly replicate in their corresponding namesake host, it now appears rather clear that some of these poxvirus names are really misnomers. Both cowpox and monkeypox viruses present good examples of misnomers. Both of these viruses infect humans in addition to their namesake hosts. However, in natural ecosystems, both viruses are persistent in specific rodent species. Cowpox virus can be found in bank voles (*Clethrionomys glareolus*), field voles (*Microtus agrestis*), and wood mice (*Apodemus sylvaticus*) and shows a high prevalence in these species in northern Europe. Infected voles, in particular, appear to be able to maintain the virus in the natural habitat. In bank voles, cowpox virus is always persistent and inapparent, and animals show few signs of infection. Also, cowpox virus transmission is not dependent on vole population density and appears to occur via close social or sexual interactions. Although mice (*Mus*) and rats (*Rattus*) are present in the same habitats and can also be infected with cowpox virus, these species develop acute disease and cannot maintain the virus in the ecosystem. Furthermore, natural transmission between species is rare in field studies; thus, interspecies transmission is not typical of the biological maintenance of cowpox virus in natural habitats. In terms of human infection with cowpox virus, humans are most commonly infected by contact with infected cats, which, being active hunters, prey on infected rats. However, cats are not highly susceptible to cowpox virus infection unless also infected by and immunosuppressed with FeLV, which renders them fivefold more susceptible to cowpox disease. Human cowpox disease is normally self-limiting and localized. However, humans also infected with HIV-1 can acquire lethal cowpox virus infections. It thus seems most likely that cowpox virus has evolved as a virus that can persist in bank voles as its native northern European host. However, the virus can also readily jump species to replicate as an acute agent in other rodents, cows, cats, and humans.

An almost identical biological pattern applies to monkeypox virus, with different host species. Monkeypox is an acute disease of central African monkeys and is most prevalent in the Democratic Republic of the Congo (formerly Zaire). Humans often acquire infection by the consumption of monkey "bush meat," which can result in an acute infection almost as lethal as smallpox. However, it is the giant Gambian pouched rat that is persistently infected with asymptomatic monkeypox virus. Although field studies have not yet been performed, it is expected that such a rodent-specific persistent infection provides for the stable maintenance of monkeypox virus in this habitat. Recently, a giant Gambian pouched rat asymptomatically infected with monkeypox virus was imported into the United States. The rat was housed with prairie dogs, which became acutely infected with monkeypox virus, leading to an outbreak of human infections. Monkeypox virus shows considerable sequence similarity to cowpox virus. Thus, monkeypox virus might in reality be considered the persistent giant Gambian pouched rat poxvirus.

Poxvirus phylogenetics: rodent cowpox virus, the ancestor of smallpox virus

The sequence relationships among the various poxviruses have recently been clarified with the completion of numerous poxvirus genomic DNA sequences. Phylogenetic analysis supports the idea that the cowpox virus represents the most basal of the mammalian poxviruses. The various host-specific poxviruses, including *Vaccinia virus* and smallpox virus, appear to have descended from an ancestor most likely represented by cowpox virus. In addition, it was noted that the cowpox virus genome is more complex than many other poxvirus genomes and contains a greater number of immuno-modulatory genes than does the smallpox virus genome, for example. These observations, along with the above discussion, now allow proposal of a scenario for the evolution of smallpox virus consistent with both the known facts and the evolutionary consequences of persistent and acute life strategies as I have outlined throughout this book. Various rodent species, especially Old World rodents, have a stable and persistent relationship with poxviruses. It is thus likely that the basal poxviruses have coevolved in a persistent life strategy with these hosts. However, these rodent-specific poxviruses have a natural tendency to replicate as acute virus infections in other vertebrate hosts, including other rodent species. In some cases, where supported by appropriate and dense host population structures, these rodent-derived poxviruses have adapted to become prevalent acute agents in these other hosts. Generally, this has involved loss of genes, presumably genes associated with persistence in the original rodent host. For example, cowpox virus has 39 ORFs that differ from or are absent from *Vaccinia virus* and smallpox virus. With sufficient selection, however, this adaptation to the acute life strategy in a new host can become irreversible, such as occurred for smallpox virus, which is now a human-specific acute virus. This proposal is very similar in strategy to that which I proposed earlier for the evolution of human-specific acute influenza B virus from persistent influenza A viruses of various waterfowl species.

Vertebrate poxvirus evolution from entomopoxviruses

The broader question of how poxviruses might have initially adapted to vertebrates, since they do not seem to persist in a large number of other vertebrate hosts, can now be considered. Poxviruses are known for shrews, voles, birds, and crocodiles, but not for vertebrate fish or shellfish. When did they first adapt to the vertebrate host? Besides the chordopoxviruses, which have been mainly discussed above, there are several other classes of poxviruses, including entomopoxviruses, capripoxviruses, parapoxviruses, and *Molluscum contagiosum virus* (MOCV) of humans. These viruses differ significantly from each other in sequence and in other important ways. However, the entomopoxviruses are clearly the most diverse, complex, and basal of all of these poxviruses, as they are notably larger and more complex and their gene order is less well conserved. These differences have led to numerous suggestions that the vertebrate poxviruses were adapted originally from the insect entomopoxviruses. In addition to strong phylogenetic support, there are also some experimental results consistent with this hypothesis. Vertebrate poxviruses often show curiously strong activity in insect cells. For example, although entomopoxviruses enter vertebrate host cells, the resulting infection is abortive and cells express only early genes. However, the converse infection of insect cells by animal poxviruses is much more efficient: not only does vaccinia virus enter insect cells successfully, but also it expresses early genes, induces increased viral DNA synthesis, and induces late gene expression. In fact, the only factor limiting a fully productive vaccinia virus infection in insect cells appears to be faults in processing of viral proteins that occur in the insect cells. Thus, vertebrate poxviruses show considerable and unexpectedly complete activity in insect cells.

If entomopoxviruses do indeed represent the predecessors to vertebrate poxviruses, then it appears that the various lineages of more distant poxviruses (e.g., leporipoxviruses, suipoxviruses, and capripoxviruses) have maintained some additional similarities with respect to biological properties, such as insect-mediated transmission and virus persistent mechanisms. Mostly, these other poxviruses appear to establish transient persistent infections in their specific hosts by inducing benign fibromatous growths in infected cells. These growths are slowly cleared by the host cellular immunity, and virus can persist during this clearance. In addition, these other poxviruses are not transmitted by respiratory routes, as is smallpox, but are mainly transmitted by biting insects (e.g., mosquitoes and fleas). However, in some hosts, these initial areas of skin hyperplasia can become necrotic pustules and lead to severe systemic infections.

Leporipoxvirus: New World and Old World rabbits and myxomatosis
The best-studied example of a persistent and virulent leporipoxvirus is *Myxoma virus*. In natural settings, this virus infects various New World rabbits, such as the California bush rabbit or the South American *Sylvilagus brasiliensis*. Infection is readily observed in field settings and results in a localized benign cutaneous fibroma that is slowly cleared. The cycle of transmission is maintained by mosquitoes feeding on infected rabbits. Similar situations apply to poxviruses of hares, squirrels, and domestic pigs. However, when myxoma virus infects the domesticated European rabbit (*Oryctolagus cuniculus*), a fulminate and highly fatal disease (myxomatosis) results. Thus, a species jump by this virus results in a switch to an acute virus life strategy and the

loss of persistence. In 1950, myxoma virus was introduced into Australia to help control the European rabbits, which had become a major pest species on that continent. Following a very high mortality sweep (99%) as a virgin soil epidemic, rabbit population densities were severally reduced and the virus had to adapt its life strategy to this sparse host population. The drop in host population density also led to adaptations in the rabbit hosts. In this new situation, the virus needed to be able to infect the less numerous and immunologically naive newborn rabbits, since most surviving adults were immune. This altered selective pressure produced what appears to be a somewhat slower-replicating acute virus variant that has a mortality rate of only about 60% in rabbits. Many evolutionary biologists have pointed to this result to argue that a highly virulent virus is not evolutionarily stable because it kills too many of its hosts and that the virus is thus constantly evolving to lose its virulence. It is worth noting a couple of points concerning this idea. First, the virus has not evolved an ability to persist in these hosts (as it does in New World rabbits), and the infection continues to be strictly acute even after adaptation to lower virulence. Second, the resulting 60% mortality rate is still considered highly virulent. To put this in a human perspective, consider that this mortality rate is higher than the 30 to 50% mortality rate of smallpox virus in humans.

A collection of early poxviruses, including MOCV

Another group of insect-transmitted poxviruses are the capripoxviruses, which are generally specific to hoofed hosts. These viruses were first observed in 1929 in South Africa, where they infected domestic cattle. Capripoxviruses are geographically restricted and induce skin growths, such as lumpy skin disease, in cattle, sheep, and goats. The specific capripoxviruses have adapted mainly to specific domestic animal hosts. Little is known, however, concerning the origin and natural biology of these viruses, although Kenya appears to be the source of multiple outbreaks and may be the homeland. Virus replication is mostly restricted to the skin, but other organs can also be infected. The initial infection results in skin hyperplasia, and as a consequence of the host immune response, these foci develop into papules and lesions that produce virus. Thus, the infection of domestic animals is mainly an acute, nonpersistent infection. As no lung transmission has been observed, these capripoxviruses are clearly unlike the chordopoxviruses of mammals. In addition, parapoxviruses are also insect transmitted (as well as being transmitted by skin abrasion) and cause widespread disease in small and large ruminants. Of these, orf virus is economically the most important infectious agent of sheep and induces benign cutaneous growths. There is one human-specific poxvirus that also fits this pattern of inducing cutaneous growths: MOCV. However, MOCV is mechanically transmitted through contact with infected skin and is not known to be transmitted by insect bites. MOCV was first reported in 1841 to be a transmissible agent that induced basal skin cell hyperplasia. Although it is classified as a chordopoxvirus, it has several distinct characteristics: it is specific to its human host, is linked to keratinocyte differentiation, and is the only poxvirus that codes for glutathione peroxidase, which protects infected cells from UV-induced DNA damage, stress, and oxidation. This virus also induces benign skin growths associated with persistent virus production. Infection is spread worldwide and tends to occur mainly in children. Although most transmission appears to be mechanical (e.g., through skin contact), sexual transmission is also known and is common in some areas. The duration of infection is a few months to years, or even longer in immunocompromised hosts. This virus appears to have an extended but limited persistence life strategy and closely resembles the keratinocyte biology of HPVs, described below. Infection is generally localized, benign, and self-limiting, although lesions sometimes develop. Acute infections are not observed, which is also similar to the case with HPV.

Broad patterns of poxvirus evolution: insect to bird to mammal

MOCV appears to represent an old infectious agent of humans that has evolved to become a human-specific virus. This virus is unusual in that no other members of the poxvirus families are very similar to it. Although MOCV has clearly discernible similarities to the other chordopoxviruses, it is phylogenetically the most distant of the chordopoxviruses and thus appears to represent a basal virus that diverged very early from the other poxvirus lineages. In terms of overall genome organization and sequence similarity, MOCV also shows clear similarity to the capripoxviruses and the leporipoxviruses, described above. Thus, it would seem to represent a virus lineage that diverged prior to these two other virus lineages. MOCV also shows clear organizational similarity to fowl poxviruses. As mentioned previously, fowl poxviruses have the largest genomes of the poxvirus family members. The next-largest genomes are those of the entomopoxviruses. However, fowl poxviruses maintain a gene order that is similar to that of MOCV, whereas the entomopoxviruses have significantly altered gene orders, including changes in the conserved gene families. Finally, phylogenetic analysis suggests that fowl poxviruses are basal to MOCV, which is basal to the other orthopoxvirus members. All these observations together allow the proposal of an evolutionary path for all of the poxviruses and how they adapted to mammals. The entomopoxviruses appear to be the earliest members of the family *Poxviridae,* as noted in chapter 7. In addition, the entomopoxviruses show significant similarity to other large insect-specific DNA viruses, such as the ascoviruses (which are not found in hosts other than insects). Thus, these large DNA entomopoxviruses most likely initially evolved in insects. It appears that at some point, likely early during vertebrate evolution, this insect virus lineage underwent a species jump into the reptilian and avian lineages, probably via biting insects. These reptilian and avian viruses adapted to persist as skin growths in specific avian hosts, but they can also induce acute disease lesions in other host species. At some point (likely after the radiation of avians), a fowl poxvirus-like progenitor virus was able to adapt to mammalian hosts, resulting in a virus that resembles MOCV. This lineage of virus underwent an adaptive radiation into various other hosts, resulting in capripoxviruses, leporipoxviruses, and swinepoxviruses. All of these viruses maintained a tendency for insect-mediated transmission and the induction of benign skin growths for persistence in their respective hosts, but they were sometimes capable of inducing acute lesions in other hosts. However, one poxvirus lineage adapted to a rodent or rodent-like (shrew) host. This was the orthopoxvirus lineage. In their native rodent hosts, these viruses establish benign persistent infections. Representatives of these viruses include cowpoxviruses (specific to bank voles) and monkeypoxviruses (specific to Gambian rats). However, this poxvirus lineage also acquired a capacity for acute systemic replication and respiratory transmission in other hosts (hence the ability of cowpoxvirus to replicate in rats, cats, and humans). Many of these viruses became adapted as acute viruses to these new hosts, resulting in species-specific acute viral agents (such as smallpox virus). Thus, in contrast to the herpesvirus lineages, which have evolved in close congruence with all vertebrate animal lineages, the poxviruses have not, for the most part, coevolved with the majority of their hosts. Rather, they appear to trace an evolutionary path originating in insects and adapting to birds and then mammals. And although they sometimes establish persistent infections, they often produce stable acute infections with high mortality rates.

Other persistent and emergent DNA viruses of mammals

Smaller DNA viruses: the adenoviruses

Unlike the poxviruses, the adenoviruses appear to have infected vertebrate hosts throughout the evolution of these hosts, as they are represented in all of these terrestrial vertebrate lineages, from fish to amphibians to avians to marsupials. Thus, adenoviruses have, in general, patterns of coevolution with their hosts. The adenoviruses have double-stranded DNA (dsDNA) genomes that usually contain about 50 genes, but only 16 genes are known to be conserved in all genera. Surprisingly, the highly studied E1A regulatory gene from adenovirus type 5 is not conserved in all adenovirus lineages (such as mouse adenovirus). The marsupial (possum) adenovirus is a member of the atadenovirus family. With respect to mammals, all examined adenovirus lineages appear to include species-specific adenoviruses. The adenovirus that infects the tree shrew is the basal member of the mastadenoviruses, and phylogenetic analysis suggests that it diverged from the atadenoviruses. This virus appears to be able to establish a persistent relationship with its host, since a mastadenovirus-like adenovirus can be isolated from the kidneys of healthy tree shrews. Consistent with a persistent life strategy, the ends of this mastadenovirus DNA are high in GC content and may undergo DNA methylation, which suggests transcriptional silencing of the DNA during the virus life cycle. In humans, about 50 types of specific adenoviruses are known. Most, but not all, of these are also known to establish prevalent persistent infections. In fact, using normal tonsils removed from children in 1959 in New York, the spontaneous reactivation of adenovirus from cells grown in culture was observed and led to the first isolation of human adenovirus. Within the human adenovirus group, the viruses are divided into respiratory and enteric families. Human adenovirus 40 appears to be the basal member of both groups according to phylogenetic analysis. A biological characteristic that has been lost in most mammalian adenoviruses is the tendency to induce host cell hyperplasia, as was seen in amphibians. Mammalian adenoviruses are also frequently associated with and coisolated with satellite viruses, such as adeno-associated viruses.

Polyomaviruses

Polyomaviruses (small circular dsDNA viruses) are known for mice, monkeys, and humans. However, the broader host distribution of these viruses has not been well studied. Simian virus 40 was the initial member of this family to be isolated from persistently infected normal monkey primary kidney cells, which were being used to grow vaccine strains of poliovirus. So far, all mammalian versions of this virus appear to have the tendency to persistently infect kidneys, but they probably initially enter their hosts via the respiratory tract. Although some diseases can occur from polyomavirus infections (such as progressive multifocal leukoencephalopathy), these are rare and normally associated with compromised immunity. The vast majority of infections are asymptomatic and persistent. Furthermore, persistence is generally lifelong, with the quantity of virus shedding into the urine increasing with age. Like the other mammalian polyomaviruses, human polyomaviruses (BK virus and JCV) show extreme host specificity. Furthermore, there is no evidence that host species jumps ever occur. These viruses are genetically stable and display a phylogenetic congruence with their hosts, characteristic of most persistent viruses. The best-studied virus in this respect is JCV. This virus exists in three stable genetic types (A, B, and C) and four subtypes that are distributed among the human populations. The distribution of these viruses can be used to trace the movement of human populations and has been used to evaluate the origins of the Asian populations in China, Korea, and Japan. Such population-based analyses are similar to those that have also been done using human T-lymphotropic virus 1, HPV-16, and HPV-18. In the case of JCV, virus transmission requires close familial contact and the virus does not readily move between mixed human populations. In natural mammalian populations, polyomaviruses seldom induce tumors, in contrast to the polyomaviruses of fish, frogs, and turtles, which all induce tumors (either epidermal hyperplasia in fish or kidney tumors in frogs).

Papillomaviruses

The papillomaviruses, like the polyomaviruses, are small circular dsDNA viruses. Many of these viruses are associated with skin growths (warts) in their hosts. Virus replication (like for MOCV) is linked to the differentiation of infected keratinocytes, but here the infected cells are stimulated to grow. These growths produce virus and can often persist for months. However, some persistent infections are much more silent and not associated with skin growth. Such infections may be the most common. Papillomaviruses are common to many mammalian species. Humans are known to host over 100 types of human papillomaviruses (HPVs), and these are classified into two large groups: cutaneous and mucosal. Most populations have a significant prevalence of these viruses (up to 40%). The two best studied of these viruses are HPV-16 and HPV-18, which have received much attention due to their clear association with cervical dysplasia. There are also a large number of primate papillomaviruses. As mentioned above, papillomavirus evolution is phylogenetically congruent with that of the host species; thus, the patterns of coevolution have been used to trace the geographical and racial patterns of human populations and their migrations. One curious question is why there are so many types of HPVs, given that human polyomaviruses exist in only two types. One factor is that competition may exist among HPV types. Also, HPVs may not persist as efficiently as do the polyomaviruses, as infections are not lifelong and either resolve or are replaced by infection with another virus type. Genetic analysis suggests that most of the HPV types are old and may have originated around the time of human divergence from other primates. Thus, the large diversity of HPV types also appears to be old on an evolutionary timescale.

Other small DNA viruses

Mammals also support infections with small DNA viruses like parvoviruses and TT virus of humans and primates. Parvoviruses show congruence with their hosts but also show rapid evolution with species jumps and associated acute disease. Four hundred seventy-two types of animal parvoviruses are known. These exist in three phylogenetic groups: group 1, which includes rodent, pig, and carnivore viruses; group 2, which includes bovine and autonomous human and primate viruses; and group 3, which includes human helper-dependent and autonomous avian viruses. Group 2 has members that are host congruent (host coevolving), but it also has members that appear to have undergone numerous species jumps and are not congruent with their hosts. Curiously, no invertebrate parvoviruses are known. The mammalian adeno-associated virus group is most similar to avian viruses and may represent a species jump between avian and mammalian hosts. A parvovirus of the tree shrew is known, but this virus is latent. Latent parvoviruses of mice, such as orphan parvovirus, are also known and prevalent in nature, as most wild mice have antibodies against these viruses. Similar latent parvoviruses are known for other wild rodent species. However, the shrew virus is lytic in guinea pigs and mouse cell lines, suggesting that a species jump by this virus could result in acute infections. Species jumping and the ability to cause acute disease also seem to be characteristics of the carnivore parvoviruses, such as canine distemper virus.

The human TT virus represents another family of small DNA viruses and was initially discovered in the blood of human patients that had hepatitis. Although it was initially suspected to be involved in human disease, subsequent analysis indicated that it did not contribute to hepatitis or any other known disease. The virus is instead maintained as a highly prevalent, inapparent, and stable persistent infection of humans. Very similar viruses can also be isolated from chimpanzees, especially after exposure to hepatitis A and C viruses. In one evaluation of human patients with hepatitis it was observed that 29 of 99 patients were infected with TT virus. Thus, TT virus often behaves like satellite viruses and is frequently associated with mixed virus infections. Four new variants of TT viruses have recently been reported: A, M1, Mz, and M3. The M3 viruses are the chimpanzee-specific variants. Like JCV, the natural TT virus variants are distributed among human populations. TT virus type 1 is found in Asia, whereas TT virus type 3 is found in Africa and not Asia. TT virus infection appears to be acquired during childhood as an upper respiratory tract infection transmitted from infected adults. The virus is persistently shed in saliva. Overall, TT viruses seems to be old and stable viral parasites of humans.

In summary, the majority of small DNA viruses (adenoviruses, papillomaviruses, polyomaviruses, and TT viruses) have persistent life strategies, are host specific, and are phylogenetically congruent with the evolution of their hosts. Most often, these viruses are benign, only occasionally associated with acute or proliferative diseases. The notable exception to this generalization is the parvoviruses. The parvoviruses include members that are both persistent and host specific, as well as members (such as the carnivore parvoviruses) that frequently jump species and cause acute disease.

Emergence of RNA viruses in mammals

Persistent RNA viruses of placental species

In contrast to the ubiquity of persistent DNA viruses in humans and other mammals, stable persistent infections with RNA viruses are much less common. Given the frequency with which such agents can cause acute human disease, persistence by RNA viruses is not as common as many might expect. Furthermore, in those cases where RNA virus persistence is prevalent, it tends to be highly restricted to specific host species, especially rodents and bats (see below). Humans are known to support a few RNA viruses that cause stable persistent infections, such as hepatitis C virus. However, the bloodborne nature of transmission of this virus and the results from phylogenetic analysis suggest that this virus may have been recently introduced into the human population (possibly from avian sources) and that hepatitis C virus did not likely evolve along with its human hosts.

Negative-strand ssRNA viruses

Negative-strand ssRNA viruses, such as the myxoviruses, paramyxoviruses, and rhabdoviruses, are known to infect almost all orders of placental mammals. For the most part, these infections cause acute disease, and the viruses do not persist in their hosts. There are, however, some clear exceptions to this situation that I examine.

Rhabdovirus and lyssavirus persistence

The interest in rhabdoviruses stems mainly from the ability of rabies virus to cause highly lethal infections in humans and domestic animals. Lethality from rabies virus is very near 100%. The virus is shed in saliva and is normally transmitted by biting, often by infected (and agitated) carnivores. Virus replication occurs in the peripheral nerves of infected animals, allowing virus to migrate up the nerves to the CNS, where infection induces aggressive behavior and leads to an inevitably lethal disease. Because of the slow pace of this transmission, postexposure immunization against rabies virus (first established by Louis Pasteur) can be lifesaving. As mentioned above, *Rabies virus* is a member of the rhabdoviruses, a large group of viruses that includes lyssaviruses. The evolution of these negative-strand ssRNA viruses, which lack recombination, is mainly due to a point mutation process, gene duplication, and gene deletion. Thus, it is possible to infer the long-term evolutionary relationships among these viruses. Recent phylogenetic analysis of the entire rhabdovirus group suggests that *Mokola virus,* which is found in African shrews, is the most basal member. Mokola virus was isolated from insectivorous shrews, and this virus has the unique distinction among the rhabdoviruses of also being able to replicate in insects (e.g., *Aedes aegypti*). This dual host biology suggests that rhabdoviruses initially adapted to mammalian hosts from insect hosts. The rhabdovirus families are organized into genus level subfamilies. The *Lyssavirus* genus consists of two phylogenetic groups (phylogroups) that are associated with their corresponding vector hosts (carnivore and chiropteran). Phylogroup I contains the African *Mokola virus* and *Lagos bat virus* (of fruit- and insect-eating bats). Phylogroup II includes the Australian bat viruses as well as the worldwide *Rabies virus*. Figure 8.4 shows a phylogenetic dendrogram that outlines rhabdovirus evolution. Sequence analysis suggests that emergence of *Rabies virus* worldwide in carnivores was a relatively recent evolutionary event and may have occurred as recently as 900 to 1,500 ybp. The larger worldwide rabies "group" of viruses appears to have originated about 4,000 ybp. If so, this would further suggest that human culture was likely involved in the origination of modern *Rabies virus*. For the most part, the virus seems to have evolved mainly by host switching, which is most often associated with glycoprotein changes. The lyssaviruses also appear to have often switched hosts during evolution. Like infections with rabies virus, other lyssavirus infections of carnivores are generally lethal. However, lyssaviruses are also known to persist as asymptomatic infections in specific bat species. This would

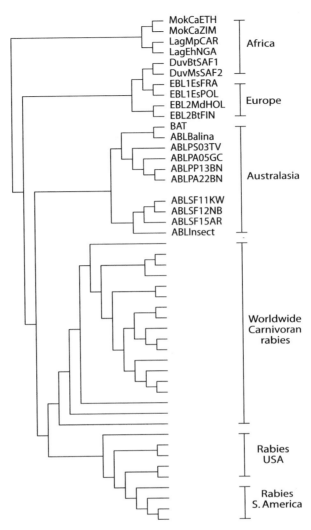

Figure 8.4 Phylogenetic relationship of rhabdoviruses. Data are from H. Badrane, C. Bahloul, P. Perrin, and N. Tordo, *J. Virol.* 75:3268–3276, 2001.

suggest that bats are the evolutionarily stable source of acute lyssavirus infections in other species.

Bats are chiropterans, which are placental species that diverged early from the other placental lineages and thus represent a rather basal placental group. Although bats are found throughout the world, in Australia they represent one of the few native placental species in an otherwise marsupial-populated habitat. Thus, it is highly interesting that all common Australian bat species appear to harbor persistent lyssaviruses. Why might there be a link between bats and lyssavirus persistence? Phylogenetic analysis indicates that Australian bat lyssaviruses are monophyletic (based on G-protein similarities) and can be separated from other lyssaviruses. These bat viruses appear to be rather stable genetically and can be grouped into host species-specific clades. This is unlike the other acute rhabdoviruses, whose clades show much greater genetic diversity. Thus, the lyssaviruses, although strictly acute and highly lethal in numerous species (especially carnivores), appear to maintain an evolutionarily stable relationship as a persistent infection of bat species.

Paramyxovirus persistence

Paramyxoviruses are another group of negative-strand ssRNA viruses that infect almost all placental species and include viruses such as measles, mumps, canine distemper, rinderpest, and respiratory syncytial viruses and the avian paramyxoviruses. Like the rhabdoviruses discussed above, these paramyxovirus infections tend to be acute and disease associated. Sequence relationships among all of these paramyxoviruses can readily be seen, especially in the polymerase genes. The most basal member of the mammal-specific members appears to be *Sendai virus*. In spite of the broad array of species that can be infected by paramyxoviruses, there are very few examples of persistent infections that are part of the normal biological strategy for any of these viruses. They almost always establish acute infections. However, there is at least one placental host that can be persistently infected with paramyxoviruses: tree shrews infected with Tupaia virus. This virus establishes a silent and persistent infection in its native host and shows no antigenic cross-reaction to the other paramyxoviruses. However, Tupaia virus does show clear resemblance to Hendra virus. Hendra virus is a paramyxovirus (a morbillivirus) that was isolated in Australia. Its genome has a nonsegmented negative-strand ssRNA template of 18.2 kbp, which is large. This is the most complex of the paramyxovirus genomes, and other paramyxoviruses have a more uniform genome size. This virus was observed to cause acute and lethal outbreaks in horses and the humans who were in contact with infected horses. It now seems that Hendra virus had jumped from other species into domestic animals and humans. Based on conservation of the L protein domain, Hendra virus is related to *Nipah virus,* another paramyxovirus that has caused a major epidemic in pigs in Malaysia. Infected pigs were able to transmit Nipah virus infections to humans, resulting in about 100 human deaths. The specific source of this outbreak has not yet been identified. However, a broader epidemic was apparently prevented by culling millions of pigs. Flying foxes (genus *Pteropus*) are a natural persistent host for Hendra virus in Australia and harbor the closely related Nipah virus in Malaysia. These two viruses resemble both the Ebola viruses (which are filoviruses) and *Measles virus* (a paramyxovirus) in their genetic organization, suggesting that all of these viruses may share a common evolutionary background. It is interesting that like the lyssaviruses, morbilliviruses (a genus of paramyxoviruses) also cause many lethal infections in carnivores by inducing a distemper-like disease, associated with increased aggressive behavior. This suggests that these viruses might have some involvement in the predator-prey relationships of their hosts.

In summary, the negative-strand ssRNA viruses include a large number of viruses that cause serious acute diseases in most mammalian species. While most of these infections are acute, some persistent host infections are also known. The rhabdoviruses all appear to have common ancestors, which may have been insect-specific viruses. The paramyxoviruses also appear to have one common ancestor. Most of the acute infections caused by these viruses appear to have adapted or evolved from persistent infections in specific hosts. Bats in particular appear to be prone to persistent infections with negative-strand ssRNA viruses.

RNA viruses of rodents: an untold story of inapparent virus

Rodent evolution, phylogenetics, and RNA viruses

Excluding humans, rodents are by far the best-studied placental species. Rodents, especially *M. musculus domesticus* (the lab mouse) and *Rattus rattus* (the lab rat), have also been extensively examined with respect to their viruses. As mentioned previously, rodents comprise about one-half of all placental species. Within the rodents, the murid family (Muridae) contains the most species. Although rodent-like mammals are very ancient, the modern rodents have evolved much more recently. Most modern rodents have adapted to eat seeds, which links them to the evolution of modern grasses. All rodents evolved from shrewlike carnivorous or insectivorous ancestors and have since diverged into various families and subfamilies. For example, the Muridae family, which is phylogenetically monophyletic, has diverged into several subfamilies (see Fig. 8.5). The subfamily Murinae represents an early branch (9.8 million ybp) and contains various Old World species, including the *M. musculus* and *Rattus* species that are the current laboratory standards. New World rodents are part of the Sigmodontinae subfamily and have more recently diverged (5.7 million ybp). This subfamily contains the *Peromyscus* species (deer mouse) studied in the Americas. The most recently diverged subfamily of the Muridae is the Arvicolinae (3.6 million ybp), which is further divided into Old World and New World lineages. Of all subfamilies of the Muridae, the Arvicolinae is the most diverse in species, including lemmings and voles. A recent radiative expansion seems to have occurred in this subfamily. This last evolution of rodent species has been especially rapid and recent compared to the evolution of the primates.

The laboratory mouse: so much from so few

The laboratory mouse, *M. musculus domesticus,* is a member of the Murinae. The *Mus* genus itself has an extensive lineage history, which originated from a very small genetic pool. There are 30 to 40 *Mus* species that originally inhabited the Old World (Europe, Asia, and Africa). The genus *Mus* is divided into four subgenera: *Coelomys* (southern Asia), *Pyromys* (southern Asia), *Nannomys* (Africa), and *Mus* (Europe, northern Africa, and Asia). The subgenus *Mus* includes *Mus caroli, Mus cervicolor, Mus cookii,* and

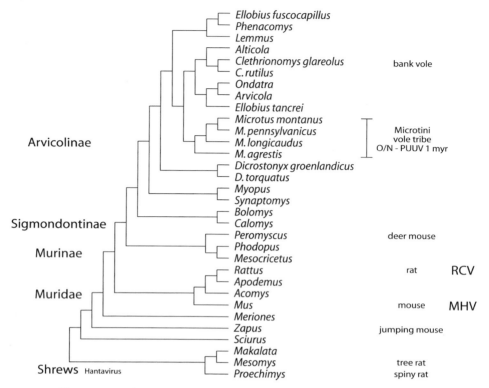

Figure 8.5 A dendrogram schematic of the evolutionary relationships of the Muridae family of rodents, based on mitochondrial DNA and cytochrome B sequence data. Data are from C. J. Conroy and J. A. Cook, *J. Mammal. Evol.* **6**:221–245, 1999. Also shown are some common names as well as patterns of hantavirus (Puumula virus [PUUV]) and coronavirus (rat coronavirus [RCV] and MHV) rodent infection.

M. musculus, as well as the aboriginal *Mus spretus*, *Mus spicilegus*, and *Mus macedonicus*, all of which live in natural field settings. The commensal, human-associated (semidomesticated) subspecies of *M. musculus* (the house or lab mouse) are *M. musculus domesticus* (from western Europe and the Middle East), *M. musculus musculus* (from Eastern Europe), and *M. musculus castaneus* (from southeast Asia). The domestic *M. musculus molossinus* (found in Japan) appears to be a permanent hybrid cross between *M. musculus musculus* and *M. musculus castaneus*. The Americas were colonized by *M. musculus musculus* and *M. musculus domesticus* with the European colonization, and these immigrants differed significantly from the native mouse species of the Sigmodontinae and Arvicolinae. All three of the domestic *M. musculus* subspecies (*M. musculus musculus*, *M. musculus domesticus*, and *M. musculus castaneus*) appear to have initially evolved in the northern Indian subcontinent and diverged about 600,000 ybp. The great bulk of scientific study has been done on these domestic *Mus* subspecies, and the current lab strains are derived mainly from a very small population of breeding stocks originating in Japan. Although these strains were quickly crossed with European mice, the extant strains now harbor mainly *M. musculus domesticus* genes, *M. musculus domesticus* mitochondrial DNA, and mixed *M. musculus musculus* and *M. musculus castaneus* Y chromosomes. In 1909, C. C. Little established the first inbred mouse strain, DBA (which has the coat color alleles dilute [*d*], brown [*b*], and nonagouti [*a*]). The BALB inbred mouse strain was established in 1913 by Columbia University graduate student H. J. Bagg from an albino line, and after generation F_{26}, the "/c" was added at Jackson Laboratories to yield the common BALB/c strain. BALB/c and DBA mice have been used, along with outcrosses, to establish most of the lab mouse lines in use today. As a consequence, feral *Mus* species have not often been systematically studied, especially with respect to virus infections. However, one major field study in Australia (where the introduced *Mus* became a pest species) showed a high prevalence of coronaviruses (such as mouse hepatitis virus), a gammaherpesvirus (mouse cytomegalovirus), and mouse parvoviruses. These infections were all asymptomatic, and acute virus infections were rare. Also, wild species of the Sigmodontinae (New World) and Arvicolinae (New World and Old World) have been extensively studied with respect to hantavirus and arenavirus infections.

Hantaviruses and persistence in rodents

Rodent field studies, although limited in number, have led to the realization that many wild rodent species tend to harbor asymptomatic persistent infections with various RNA viruses. Two virus families in particular have been studied in this regard. Hantaviruses and arenaviruses (negative-strand ssRNA viruses) are both, in their native rodent hosts, generally maintained as persistent asymptomatic infections. For the hantaviruses, many virus types have been characterized. These viruses can be broadly classified into New World and Old World groups, in keeping with the classification of their rodent hosts. In all cases examined, each particular virus is highly specific to its host, is genetically stable, and is distributed in a geographically restricted way. Furthermore, the virus and host rodent clades are congruent, and the viruses and hosts are coevolving with essentially the same slow molecular clock (which is much slower than seen for acute RNA viruses). This tight link between virus and host compels the hantaviruses that infect genera of the Arvicolinae (such as *Microtus* [voles]) to be distinct from those that infect genera of the Sigmodontinae (such as *Peromyscus* [deer mice]). Since hantaviruses are RNA viruses with the established potential for very fast RNA variation, it seems that the persistent state imposes a restriction on the rate of virus evolution. Northern European hantaviruses can be classified into two groups, both infecting members of the host vole group. Although these viruses and their rodent hosts are generally coevolving, there is also evidence that occasional host species jumps have occurred, establishing hantavirus infections in new rodent lineages. In humans, hantaviruses tend to cause serious respiratory and renal diseases. However, human-to-human transmission has been very inefficient, so humans tend to be dead-end hosts for these viruses. Table 8.3 summarizes hantaviruses of rodents and identifies those that have established an ability to cause human disease. Although hantaviruses are also known for some shrew species, these viruses have not been well studied, nor is it clear if these viruses are basal to the hantaviruses of other rodents. Hantaviruses are not known to establish persistent infections in hosts other than rodents.

Arenaviruses

Arenaviruses are ambisense (having both positive- and negative-strand segments) ssRNA viruses that are responsible for various types of hemorrhagic fevers in humans. These human infections are generally disease associated and strictly acute. However, in rodents arenaviruses are known for their ability to establish asymptomatic persistent infections, and persistently infected animals can be the source of acute infections of other species. In Sierra Leone, for example, the consumption of asymptomatically infected rats has been a major source of human infection with Lassa virus. The best-studied member of this family is *Lymphocytic choriomeningitis virus* (LCMV) of *M. musculus*. This virus is ubiquitous in natural mouse populations and establishes lifelong persistent infections. Similar to the hantaviruses, the arenaviruses are classified into two groups. One, the *Lymphocytic choriomeningitis virus-Lassa virus* group, is monophyletic, with three distinct lineages. The other group, the Tacaribe viruses, is also monophyletic, with three lineages. These groups correspond to respective New World and Old World rodent hosts. And reports show that the arenaviruses are coevolving with their native rodent hosts. In contrast, the pathogenic arenaviruses are not monophyletic, show multiple independent origins, and have high rates of genetic change. Thus, although we see a New World-Old World congruence between virus and host, there is also evidence that some stable species jumps between rodents have occurred. However, there is no evidence that arenaviruses have established stable persistent infections in nonrodent hosts, and even the acute infections in nonrodent hosts appear to be unstable, requiring rodents as reservoirs to continue the chain of transmission. There

Table 8.3 Major examples of persistence emergence

Virus	Comment(s)
Smallpox virus	Emerged following development of agriculture (Indian subcontinent; written documentation in Sanskrit)
	An acute-only, human-specific virus; no known reservoir in other species
	Basal virus is cowpox virus (based on accessory genes). Cowpox virus has stable, asymptomatic persistence in rodents (e.g., bank voles); causes acute infection in rats and mice; is often transmitted to humans via cats (FeLV); and can be fatal to HIV-infected humans.
Influenza virus	Reservoir is avian (waterfowl) or domestic animals.
Hantavirus	Old World and New World rodents (e.g., bank voles) are hosts.

are no primate versions of arenaviruses known, for example. We have no explanation for the restriction of both the persistent hantaviruses and the arenaviruses to rodents.

Current issues in emergence, persistence, and human evolution

Recent examples of human viral epidemics resulting from persistent infections in other species

The epidemiological concept of a reservoir species has been used for many decades to help explain the recurrence of viral disease after it has apparently been eliminated from a population or habitat. Although this has been a "practically" helpful concept, it has also caused confusion in its oversimplification. Because this concept fails to indicate a distinct difference in the persistent virus life strategy and the acute life strategy, with correspondingly distinct differences between the viral fitnesses and the evolutionary relationships between virus and host, it has led generations of infectious-disease students to think of viral fitness and evolution simply from the context of acute disease. Thus, when a new disease emerges, it is simply assigned to having emerged from some "reservoir animal," as if that is sufficient to explain the origin of a fully intact virus system. In this book I have attempted, through numerous examples spanning all classes of virus and host, to correct this misunderstanding and establish the central importance of persistent infections to the evolution of viruses, their stability in their hosts, and how persistence contributes to the emergence of a more simple virus-host relationship—acute replication. Because stable viral persistence generally imparts some fitness consequence to its infected host, persistence is not evolutionarily neutral. The virus is not simply a selfish replicator. Persistence matters to host survival. It now appears that we can account for almost all examples of recent human (and domestic animal) viral epidemics as the results of species jumps and adaptations of viruses, which were and are stable persistent viruses in other species, to humans or domestic animals. These are not simply isolated events. The transmission of these agents to new species appears to be an ongoing and never-ending process. Even within the same host species, if two populations become sufficiently isolated, they can and often do acquire distinct persistent viral agents. And this presents the opportunity to initiate acute viral infections when these isolated populations reconnect.

Thus, acute viral infections, especially of humans, originate from persistent infections. As I have shown, with close examination there seem to be few, if any, exceptions to this persistent origin. Yet, there are, however, some seemingly clear examples of acute human viral infections for which we have no obvious source of persistent virus from some other host. The human cold viruses, such as rhinoviruses (a type of picornavirus), appear to be one such example. Given that there are about 100 types of human rhinovirus alone, how could all of these have derived from other hosts? And yet, closer examination of existing sequence data can still identify potential persistent sources for these viruses. For example, Theiler's virus (a picornavirus) causes a ubiquitous, generally asymptomatic and naturally persistent enteric infection of mice. The RNA sequence of this virus appears to represent the basal member of the entire rhinovirus family, leaving open the possibility that this mouse virus also represents the ancestor that originated human rhinoviruses.

I have already examined the major emergent human viral epidemics of the last century from the perspective of virus persistence. Thus, I can propose, with good experimental support, the idea that the influenza pandemic of 1918 derived from a persistent avian infection having the H1 genome segment, following reassortment. A similar scenario can be proposed for the source of the 1957 influenza pandemic (H2N2) and also the Hong Kong influenza outbreak of 1997 (H5N1). In both cases, the persistent "reservoir" for the genes was waterfowl (geese). Given that waterfowl harbor 15 H types and 9 N types, whereas humans harbor only 3 types of each gene, we can expect that persistently infected waterfowl will continue to be the major source of influenza virus genes that can adapt to infect humans. This appears to be an almost inevitable consequence of humans, domestic animals, and waterfowl living in close proximity. Similarly, as outlined above, both HIV-1 and HIV-2 are now well supported as having originated from retroviral genomes persisting in other (monkey and then chimpanzee) hosts. Very recently, there was an outbreak of monkeypoxvirus in the United States, causing disease in humans and prairie dogs. Even this outbreak could be traced to the importation of an asymptomatically and persistently infected giant Gambian pouched rat from Africa. Table 8.4 summarizes some major

Table 8.4 Hantavirus emergence from rodent- or shrew-specific persistence[a]

Common name of animal	Subfamily	Persistent hantavirus (rodent or shrew specific)	Human disease	Human-to-human transmission
Striped field mouse	M	Hantaan virus	HFRS	NA
Yellow-neck mouse	M	Dobrava-Belgrade virus	HFRS	NA
Norway rat	M	Seoul virus	HFRS	NA
Bank vole	A	Puumala virus	HFRS	NA
Deer mouse	S	Sin Nombre virus	HFS	NA
White-footed mouse	S	New York virus	HFS	NA
Cotton rat	S	Black Creek Canal virus	HFS	NA
Rice rat	S	Bayou virus	HFS	NA
Long-tailed pygmy rice rat	S	Andes virus	HFS	Yes
Meadow vole	A	Prospect Hill virus	NA	NA
California vole	A	Isla Vista virus	NA	NA
Western harvest mouse	S	El Moro Canyon virus	NA	NA
Musk shrew	So	Thottapalayam virus	NA	NA
Bandicoot rat	M	Thailand virus	NA	NA

[a]M, Murinae; A, Arvicolinae; S, Sigmodontinae; So, Soricidae; HFRS, hemorrhagic fever with renal syndrome; HFS, hemorrhagic fever syndrome; NA, not applicable.

examples of emergent human diseases originating from stable persistent infections of other species.

Persistence misstudied

Unfortunately, because the above cases have all been so focused on the acute pattern of virus replication, we still essentially lack any understanding of the mechanisms by which these viruses persist in their natural hosts. Our general operating assumption is that viral genes have, for the most part, evolved to function in an acute mode, that is, to make more virus and not to dampen the virus replication in order to persist. Thus, we are often surprised to learn that acute viral diseases, recently introduced from reservoir hosts, are often associated with an overactive or toxic host immune response and not with toxicity resulting directly from virus replication. Hantavirus severe respiratory syndrome appears to be a good example of exactly this situation. Hantavirus replicates poorly in human lung tissue, but it induces extensive cytokine-mediated inflammatory or oxidative damage. In fact, there may be some general mechanistic principles of disease induction that apply when a virus, evolved to persist in one host, finds itself able to replicate acutely in a new host. It may well be that the normal dampening mechanisms of persistence (such as signal transduction or immune modulation) are often species specific, selected to dampen virus replication in the natural species but no longer functional in the new host species. After all, if a virus fails to grow well in culture, virologists generally assume that they have failed to find the right conditions for virus growth, as opposed to assuming that persistence or silence is in fact the expected and normal outcome of the infection. A process will then be undertaken to select for the version of the virus, or host cell, that will coax the virus to replicate efficiently and kill host cells, thereby losing the very gene functions that were necessary for persistence. The isolation of CMVs bearing the UL144 gene (a mini-tumor necrosis factor receptor sequence) from clinical, but not lab, sources may be an appropriate example of just such an inadvertent selection. Such dispensable genes are generally considered "accessory," as they do not contribute to acute virus replication, although they can be highly conserved in nature. This observation may also apply to explain the current dilemma of why the poxviruses that persist in rodents (such as cowpoxvirus or MOCV) have so many additional immune modulators compared to the strictly acute viruses, like smallpoxvirus of humans. Persistence is not an accessory function. It is a most basic virus life strategy and deserves to be studied as such.

SARS: a most recent example of emergence, and the future

Prior to 2003, the virologists who study coronaviruses had a real problem with obtaining federal grant monies. They needed to carefully justify why anyone would fund the study of a family of viruses that cause little human disease (aside from colds). For the most part, they struggled to make this justification. The best-studied member of this family, Mouse hepatitis virus, caused little disease in natural *M. musculus* populations, but it could be coaxed by virologists with some mutation of the virus into causing neurological disease. All of this changed in March 2003, when, in southern China, a new coronavirus emerged for the first time to infect humans and cause a serious, often lethal disease of the human respiratory system, severe acute respiratory syndrome (SARS). SARS was quickly established to be caused by a family of coronaviruses that had not previously been known. Due to rapid and effective quarantine measures, a SARS pandemic was prevented. However, the question became, Where did this fully formed and new fully intact member of the coronavirus family come from? As all the genes appeared to be equally different from those of existing coronavirus strains, the possibility that the SARS virus either represented a recombinant virus or had acquired a set of mutations from a known coronavirus was quickly dismissed. The virus seemed to have appeared relatively intact from an unknown source. The search was then initiated for an animal reservoir of SARS virus, and several species of carnivores inhabiting the region (e.g., civets and raccoon dogs) were identified to be asymptomatically infected with a SARS-like virus that was very similar to the one infecting humans. It has yet to be established if these carnivores are the stable source of SARS infection, or if they too acquired the infection from some prey. However, it is clear that the virus is in the ecosystem.

Coronaviruses are likely to be an ancient family of virus. Gill-associated virus of prawns shows limited similarity to the SARS virus replicase but is clearly a member of the same general virus family. This virus of shrimp is the simplest member of this family, the nidoviruses. Although gill-associated virus is highly pathogenic in shrimp farms, it is found as a persistent asymptomatic infection in wild shrimp populations. It thus seems likely that the nidoviruses are ancient viruses that have infected vertebrates for a long time. However, if this evolutionary stability has been maintained by persistently infected hosts, we know little of what those hosts might be. Coronaviruses are common in feral *Mus* species but do not appear to infect natural populations of New World rodents. But this issue is poorly studied. Most of the known nidoviruses (coronaviruses) do not establish persistent infections in their corresponding hosts (e.g., humans and turkeys), but some of these viruses (e.g., those of cats) do. Thus, it seems almost certain that the SARS virus is an ancient persistent virus (possibly specific for a carnivore or its prey) that has adapted as an acute agent of the human respiratory tract. Thus, the emergence of SARS virus appears to fit well with the pattern of acute adaptation from a persistent virus life strategy.

Factors that favor emergent viruses

The capacity of a persistent virus to adapt to become an acute viral agent in another host seems to be a never-ending ability. As long as there are multiple species that can share viruses, the possibility of virus adaptation appears ever present. Related species would almost always seem at the highest risk of such virus exchange, since the new host habitat would be close to what the jumping virus needs. If this situation is extrapolated, it suggests that virus persistence may well affect competition between host species. It may be that the most successful species could be the ones colonized with the appropriate persistent viruses, which could acutely infect competing species. This idea brings to mind the observations presented early in this book concerning the mixing of bacterial strains harboring different prophages, leading to the lysogenic death of uninfected bacterial strains. Clearly, virus persistence can affect host species competition.

Host populations

The successful growth, or overgrowth, of any one population of host species appears to increase the probability of acquiring acute viral agents. Large host populations create a population dynamic with increased contact rates that predispose this successful population to the possible adaptation of and infection by a persistent virus in another species. The more successful the host, the greater the number of acute viral agents that are likely to adapt. Thus, we expect that success itself of a population brings with it an increased risk of contact with other species harboring persistent viruses. The current human population, with its highly dense distributions throughout the world and high migration rates, is likely to have high rates of interspecies contact and acute viral agent adaptation. Also, any situations that bring together previously separated species or situations that lower biological barriers will increase the likelihood that a persistent virus of one species will adapt to cause an acute infection of another species. Thus, the placement of African and Asian elephants adjacent to one another in zoos can be expected to allow the exchange of their corresponding persistent herpesviruses. Similarly, the mixed farming of related shrimp species might also result in virus transmission.

Biological barriers

Besides changes in population structure and exchange rates, which are mainly social, changes in biological barriers can also contribute to disease emergence. Factors that lower immunological status or physical biological barriers are expected to increase the probability of adaptation of a persistent virus of one host to become an acute agent of the altered host. For example, the human population now contains a large number of HIV-infected people. In HIV-induced AIDS, HIV lowers cellular immunity, thereby increasing the likelihood of reactivation of many persistent human viral agents and increasing the probability that these agents will adapt to become acute agents in HIV-infected individuals. The increased rate of human infections with monkeypoxvirus in the Democratic Republic of the Congo (where there are high rates of HIV infection) may be exactly such a situation. Lowered immunity also allows reactivation of persistent agents. Indeed, such latent virus reactivation is one of the main clinical problems faced by AIDS patients. Besides AIDS, the transplantation of organs can also lead to latent virus reactivation, as was seen for the reactivation of BK virus in transplanted kidneys or hepatitis C virus in transplanted livers. Also, persistent asymptomatic VZV in skin and thymus can reactivate in xenografts. Similar kinds of tissue transplants in mice have also resulted in reactivation of silent persistent infections. For example, the transplantation of mouse brain slices resulted in the reactivation of mouse CMV in 75% of the subjects, most likely due to stem or progenitor cell proliferation. Xeno-transplantation can also lead to persistence reactivation and acute transmission to other species. For example, pig kidneys reactivate CMV when implanted into baboons. In addition to these exogenous viral agents, most mammalian tissues also have the possibility of the reactivation of ERVs in the germ line, such as PERVs in pig tissues. Thus, placement of tissues into new hosts represents a novel avenue for the transmission of persistent virus that was not available prior to human intervention. It should be noted that even if no additional or outside species were potential sources of emergent viral agents, in some cases an organism's genome itself can provide the virus. I have mentioned before the evolution of leukemogenic retroviruses from an ERV in the AKR mouse line. Similarly, for the Lake Casitas mice I mentioned another emergent virus from the germ line. Thus, emergent acute viruses can develop through reactivation of persistent viruses from both exogenous and endogenous sources.

Factors affecting persistent virus colonization

The main concern expressed above is the occurrence of new acute human epidemic viral agents. The attention of virologists has long been focused on the emergence of just such agents. What about the occurrence of new persistent agents? What do we know about conditions or viral phenotypes that favor their emergence? For the most part, persistent viruses tend to evolve along with their host lineages. Hence, these viruses have mainly previously existing virus-host relationships and tend not to represent new viral agents. Yet, in various orders of organisms there is a characteristic pattern of harboring a particular type of persistent infection. How did these persistent viral infections come about? As persistence often requires specific mechanisms and gene functions to avoid immune elimination, maintain persistence, or allow virus reactivation, this life strategy represents a more complex phenotype than does the acute life strategy, generally involving multiple genes. Consistent with this idea is the fact that the persistent DNA viruses often have a large number of disposable accessory genes that can be deleted in viruses propagated in culture. The complexity of this situation might lead one to expect that colonization of a host by new persistent viral agents is a relatively uncommon and perhaps evolutionarily important event. However, genetic evidence suggests that such new colonization does occasionally happen. In "RNA Viruses of Rodents: an Untold Story of Inapparent Virus" above, I noted evidence that both hantaviruses and arenaviruses have on occasion jumped into other related rodent species and established new lines of virus that persistently infected the hosts. Can anything be said about the factors that might affect this situation? Since there are few experimental results that are relevant to this issue, much of the consideration of this issue must be from a more theoretical perspective. In chapter 7, I mentioned the stable colonization of wild American *Drosophila melanogaster* with retroviral agents. This process was associated with sexual compatibility (hybrid dysgenesis) between two previously isolated species of *Drosophila* and also appeared to have elements of an addiction module (i.e., protected if infected, harmed if not infected). However, for the vertebrate animals, I am hardpressed to point to any experimental observations that demonstrate new colonization by a persistent virus. One possibly related example would be the Lake Casitas feral mice and the role of the ERV in the emergence of the autonomous retrovirus in the F_1 cross of two subspecies. However, this example represents a loss of protection from acute disease through the corresponding sex-mediated loss of the persistent and protecting ERV, not a gain of a new persistent infection. The limited populations of chimpanzees that harbor SIVcpz in Africa might be one example of a new persistent infection.

Although we lack clear experimental examples of the acquisition of new persistent viruses in mammals, this possibility can still be considered from a theoretical perspective. The issues to examine include what conditions favor the acquisition of new persistent agents, what conditions favor the transmission of persistence, and what are the consequences of acquisition to the colonized host. One theoretical conclusion would be that if versions of an acute virus infecting a particular species were highly prevalent, then this would favor the establishment and survival of a persistent virus (generally derived from the variants or defectives of the same acute virus). And this would bestow immunity on the persistently infected host. Since by definition persistence requires the regulation of acute viral replication, in this situation the persistently infected host would be more likely to survive. Thus, persistence acquisition is favored when highly prevalent acute viral agents infect a specific host. The transmission of persistence to uninfected members of the same or related species has often been seen to involve an addiction module- or persistence module-like process. The *D. melanogaster* gypsy retrovirus mentioned previously, for which viral transmission was also sex associated, is an example of this situation. However, as also mentioned, we have no examples of this situation occurring in mammalian species, despite the high-level colonization of mammalian genomes by ERVs. As I have mentioned, the acquisition of ERVs as a way to protect the host against autonomous versions of the viruses has often been invoked to explain the presence of the many ERVs in the genomes of mammals.

Let us consider this ERV acquisition issue as a theoretical extrapolation of the ongoing human pandemic with HIV-1 to see how acquisition of such ERVs and defective ERVs might have come about. If one assumes that there is no intervention by human activities (cultural or scientific), one can envision a resulting HIV-1 epidemic that is not controlled and could even lead to increased rates of HIV-1 transmission. In this circumstance, one can easily imagine how essentially the entire communicating human population might become infected with HIV. As most infected people would be expected to succumb to the infection, those few survivors would be disease nonprogressors that still maintained the virus. That is, the resulting human population would be persistently infected with HIV, and these infections would be asymptomatic. However, a persistently infected mother would still present the risk of infection to her uninfected offspring. This would provide a selective pressure that would favor germ line colonization by HIV and HIV defectives of the offspring's genome, which would establish a persistence module in this child. In order to stabilize this persistence against all the genetic variants of HIV that might prevail, this persistence module would itself need to inhibit the population of HIV variants; hence, multiple germ line colonizations with many slight variants of HIV defectives might be needed to stabilize the state of persistence. The resulting new human species would be inert to HIV infection. In summary, in the situation of a prevalent and lethal viral infection, selective conditions would favor the survival of hosts carrying the virus or its defectives in a persistent state in their genomes. Furthermore, this surviving population would pose a risk of HIV infection to other, naive human populations that had not undergone HIV colonization and selection. Finally, it is worth noting that if such a human species were to evolve, it would have gained various new molecular circuits, such as *vpu, nef,* and *tat,* plus a new *env* gene, all of which would be available for additional Darwinian evolutionary selection in the evolution of the human genome. Do such events really happen? There are some reasons to question this scenario. For example, HIV-1 is a lentivirus, and all lentiviruses have distinct RT sequences in addition to various accessory genes. Although lentiviruses are not highly prevalent in most natural populations, species-specific versions are known. However, there do not seem to be any examples of ERVs that have the lentivirus RT sequence. Most ERVs have MLV-like RT sequences. Perhaps germ line lentivirus integration is precluded for some unknown reason. Yet it is known that lentivirus-based gene therapy vectors are very efficient at germ line integration in mouse transgenic studies. Thus, the question remains, Can one see examples of ERV colonization events that occurred at the origination of the various mammalian lineages that resemble this hypothetical scenario?

Human and primate evolution and ERV acquisition

About a million years ago, the human and chimpanzee primate lineages diverged from a common ancestor somewhere in Africa. With the completion of the sequencing of the human genome, we are now better able to consider what types of genetic changes were associated with this recent stage in human evolution. One of the surprises of the Human Genome Project was learning how relatively little difference there is between the coding sequences of the human genome and the coding sequences of the chimpanzee genome. Based on coding sequences alone, one would be hard-pressed to differentiate human DNA from chimpanzee DNA, as they are are 99% similar. Yet humans and their primate relatives differ in some rather major and complex ways, especially with respect to cognition and use of language. Can we relate any differences we see in their respective genomes to corresponding differences in their primate phenotypes? What distinguishes human from chimpanzee, and do their respective ERVs have any role in this distinction? Which of these differences are the most recent changes?

The largest and most distinct differences between human and chimpanzee genomes are found in their corresponding Y chromosomes. The recently sequenced human Y chromosome contains a surprisingly small number of genes (about 70 possible ORFs, but perhaps as few as 20 authentic protein genes), and half of these are retroposon associated. The biggest difference between the human and chimpanzee Y chromosomes occurs mainly in the pattern of retroposon or ERV colonization, not in the coding sequences. This ERV difference also applies to the autosomes, in that all other human chromosomes also show evidence of recent HERV colonization. Overall, the human genome has about twice the content of HERVs and LINEs as does the chimpanzee genome. In addition, the highly numerous human *Alu* elements appear to have undergone an expansion corresponding to (and probably synchronized with) the divergence of human and chimpanzee lineages. This expansion most likely resulted from the transposition-mediated action of LINEs. Thus, in the African primates there must have occurred a major event leading to human ERV colonization. And it is the HERV-K family of viruses (and their SINE derivatives) that most distinguishes the human genome from the chimpanzee genome. What evolutionary forces might have caused this major colonization event in the genomes of the African primates?

In earlier chapters, it was argued that hosts which are susceptible to viral parasites tend also to be more dynamic and more able to generate new host species. Is there any evidence that at some point during their evolution the African primates became more open to viral parasites, which could have increased their rates of ERV colonization? There are some distinct viruses in African monkey species that should have affected all African primates. The most obvious of these, from the perspective of retroviruses, are the lentiviruses of African monkey species, the SIVs. Most wild African (but not Asian or New World) monkey species appear to harbor prevalent infections with species-specific versions of SIV. These include the African green monkey, the sooty mangabey, the L'Hoest's monkey, and the nose-spotted monkey. All of these species are chronically infected with their own phylogenetically stable versions of SIV and produce these SIVs at high levels without showing signs of disease. In the case of the African green monkeys, three monkey species, all of which support their own versions of SIV in an inapparent manner, are known. Yet these viruses, when used to infect the closely related Asian macaques, produce an AIDS-like disease. The relationship between the SIVs and their African monkey hosts appears to identify an important new development in retrovirus-host interactions. In sharp contrast to the other retroviruses discussed so far, these chronic SIV infections are highly productive and are not silent persistent infections. Thus, these monkeys are like lentivirus factories, making high quantities of virus throughout their lives. We now know that these SIVs have frequently undergone genetic alterations and have jumped species to infect various other African primates, including humans (i.e., HIV), in which they often cause disease.

There is reason to believe that the monkey lentiviruses may have been involved in the HERV-K colonization that is characteristic of all African primates. In contrast to the ERVs of other host lineages, primate HERV-Ks encode a functional dUTPase. Thus, primates can express a cytoplasmic dUTPase activity. It is therefore highly curious that the primate lentiviruses can be distinguished from the nonprimate lentiviruses (e.g., feline or small-ruminant viruses) based on their lack of dUTPase genes. That is, the primate lentiviruses alone do not code for a viral dUTPase. What might be the role of dUTPase with respect to retrovirus infection? In 1996, E. M. McIntosh and R. H. Haynes proposed a most intriguing idea along these lines. They suggested that the cytoplasmic synthesis of cDNA by the lentiviral RT will normally be poisoned by the presence of dUTP. If so, then cytoplasmic expression of a dUTPase would be crucial for the high-level virus replication and chronic infection seen for the primate lentiviruses. HERV-K colonization may have been the genetic event that allowed such high-level SIV replication to occur. Thus, there may have been a type of complementation between HERV-Ks and autonomous retroviruses such as SIV. The result was a new lineage of African monkeys that were highly prone to chronic retroviral infections, thereby driving the evolution and speciation of their African primate hosts.

Human cognition and ERVs

The acquisition of human language and the associated changes in human cognition are the most complex distinctions between humans and chimpanzees. It is thought that brain lateralization is involved in this processing of language, as language is clearly processed in an asymmetric manner by the human brain. Brain lateralization and language acquisition are also believed to be related to the acquisition of abstract thought. Currently, this is a very poorly understood process, so little can be said about its underlying mechanisms or how it might have come about during human evolution. However, it also appears that abstraction and brain lateralization are related to a uniquely human disease, schizophrenia (which appears to be absent from other primates). Some feel that schizophrenia and psychosis are associated with brains that are closer to symmetry and thus associated with less lateralization. All human populations are prone to schizophrenia at similar rates, so no environmental agents appear to be involved. Schizophrenia strongly reduces attachment behavior. Although there is a clear genetic component to the disease, its nature is not known. Lateralization is developmentally controlled, but how this might be perturbed to produce schizophrenia is not clear. Its onset postpuberty demonstrates a clear association of schizophrenia with sex development, and schizophrenia can be seen in XO and XXY humans. There is, however, also a curious link between schizophrenia and human-specific HERV-K family sequences. Affected regions of schizophrenic brains have been shown to express HERVs and LINEs (such as MSRV and SINE-R.C2) in the frontal lobes at high levels (10- to 50-fold above the levels in similar regions in healthy control brains). There is a human-specific HERV-K(10) family agent found at high copy level on the X and Y chromosomes. Figure 8.6 presents a phylogenetic tree that shows the relationships of SINE-R.C2, Schizo cDNA, and HERV-K(10). As the role of HERV or SINE-R.C2 expression in brain function is not known, a causal or mechanistic role for the ERVs in human cognition cannot now be proposed. That is, HERV expression might be the result, not the cause, of the disease. The congruence of these circumstances is highly intriguing.

Complex behavior and viruses

Other aspects of complex human behavior also pose a major evolutionary puzzle. However, to most readers, the idea that viral genetic parasites might be involved in the evolution of such complex human behavior may seem ludicrous. Nonetheless, I will discuss some of the most human traits from the perspective of possible virus involvement. One such puzzle is the acquisition of associative learning, which relates to the learning and development of social attachments and the development of human cooperation and society. Such cooperation underlies familial, tribal, cultural, religious, and social cohesion, as well as altruistic behavior. It has recently been proposed that other extinct human lineages (e.g., the Neanderthals) may have lacked the high degree of social cooperation present in modern humans. What might be the genetic changes that predisposed modern humans to such behaviors? All of the mentioned social situations represent cooperative group activities or systems that are supportive of, pleasurable for, and/or beneficial to participants of the group, whereas they are harmful, painful, or punishing to

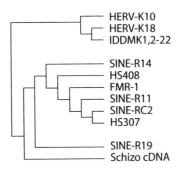

Figure 8.6 Dendrogram schematic of the HERV elements associated with schizophrenia by subtractive cDNA analysis. Relationship to the HERV SINE-R.C2 is shown. Data are from H.-S. Kim, R. V. Wadekar, O. Takenaka, C. Winstanley, F. Mitsunaga, T. Kageyama, B.-H. Hyun, and T. J. Crow, *Am. J. Med. Genet.* **88:**560–566, 1999.

individuals outside the group (features reminiscent of addiction modules). In fact, it has recently been reported that social rejection activates the same region of the brain as does physical pain. What is the evolutionary force that drives the acquisition of such behavioral complexity? One experimental model that seeks to understand the mechanisms of associative learning (via attachment behavior) is the study of partner preference formation in male prairie voles. Mating-dependent learning by these voles results in lifelong monogamous bonding between mates. Vasopressin receptors and dopamine are known to be involved in this learning, possibly affecting both reward and anxiety circuits. Brain imaging studies indicate that this attachment learning affects the same region of the brain as is affected by drug addiction. Intriguingly, infection of the affected region of vole brains with a recombinant adenovirus expressing the vasopressin receptor results in the same attachment behavior. Clearly, humans are not voles, nor is human sexual behavior so simple. Yet the human genome has, in recent evolutionary times, been colonized by various viral or virally derived agents. I have already noted situations in which viruses can colonize and manipulate the complex sexual behaviors of their hosts, sometimes via hormonal control of host nervous tissue (e.g., the very successful polydnaviruses and other viruses of parasitoid wasps). Because persistent viruses can be highly dependent on specific sexual behaviors of their hosts, there is an inherent link between virus survival and host sexual behavior. It thus seems plausible that an ancient virus could also have affected the evolution of human behavior by contributing addictive behavioral elements that were responsive to specific environmental stimuli. However, although the human CNS has and expresses many such elements, I know of no specific viral agent that can currently be identified as a candidate that could be used to experimentally evaluate this idea.

With respect to the general consequences of viral agents to human survival, human social behavior matters greatly because survival of the social group is strongly affected by altruistic group behavior. It is the social or group behavioral responses to the HIV and SARS epidemics that have mainly been responsible for limiting human deaths due to these agents. In a sense, this behavior is a successful "persistence phenotype" that has precluded other competing viral agents from succeeding in human hosts. Thus, human behavior is a most powerful and adaptive phenotype with respect to the unending threats posed by emergent human viruses.

Role of viruses in the tree of life: a never-ending story of creation

This book has examined how viruses influence the evolution of life from the perspective of acute and persistent viruses. From the prebiotic beginnings of replicator molecules and replicating programs to the first cells, from the origin of the eukaryotic nucleus to multicellularity, from worms to humans, I have surveyed how viruses influence their hosts and provide a never-ending and dynamic environment for the weaving of life. All life has been touched by viral influences, and most genomes clearly show the lasting evidence of viral footprints. These footprints represent those viral genomes that have persisted, and new footprints continue to be left on life's genomes, as our own human DNA can attest. Yet most evolutionary biologists continue to disregard viruses as an element of the tree of life. Viruses were not perceived to be living entities. Yet viruses have all the characteristics needed in order to be subjected to the laws of evolution and to contribute to the fabric of life. And now with the known genomic sequences and the well-documented presence of viruses, we can no longer deny their important role in the evolution of life. However, we still have a long way to go towards understanding the network of life that includes the evolutionary history of viruses. In this regard, viruses confuse us with their multiple identities, mixed lineages, mixed clock rates, episodic or punctuated emergence, and dizzying diversity. We prefer to think of the tree of life as having a coherent and congruent topology. If these viral characteristics are overlaid onto the tree of life, we might be placing a shroud on this tree that confuses us rather than clarifies our understanding. There is clearly a discernible ancestry in the lineage of living species, and the tree-of-life analogy reflects what relationships we can see in living organisms. Perhaps that is why there remains a significant reluctance to include viruses in the tree of life: although their origins and lineages are old and their influence on all life is major, their evolution is complex and not consistent with our accepted tree topology. The inclusion of virus would force us away from the comfortable and useful analogy provided by the tree of life into new, undefined, and seemingly incomprehensible paradigms. However, we can now begin to see that viruses can be involved in and contribute to the origination of species. Viruses, in their competition with themselves, provide a selective pressure that differentiates host lineages and can make previously compatible host populations become incompatible. They can also add new layers of complexity and genetic identity to their hosts in episodic events through persistence and genome colonization. This process would appear to be cumulative, resulting in ever-greater host complexity. Since their discovery over 100 years ago, viruses have been inherently inscrutable, and aside from causing diseases, they remain almost invisible to most biologists. It is time to acknowledge these unseen creators that can and do explore the vastness of sequence space at previously unimagined rates.

An inherently invisible nature of viruses persists to this day. Consider the oceans, the vast ancient cauldron that gave birth to all life on this planet. Few realize that the oceans are also vast cauldrons of viruses with unending, hyperaccelerated rates of evolution. Measurements indicate that the oceans contain about 10^{31} virions or phage particles in total. The great majority of these viruses correspond to large DNA viruses and phages that can acutely and lysogenically infect most bacteria and algae. To get a sense of the vastness of this number, we can estimate that most of these virions will have a diameter of about 100 nm. If laid end to end, these viruses would span the length of the observable universe (about 10^{24} m)! In addition, the great majority of these viruses turn over every day. If this pattern has persisted for the last 3 billion years, which appears likely, then there have been about 10^{43} generations of individual viruses during this period. Given the high rates of recombination and genetic variation inherent in DNA virus replication, we can see that there has been a vast exploration of the sequence space by this viral process of turnover. Furthermore, the successful genomes that have colonized host cells might be expected to persist in the ecosystem and contribute this vast genetic creativity to the tree of life. Perhaps from such a perspective, we can better appreciate why our very own human genomes appear to represent an ocean of ancient retroviral elements.

Recommended Reading

Mammalian Evolution

Benit, L., A. Calteau, and T. Heidmann. 2003. Characterization of the low-copy HERV-Fc family: evidence for recent integrations in primates of elements with coding envelope genes. *Virology* **312**:159–168.

Conroy, C. J., and A. J. Cook. 1999. MtDNA evidence for repeated pulses of speciation within arvicoline and murid rodents. *J. Mamm. Evol.* **6**:221–245.

Dewannieux, M., C. Esnault, and T. Heidmann. 2003. LINE-mediated retrotransposition of marked Alu sequences. *Nat. Genet.* **35**:41–48.

Murphy, W. J., E. Eizirik, W. E. Johnson, Y. P. Zhang, O. A. Ryder, and S. J. O'Brien. 2001. Molecular phylogenetics and the origins of placental mammals. *Nature* **409**:614–618.

Murphy, W. J., E. Eizirik, S. J. O'Brien, O. Madsen, M. Scally, C. J. Douady, E. Teeling, O. A. Ryder, M. J. Stanhope, W. W. de Jong, and M. S. Springer. 2001. Resolution of the early placental mammal radiation using Bayesian phylogenetics. *Science* **294**:2348–2351.

Nonplacental ERVs

Dimcheff, D. E., S. V. Drovetski, M. Krishnan, and D. P. Mindell. 2000. Cospeciation and horizontal transmission of avian sarcoma and leukosis virus *gag* genes in galliform birds. *J. Virol.* **74**:3984–3995.

Dimcheff, D. E., M. Krishnan, and D. P. Mindell. 2001. Evolution and characterization of tetraonine endogenous retrovirus: a new virus related to avian sarcoma and leukosis viruses. *J. Virol.* **75**:2002–2009.

Fadly, A. M., and E. J. Smith. 1997. Role of contact and genetic transmission of endogenous virus-21 in the susceptibility of chickens to avian leukosis virus infection and tumors. *Poult. Sci.* **76**:968–973.

Hanger, J. J., L. D. Bromham, J. J. McKee, T. M. O'Brien, and W. F. Robinson. 2000. The nucleotide sequence of koala (*Phascolarctos cinereus*) retrovirus: a novel type C endogenous virus related to gibbon ape leukemia virus. *J. Virol.* **74**:4264–4272.

Huder, J. B., J. Böni, J.-M. Hatt, G. Soldati, H. Lutz, and J. Schüpbach. 2002. Identification and characterization of two closely related unclassifiable endogenous retroviruses in pythons (*Python molurus* and *Python curtus*). *J. Virol.* **76**:7607–7615.

Iraqi, F., and E. J. Smith. 1995. Organization of the sex-linked late-feathering haplotype in chickens. *Anim. Genet.* **26**:141–146.

Leblanc, P., S. Desset, F. Giorgi, A. R. Taddei, A. M. Fausto, M. Mazzini, B. Dastugue, and C. Vaury. 2000. Life cycle of an endogenous retrovirus, *ZAM*, in *Drosophila melanogaster*. *J. Virol.* **74**:10658–10669.

O'Neill, R. J., M. J. O'Neill, and J. A. Graves. 1998. Undermethylation associated with retroelement activation and chromosome remodeling in an interspecific mammalian hybrid. *Nature* **393**:68–72.

Smith, E. J., and A. M. Fadly. 1994. Male-mediated venereal transmission of endogenous avian leukosis virus. *Poult. Sci.* **73**:488–494.

Tristem, M., E. Herniou, K. Summers, and J. Cook. 1996. Three retroviral sequences in amphibians are distinct from those in mammals and birds. *J. Virol.* **70**:4864–4870.

ERVs and the Placenta

Jurka, J., E. Zietkiewicz, and D. Labuda. 1995. Ubiquitous mammalian-wide interspersed repeats (MIRs) are molecular fossils from the Mesozoic era. *Nucleic Acids Res.* **23**:170–175.

Levy, J. A. 1977. Endogenous C-type viruses in normal and "abnormal" cell development. *Cancer Res.* **37**:2957–2968.

Levy, J. A. 1975. Host range of murine xenotropic virus: replication in avian cells. *Nature* **253**:140–142.

Levy, J. A., J. Joyner, and E. Borenfreund. 1980. Mouse sperm can horizontally transmit type C viruses. *J. Gen. Virol.* **51**:439–443.

Levy, J. A., O. Oleszko, J. Dimpfl, D. Lau, R. H. Rigdon, J. Jones, and R. Avery. 1982. Murine xenotropic type C viruses. IV. Replication and pathogenesis of ducks. *J. Gen. Virol.* **61**(Pt. 1):65–74.

Nelson, J. A., J. A. Levy, and J. C. Leong. 1981. Human placentas contain a specific inhibitor of RNA-directed DNA polymerase. *Proc. Natl. Acad. Sci. USA* **78**:1670–1674.

Revoltella, R. P., and Consiglio Nazionale delle Ricerche (Italy). 1982. *Expression of Differentiated Functions in Cancer Cells*. Raven Press, New York, N.Y.

Seman, G., B. M. Levy, M. Panigel, and L. Dmochowski. 1975. Type-C virus particles in placenta of the cottontop marmoset (Saguinus oedipus). *J. Natl. Cancer Inst.* **54**:251–252.

ERVs and the Live-Birth Hypothesis

Blaise, S., N. De Parseval, L. Benit, and T. Heidmann. 2003. Genomewide screening for fusogenic human endogenous retrovirus envelopes identifies syncytin 2, a gene conserved on primate evolution. *Proc. Natl. Acad. Sci. USA* **100**:13013–13018.

Bromham, L. 2002. The human zoo: endogenous retroviruses in the human genome. *Trends Ecol. Evol.* **17**:91–97.

de Parseval, N., J. Casella, L. Gressin, and T. Heidmann. 2001. Characterization of the three HERV-H proviruses with an open envelope reading frame encompassing the immunosuppressive domain and evolutionary history in primates. *Virology* **279**:558–569.

Espinosa, A., and L. P. Villarreal. 2000. T-Ag inhibits implantation by EC cell derived embryoid bodies. *Virus Genes* **20**:195–200.

Harris, J. R. 1998. Placental endogenous retrovirus (ERV): structural, functional, and evolutionary significance. *Bioessays* **20**:307–316.

Hohenadl, C., C. Leib-Mosch, R. Hehlmann, and V. Erfle. 1996. Biological significance of human endogenous retroviral sequences. *J. Acquir. Immune Defic. Syndr. Hum. Retrovirol.* **13**(Suppl. 1):S268–S273.

Larsson, E., and G. Andersson. 1998. Beneficial role of human endogenous retroviruses: facts and hypotheses. *Scand. J. Immunol.* **48**:329–338.

Mi, S., X. Lee, X. Li, G. M. Veldman, H. Finnerty, L. Racie, E. LaVallie, X. Y. Tang, P. Edouard, S. Howes, J. C. Keith, Jr., and J. M. McCoy. 2000. Syncytin is a captive retroviral envelope protein involved in human placental morphogenesis. *Nature* **403**:785–789.

Nakagawa, K., and L. C. Harrison. 1996. The potential roles of endogenous retroviruses in autoimmunity. *Immunol. Rev.* **152**:193–236.

Nilsson, B. O., M. Jin, A. C. Andersson, P. Sundstrom, and E. Larsson. 1999. Expression of envelope proteins of endogenous C-type retrovirus on the surface of mouse and human oocytes at fertilization. *Virus Genes* **18**:115–120.

Stoye, J. P., and J. M. Coffin. 2000. A provirus put to work. *Nature* **403**:715, 717.

Villarreal, L. P. 1997. On viruses, sex, and motherhood. *J. Virol.* **71**:859–865.

Mouse Evolution and Viruses

Charrel, R. N., H. Feldmann, C. F. Fulhorst, R. Khelifa, R. de Chesse, and X. de Lamballerie. 2002. Phylogeny of New World arenaviruses based on the complete coding sequences of the small genomic segment identified an evolutionary lineage produced by intrasegmental recombination. *Biochem. Biophys. Res. Commun.* **296**:1118–1124.

Gottlieb, K. A. 2001. Polyomavirus replication in the lungs of mice: link to host cell differentiation and the role of the early proteins. Ph.D. dissertation. University of California, Irvine.

Hart, C. A., and M. Bennett. 1999. Hantavirus infections: epidemiology and pathogenesis. *Microbes Infect.* **1**:1229–1237.

Hook, L. M., B. A. Jude, V. S. Ter-Grigorov, J. W. Hartley, H. C. Morse, III, Z. Trainin, V. Toder, A. V. Chervonsky, and T. V. Golovkina. 2002. Characterization of a novel murine retrovirus mixture that facilitates hematopoiesis. *J. Virol.* **76**:12112–12122.

Hughes Austin, L., and R. Friedman. 2000. Evolutionary diversification of protein-coding genes of hantaviruses. *Mol. Biol. Evol.* **17**:1558–1568.

Monroe, M. C., S. P. Morzunov, A. M. Johnson, M. D. Bowen, H. Artsob, T. Yates, C. J. Peters, P. E. Rollin, T. G. Ksiazek, and S. T. Nichol. 1999. Genetic diversity and distribution of Peromyscus-borne hantaviruses in North America. *Emerg. Infect. Dis.* **5**:75–86.

Nemirov, K., H. Henttonen, A. Vaheri, and A. Plyusnin. 2002. Phylogenetic evidence for host switching in the evolution of hantaviruses carried by Apodemus mice. *Virus Res.* **90**:207–215.

Singleton, G. R., A. L. Smith, G. R. Shellam, N. Fitzgerald, and W. J. Muller. 1993. Prevalence of viral antibodies and helminths in field populations of house mice (Mus domesticus) in southeastern Australia. *Epidemiol. Infect.* **110**:399–417.

Negative-Strand RNA Viruses

Badrane, H., C. Bahloul, P. Perrin, and N. Tordo. 2001. Evidence of two *Lyssavirus* phylogroups with distinct pathogenicity and immunogenicity. *J. Virol.* **75**:3268–3276.

Badrane, H., and N. Tordo. 2001. Host switching in *Lyssavirus* history from the Chiroptera to the Carnivora orders. *J. Virol.* **75**:8096–8104.

Davis, I. C., A. J. Zajac, K. B. Nolte, J. Botten, B. Hjelle, and S. Matalon. 2002. Elevated generation of reactive oxygen/nitrogen species in hantavirus cardiopulmonary syndrome. *J. Virol.* **76**:8347–8359.

Guyatt, K. J., J. Twin, P. Davis, E. C. Holmes, G. A. Smith, I. L. Smith, J. S. Mackenzie, and P. L. Young. 2003. A molecular epidemiological study of Australian bat lyssavirus. *J. Gen. Virol.* **84**:485–496.

Le Mercier, P., Y. Jacob, K. Tanner, and N. Tordo. 2002. A novel expression cassette of lyssavirus shows that the distantly related Mokola virus can rescue a defective rabies virus genome. *J. Virol.* **76**:2024–2027.

Tidona, C. A., H. W. Kurz, H. R. Gelderblom, and G. Darai. 1999. Isolation and molecular characterization of a novel cytopathogenic paramyxovirus from tree shrews. *Virology* **258**:425–434.

Wang, L., B. H. Harcourt, M. Yu, A. Tamin, P. A. Rota, W. J. Bellini, and B. T. Eaton. 2001. Molecular biology of Hendra and Nipah viruses. *Microbes Infect.* **3**:279–287.

Wang, L.-F., M. Yu, E. Hansson, L. I. Pritchard, B. Shiell, W. P. Michalski, and B. T. Eaton. 2000. The exceptionally large genome of Hendra virus: support for creation of a new genus within the family *Paramyxoviridae*. *J. Virol.* **74**:9972–9979.

The Influenza A Story: Emergent Diseases

Brownlee, G. G., and E. Fodor. 2001. The predicted antigenicity of the haemagglutinin of the 1918 Spanish influenza pandemic suggests an avian origin. *Philos. Trans. R. Soc. Lond. B* **356**:1871–1876.

Fanning, T. G., R. D. Slemons, A. H. Reid, T. A. Janczewski, J. Dean, and J. K. Taubenberger. 2002. 1917. Avian influenza virus sequences suggest that the 1918 pandemic virus did not acquire its hemagglutinin directly from birds. *J. Virol.* **76**:7860–7862.

Gammelin, M., A. Altmuller, U. Reinhardt, J. Mandler, V. R. Harley, P. J. Hudson, W. M. Fitch, and C. Scholtissek. 1990. Phylogenetic analysis of nucleoproteins suggests that human influenza A viruses emerged from a 19th-century avian ancestor. *Mol. Biol. Evol.* **7**:194–200.

Laver, W. G., N. Bischofberger, and R. G. Webster. 2000. The origin and control of pandemic influenza. *Perspect. Biol. Med.* **43**:173–192.

Makarova, K. S., Y. Wulf, E. P. Tereza, and V. A. Ratner. 1998. Different patterns of molecular evolution of influenza A viruses in avian and human population. *Genetika* **34**:890–896. (In Russian.)

Schafer, J. R., Y. Kawaoka, W. J. Bean, J. Suss, D. Senne, and R. G. Webster. 1993. Origin of the pandemic 1957 H2 influenza A virus and the persistence of its possible progenitors in the avian reservoir. *Virology* **194**:781–788.

Small DNA Viruses

Antonsson, A., O. Forslund, H. Ekberg, G. Sterner, and B. G. Hansson. 2000. The ubiquity and impressive genomic diversity of human skin papillomaviruses suggest a commensalic nature of these viruses. *J. Virol.* **74**:11636–11641.

Bahr, U., E. Schondorf, M. Handermann, and G. Darai. 2003. Molecular anatomy of Tupaia (tree shrew) adenovirus genome; evolution of viral genes and viral phylogeny. *Virus Genes* **27**:29–48.

Ikegaya, H., H. Iwase, C. Sugimoto, and Y. Yogo. 2002. JC virus genotyping offers a new means of tracing the origins of unidentified cadavers. *Int. J. Leg. Med.* **116:** 242–245.

Sugimoto, C., T. Kitamura, J. Guo, M. N. Al-Ahdal, S. N. Shchelkunov, B. Otova, P. Ondrejka, J. Y. Chollet, S. El-Safi, M. Ettayebi, G. Gresenguet, T. Kocagoz, S. Chaiyarasamee, K. Z. Thant, S. Thein, K. Moe, N. Kobayashi, F. Taguchi, and Y. Yogo. 1997. Typing of urinary JC virus DNA offers a novel means of tracing human migrations. *Proc. Natl. Acad. Sci. USA* **94:**9191–9196.

Herpesviruses

Bahr, U., and G. Darai. 2001. Analysis and characterization of the complete genome of tupaia (tree shrew) herpesvirus. *J. Virol.* **75:**4854–4870.

Darai, G., H. G. Koch, R. M. Flugel, and H. Gelderblom. 1982. Tree shrew (Tupaia) herpesviruses. *Dev. Biol. Stand.* **52:**39–51.

Gentry, G. A., M. Lowe, G. Alford, and R. Nevins. 1988. Sequence analyses of herpesviral enzymes suggest an ancient origin for human sexual behavior. *Proc. Natl. Acad. Sci. USA* **85:**2658–2661.

Huff, J. L., and P. A. Barry. 2003. B-virus (Cercopithecine herpesvirus 1) infection in humans and macaques: potential for zoonotic disease. *Emerg. Infect. Dis.* **9:** 246–250.

Lacoste, V., P. Mauclere, G. Dubreuil, J. Lewis, M. C. Georges-Courbot, and A. Gessain. 2000. KSHV-like herpesviruses in chimps and gorillas. *Nature* **407:**151–152.

McGeoch, D. J., A. Dolan, and A. C. Ralph. 2000. Toward a comprehensive phylogeny for mammalian and avian herpesviruses. *J. Virol.* **74:**10401–10406.

Zong, J., D. M. Ciufo, R. Viscidi, L. Alagiozoglou, S. Tyring, P. Rady, J. Orenstein, W. Boto, H. Kalumbuja, N. Romano, M. Melbye, G. H. Kang, C. Boshoff, and G. S. Hayward. 2002. Genotypic analysis at multiple loci across Kaposi's sarcoma herpesvirus (KSHV) DNA molecules: clustering patterns, novel variants and chimerism. *J. Clin. Virol.* **23:**119–148.

Poxviruses

Afonso, C. L., E. R. Tulman, Z. Lu, L. Zsak, N. T. Sandybaev, U. Z. Kerembekova, V. L. Zaitsev, G. F. Kutish, and D. L. Rock. 2002. The genome of camelpox virus. *Virology* **295:**1–9.

Begon, M., S. M. Hazel, D. Baxby, K. Bown, R. Cavanagh, J. Chantrey, T. Jones, and M. Bennett. 1999. Transmission dynamics of a zoonotic pathogen within and between wildlife host species. *Proc. R. Soc. Lond. B* **266:**1939–1945.

Chantrey, J., H. Meyer, D. Baxby, M. Begon, K. J. Bown, S. M. Hazel, T. Jones, W. I. Montgomery, and M. Bennett. 1999. Cowpox: reservoir hosts and geographic range. *Epidemiol. Infect.* **122:**455–460.

Feore, S. M., M. Bennett, J. Chantrey, T. Jones, D. Baxby, and M. Begon. 1997. The effect of cowpox virus infection on fecundity in bank voles and wood mice. *Proc. R. Soc. Lond. B* **264:**1457–1461.

Hazel, S. M., M. Bennett, J. Chantrey, K. Bown, R. Cavanagh, T. R. Jones, D. Baxby, and M. Begon. 2000. A longitudinal study of an endemic disease in its wildlife reservoir: cowpox and wild rodents. *Epidemiol. Infect.* **124:**551–562.

Sandvik, T., M. Tryland, H. Hansen, R. Mehl, U. Moens, O. Olsvik, and T. Traavik. 1998. Naturally occurring orthopoxviruses: potential for recombination with vaccine vectors. *J. Clin. Microbiol.* **36:**2542–2547.

Seman, G., B. M. Levy, M. Panigel, and L. Dmochowski. 1975. Type-C virus particles in placenta of the cottontop marmoset (Saguinus oedipus). *J. Natl. Cancer Inst.* **54**:251–252.

X, Y, and ERVs

Dimitri, P., and N. Junakovic. 1999. Revising the selfish DNA hypothesis: new evidence on accumulation of transposable elements in heterochromatin. *Trends Genet.* **15**:123–124.

Jones, S. 2003. Y: *the Descent of Men*, p. xvii. Houghton Mifflin, Boston, Mass.

Zsiros, J., M. F. Jebbink, V. V. Lukashov, P. A. Voute, and B. Berkhout. 1999. Biased nucleotide composition of the genome of HERV-K related endogenous retroviruses and its evolutionary implications. *J. Mol. Evol.* **48**:102–111.

Zsiros, J., M. F. Jebbink, V. V. Lukashov, P. A. Voute, and B. Berkhout. 1998. Evolutionary relationships within a subgroup of HERV-K-related human endogenous retroviruses. *J. Gen. Virol.* **79**(Pt. 1):61–70.

Human and Primate ERVs

Barbulescu, M., G. Turner, M. I. Seaman, A. S. Deinard, K. K. Kidd, and J. Lenz. 1999. Many human endogenous retrovirus K (HERV-K) proviruses are unique to humans. *Curr. Biol.* **9**:861–868.

Barbulescu, M., G. Turner, M. Su, R. Kim, M. I. Jensen-Seaman, A. S. Deinard, K. K. Kidd, and J. Lenz. 2001. A HERV-K provirus in chimpanzees, bonobos and gorillas, but not humans. *Curr. Biol.* **11**:779–783.

Benit, L., A. Calteau, and T. Heidmann. 2003. Characterization of the low-copy HERV-Fc family: evidence for recent integrations in primates of elements with coding envelope genes. *Virology* **312**:159–168.

de Parseval, N., V. Lazar, J.-F. Casella, L. Benit, and T. Heidmann. 2003. Survey of human genes of retroviral origin: identification and transcriptome of the genes with coding capacity for complete envelope proteins. *J. Virol.* **77**:10414–10422.

Lower, R., R. R. Tonjes, C. Korbmacher, R. Kurth, and J. Lower. 1995. Identification of a Rev-related protein by analysis of spliced transcripts of the human endogenous retroviruses HTDV/HERV-K. *J. Virol.* **69**:141–149.

McIntosh, E. M., and R. H. Haynes. 1996. HIV and human endogenous retroviruses: an hypothesis with therapeutic implications. *Acta Biochim. Pol.* **43**:583–592.

Menin, C., S. Indraccolo, M. Montagna, B. Corneo, L. Bonaldi, C. Leib-Mosch, L. Chieco-Bianchi, and E. D'Andrea. 1996. Identification of a human endogenous LTR-like sequence using HIV-1 LTR specific primers. *Mol. Cell. Probes* **10**:443–451.

Sverdlov, E. D. 2000. Retroviruses and primate evolution. *Bioessays* **22**:161–171.

Tonjes, R. R., F. Czauderna, and R. Kurth. 1999. Genome-wide screening, cloning, chromosomal assignment, and expression of full-length human endogenous retrovirus type K. *J. Virol.* **73**:9187–9195.

Turner, G., M. Barbulescu, M. Su, M. I. Jensen-Seaman, K. K. Kidd, and J. Lenz. 2001. Insertional polymorphisms of full-length endogenous retroviruses in humans. *Curr. Biol.* **11**:1531–1535.

Yang, J., H. P. Bogerd, S. Peng, H. Wiegand, R. Truant, and B. R. Cullen. 1999. An ancient family of human endogenous retroviruses encodes a functional homolog of the HIV-1 Rev protein. *Proc. Natl. Acad. Sci. USA* **96**:13404–13408.

Expression of HERV *env* Genes

Andersson, A. C., A. C. Svensson, C. Rolny, G. Andersson, and E. Larsson. 1998. Expression of human endogenous retrovirus ERV3 (HERV-R) mRNA in normal and reoplastic tissues. *Int. J. Oncol.* **12:**309–313.

Blond, J. L., F. Beseme, L. Duret, O. Bouton, F. Bedin, H. Perron, B. Mandrand, and F. Mallet. 1999. Molecular characterization and placental expression of HERV-W, a new human endogenous retrovirus family. *J. Virol.* **73:**1175–1185.

Blond, J. L., D. Lavillette, V. Cheynet, O. Bouton, G. Oriol, S. Chapel-Femanders, B. Mandrand, F. Mallet, and F. L. Cosset. 2000. An envelope glycoprotein of the human endogenous retrovirus HERV-W is expressed in the human placenta and fuses cells expressing the type D mammalian retrovirus receptor. *J. Virol.* **74:**3321–3329.

Herbst, H., M. Sauter, C. Kuhler-Obbarius, T. Loning, and N. Mueller-Lantzsch. 1998. Human endogenous retrovirus (HERV)-K transcripts in germ cell and trophoblastic tumours. *APMIS* **106:**216–220.

Herbst, H., M. Sauter, and N. Mueller-Lantzsch. 1996. Expression of human endogenous retrovirus K elements in germ cell and trophoblastic tumors. *Am. J. Pathol.* **149:**1727–1735.

Kjellman, C., H. O. Sjogren, L. G. Salford, and B. Widegren. 1999. HERV-F (XA34) is a full-length human endogenous retrovirus expressed in placental and fetal tissues. *Gene* **239:**99–107.

Langat, D. K., P. M. Johnson, N. S. Rote, E. O. Wango, G. O. Owiti, and J. M. Mwenda. 1998. Immunohistochemical localization of retroviral-related antigens expressed in normal baboon placental villous tissue. *J. Med. Primatol.* **27:**278–286.

Mi, S., X. Lee, X. Li, G. M. Veldman, H. Finnerty, J. Racie, E. LaVallie, X. Y. Tang, P. Edouard, S. Howes, J. C. Keith, Jr., and J. M. McCoy. 2000. Syncytin is a captive retroviral envelope protein involved in human placental morphogenesis. *Nature* **403:**785–789.

Turbeville, M. A., J. C. Rhodes, D. M. Hyams, C. M. Distler, and P. E. Steele. 1997. Characterization of a putative retroviral env-related human protein. *Pathobiology* **65:**123–128.

Lentiviruses and HIV

Allison, R. W., and E. A. Hoover. 2003. Covert vertical transmission of feline immunodeficiency virus. *AIDS Res. Hum. Retrovirol.* **19:**421–434.

Ansari, A. A., N. Onlamoon, P. Bostik, A. E. Mayne, L. Gargano, and K. Pattanapanyasat. 2003. Lessons learnt from studies of the immune characterization of naturally SIV infected sooty mangabeys. *Front. Biosci.* **8:**s1030–s1050.

Bailes, E., F. Gao, F. Bibollet-Ruche, V. Courgnaud, M. Peeters, P. A. Marx, B. H. Hahn, and P. M. Sharp. 2003. Hybrid origin of SIV in chimpanzees. *Science* **300:**1713.

Biek, R., A. G. Rodrigo, D. Holley, A. Drummond, C. R. Anderson, Jr., H. A. Ross, and M. Poss. 2003. Epidemiology, genetic diversity, and evolution of endemic feline immunodeficiency virus in a population of wild cougars. *J. Virol.* **77:**9578–9589.

Courgnaud, V., M. Salemi, X. Pourrut, E. Mpoudi-Ngole, B. Abela, P. Auzel, F. Bibollet-Ruche, B. Hahn, A.-M. Vandamme, E. Delaporte, and M. Peeters. 2002. Characterization of a novel simian immunodeficiency virus with a *vpu* gene from greater spot-nosed monkeys (*Cercopithecus nictitans*) provides new insights into simian/human immunodeficiency virus phylogeny. *J. Virol.* **76:**8298–8309.

Courgnaud, V., W. Saurin, F. Villinger, and P. Sonigo. 1998. Different evolution of simian immunodeficiency virus in a natural host and a new host. *Virology* **247:**41–50.

Patterson, B. K., H. Behbahani, W. J. Kabat, Y. Sullivan, M. R. O'Gorman, A. Landay, Z. Flener, N. Khan, R. Yogev, and J. Andersson. 2001. Leukemia inhibitory factor inhibits HIV-1 replication and is upregulated in placentae from non-transmitting women. *J. Clin. Investig.* **107:**287–294.

Rey-Cuille, M. A., J. L. Berthier, M. C. Bomsel-Demontoy, Y. Chaduc, L. Montagnier, A. G. Hovanessian, and L. A. Chakrabarti. 1998. Simian immunodeficiency virus replicates to high levels in sooty mangabeys without inducing disease. *J. Virol.* **72:**3872–3886.

Salemi, M., T. De Oliveira, V. Courgnaud, V. Moulton, B. Holland, S. Cassol, W. M. Switzer, and A. M. Vandamme. 2003. Mosaic genomes of the six major primate lentivirus lineages revealed by phylogenetic analyses. *J. Virol.* **77:**7202–7213.

Wang, B., M. Mikhail, W. B. Dyer, J. J. Zaunders, A. D. Kelleher, and N. K. Saksena. 2003. First demonstration of a lack of viral sequence evolution in a nonprogressor, defining replication-incompetent HIV-1 infection. *Virology* **312:**135–150.

Human Language, Schizophrenia, and ERVs

Berlim, M. T., B. S. Mattevi, P. Belmonte-de-Abreu, and T. J. Crow. 2003. The etiology of schizophrenia and the origin of language: overview of a theory. *Compr. Psychiatry* **44:**7–14.

Crow, T. J. 1997. Is schizophrenia the price that Homo sapiens pays for language? *Schizophr. Res.* **28:**127–141.

Crow, T. J. 1997. Aetiology of schizophrenia: an echo of the speciation event. *Int. Rev. Psychiatry* **9:**321–330.

Crow, T. J. 1999. Commentary on Annett, Yeo et al., Klar, Saugstad and Orr: cerebral asymmetry, language and psychosis—the case for a Homo sapiens-specific sex-linked gene for brain growth. *Schizophr. Res.* **39:**219–231.

Crow, T. J. 2000. Schizophrenia as the price that Homo sapiens pays for language: a resolution of the central paradox in the origin of the species. *Brain Res. Rev.* **31:**118–129.

Highley, J. R., B. McDonald, M. A. Walker, M. M. Esiri, and T. J. Crow. 1999. Schizophrenia and temporal lobe asymmetry. A post-mortem stereological study of tissue volume. *Br. J. Psychiatry* **175:**127–134.

Karlsson, H., S. Bachmann, J. Schroder, J. McArthur, E. F. Torrey, and R. H. Yolken. 2001. Retroviral RNA identified in the cerebrospinal fluids and brains of individuals with schizophrenia. *Proc. Natl. Acad. Sci. USA* **98:**4634–4639.

Kim, H. S., O. Takenaka, and T. J. Crow. 1999. Isolation and phylogeny of endogenous retrovirus sequences belonging to the HERV-W family in primates. *J. Gen. Virol.* **80**(Pt. 10):2613–2619.

Kim, H.-S., R. V. Wadekar, O. Takenaka, C. Winstanley, F. Mitsunaga, T. Kageyama, B.-H. Hyun, and T. J. Crow. 1999. SINE-R. C2 (a Homo sapiens specific retroposon) is homologous to cDNA from postmortem brain in schizophrenia and to two loci in the Xq21.3/Yp block linked to handedness and psychosis. *Am. J. Med. Genet.* **88:**560–566.

Yolken, R. H., H. Karlsson, F. Yee, N. L. Johnston-Wilson, and E. F. Torrey. 2000. Endogenous retroviruses and schizophrenia. *Brain Res. Rev.* **31:**193–199.

Sexual Behavior of Voles and Wasps

Aragona, B. J., Y. Liu, J. T. Curtis, F. K. Stephan, and Z. Wang. 2003. A critical role for nucleus accumbens dopamine in partner-preference formation in male prairie voles. *J. Neurosci.* **23:**3483–3490.

Ayasse, M., F. P. Schiestl, H. F. Paulus, F. Ibarra, and W. Francke. 2003. Pollinator attraction in a sexually deceptive orchid by means of unconventional chemicals. *Proc. R. Soc. Lond. B* **270**:517–522.

Pitkow, L. J., C. A. Sharer, X. Ren, T. R. Insel, E. F. Terwilliger, and L. J. Young. 2001. Facilitation of affiliation and pair-bond formation by vasopressin receptor gene transfer into the ventral forebrain of a monogamous vole. *J. Neurosci.* **21**: 7392–7396.

Schiestl, F. P., R. Peakall, J. G. Mant, F. Ibarra, C. Schulz, S. Franke, and W. Francke. 2003. The chemistry of sexual deception in an orchid-wasp pollination system. *Science* **302**:437–438.

Varaldi, J., P. Fouillet, M. Ravallec, M. Lopez-Ferber, M. Bouletreau, and F. Fleury. 2003. Infectious behavior in a parasitoid. *Science* **302**:1930.

Index